AF544819

# Greywater Reuse

**Co-published by IWA Publishing**
Alliance House, 12 Caxton Street, London SW1H 0QS, UK
Tel. +44 (0) 20 7654 5500, Fax +44 (0) 20 7654 5555
publications@iwap.co.uk
www.iwapublishing.com
IWA Publishing ISBN: 9781498733519

# Greywater Reuse

Amit Gross
Adi Maimon
Yuval Alfiya
Eran Friedler

CRC Press is an imprint of the
Taylor & Francis Group, an **informa** business

CRC Press
Taylor & Francis Group
6000 Broken Sound Parkway NW, Suite 300
Boca Raton, FL 33487-2742



Printed on acid-free paper
Version Date: 20150126

International Standard Book Number-13: 978-1-4822-5504-1 (Hardback)



**Visit the Taylor & Francis Web site at**
**http://www.taylorandfrancis.com**

**and the CRC Press Web site at**
**http://www.crcpress.com**

Printed and bound in Great Britain by
TJ International Ltd, Padstow, Cornwall

# Contents

# Preface

Both water scarcity and the desire to increase the sustainability of domestic water resources have stimulated the search for efficient water use practices. These reasons drove our exploration of greywater—its characteristics and potential uses—for over a decade. In our research, we examined treatment facilities in urban and rural environments, developed greywater treatment systems, and quantified potential environmental and health risks posed by greywater at different treatment levels.

Interestingly, millions of people worldwide recycle greywater to irrigate their gardens and flush their toilets. Many companies market recycling schemes for greywater, but some of these systems produce low-quality reclaimed water due to improper treatment. Despite its prevalence, public information on greywater reuse is sparse outside of that provided by interested parties such as greywater systems companies. Information on the Internet is often inaccurate or contradictory.

To help us address the dearth of readily available information, we secured support from the Israeli Water Authority to write this book, which is the first in Israel, and the world, to thoroughly describe the features and implications of greywater reuse scientifically and quantitatively.

This book reviews scores of studies in the field of greywater from around the globe. It is the result of over ten years of research, and contributions from research assistants and colleagues in Israel and abroad. The project included more than 20 postgraduate students from two institutions: Ben Gurion University in the Negev and the Technion—Israel Institute of Technology. Special thanks to Professor Alon Tal and Dr. Adi Inbar for their contributions to the chapters dealing with policy issues and public perceptions, to Rifi Ron for the English translation, Vivian Futran Fuhrman and Clara Wool for their assistance with the proofreading and final editing, Sharon Ychie from Studio Koobeeyaa and Eyal Unger for the graphic design and illustrations, the Technion—Israel Institute of technology and Ben Gurion University of the Negev for the generous support, and of course to Adi Maimon and Yuval Alfiya, who worked tirelessly in preparing this book.

Finally, we extend our thanks to Maccabi Carasso for his research support and the admirable determination with which he promotes the safe use of greywater in Israel.

We hope that this work will be a good resource to professionals and decision makers and to students who wish to enter this field.

**Amit Gross**
**Eran Friedler**

# Authors

**Amit Gross** completed his undergraduate studies in the Faculty of Agriculture of the Hebrew University of Jerusalem, Jerusalem, Israel in 1993. He earned his MSc (1996) and PhD (1999) from Auburn University, Auburn, Alabama, studying nutrient cycles in earthen ponds. During his postdoctoral training in Australia and at Ben Gurion University of the Negev, Israel, he studied various environmental issues related to water treatment and reuse and was recruited, in 2003, as a faculty member. He is currently an associate professor in the Department of Environmental Hydrology and Microbiology, Zuckerberg Institute for Water Research, Jacob Blaustein Institutes for Desert Research, Ben Gurion University of the Negev, Israel.

Dr. Gross's research areas include treatment and efficient use of marginal water and the environmental risks associated with contaminated water resources such as greywater and wastewater. He is an associate editor, is on the editorial boards of several journals, serves on the international committees of various national and international conferences, and is the coauthor of over 100 professional publications in his field.

**Adi Maimon** completed her undergraduate studies in Rupin College in 2006, focusing on marine sciences. She earned her MSc (2010) and then undertook a PhD program with Prof. Gross, studying various aspects of greywater reuse.

**Yuval Alfiya** completed his undergraduate studies in environmental engineering (2002) in the Technion—Israel Institute of Technology. He earned an MSc in agricultural engineering (2005) from the Technion. From 2004 to 2008, he worked as a research assistant in the Israel National Center for Mariculture and then for five years as a research engineer with Prof. Friedler. Toward the end of 2013, he started his PhD studies with Prof. Friedler, studying various aspects of greywater reuse.

**Eran Friedler** completed his undergraduate studies in the Faculty of Agriculture at the Hebrew University of Jerusalem, Jerusalem, Israel in 1986. He earned his PhD (1993) from the Technion—Israel Institute of Technology, studying wastewater stabilization reservoirs. During his postdoctoral training at the Imperial College, United Kingdom, he studied issues related to urban water (water and wastewater). In 1995, he returned to Israel and became a partner in an environmental consultancy, where he worked mainly on water and wastewater issues. In 2001, he was recruited as a faculty member in the Faculty of Civil and Environmental Engineering at the Technion, where he is currently an associate professor in the Department of Environmental, Water, and Agricultural Engineering. He is also a member of the Grand Water Research Institute in the Technion and a senior research fellow in the Samuel Neaman Institute for Advanced Studies in Science & Technology (Technion).

Dr. Friedler's research areas include the development of alternative water sources and their influence on sustainable urban water use, health and the environmental risks associated with reusing various types of water, and the interaction between water saving and water reuse and sewerage systems.

Friedler is an associate editor of the *Urban Water Journal*, serves on scientific committees of various national and international conferences, and is an author/coauthor of over 100 professional publications in his field.

# Introduction

Water is a basic resource for life. It is used directly or indirectly in every domain: domestic consumption, urban endeavors, industry, and agriculture. In the natural environment, the diversity and health of ecosystems depend on water. However, according to a UN estimate in 2007, about one-fifth of the world's population is facing water shortage, and this number is expected to grow. Four main drivers affect the expected growth of water shortage: population growth, urbanization, increased personal consumption due to the rising standard of living, and climatic changes (UN Water, 2007). Consequently, effective and sustainable use of water resources is a global challenge that is garnering increasing attention from various international institutions.

The need to save water and use water sources effectively is of particular importance in semiarid and arid regions, where water sources are scant and the precipitation volume is low. For example, a state of emergency was declared in 1999 in the Israeli water sector, which is still in effect to this day. Water shortage in Israel stems from excessive exploitation of a regionally limited resource accumulated over many years.

Until now, solutions to Israel's water scarcity have focused on mitigating agricultural water consumption. Over the years, innovative water-saving irrigation technologies and cultivation methods have helped increase agricultural productivity per unit of water. However, it was a change in Israel's supply strategy that really tipped the country's water use balance: moving from potable water to treated water for agricultural uses. Since 1998, Israel's water sector has been relying increasingly on marginal water (treated and brackish); meanwhile, the consumption of potable water for agricultural purposes has been declining (Tal, 2008; Israel Water Authority, 2012) (Figure I.1). Today, Israel reuses over 80% of its wastewater for agricultural irrigation, a much higher rate than any other country (e.g., Spain is in second place with about 35% of its water being recycled for agriculture).

Despite the dramatic reduction in its use of potable water for agriculture, Israel's water crisis has not been resolved. The country's total water consumption is expected to continue increasing predominantly due to population growth. To meet the growing demand, the focus has been directed to increasing water supply through desalination. While desalination is widely considered a necessity in Israel, this process has high economic and environmental costs that are still not well quantified, but include energy input, required beach area, pollutants emitted into the air that contribute to the greenhouse effect, and the discharge of the brine back into the ocean.

In 2010, potable water consumption in Israel was about 1260 million cubic meters (MCM) of which domestic consumption made up approximately 690 MCM, accounting for over 50% of Israel's potable water consumption (Israel Water Authority website). Nonetheless, efforts to save water in the domestic sector are minimal, relying mainly on public awareness campaigns and the installation of faucet aerators.

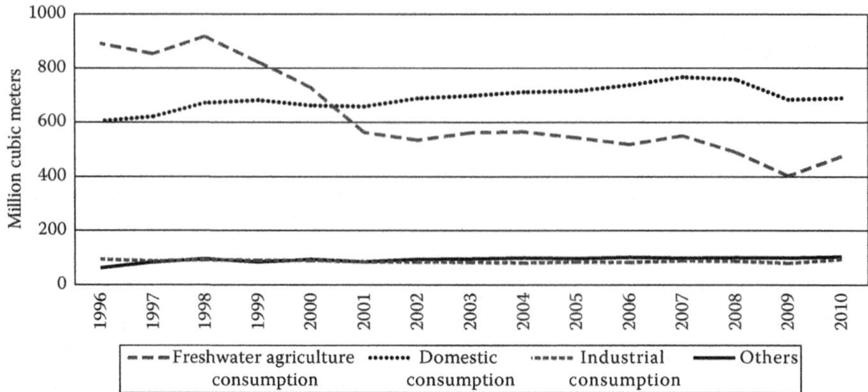


**FIGURE I.1** Distribution of potable water consumption in Israel by sector between 1996 and 2010 according to data from the Israel Water Authority.

Since domestic consumption is steadily increasing due to population growth, reduction in per capita consumption could reduce the need for desalinated water without affecting the water available for agriculture.

One way to reduce the impact of domestic consumption is with the treatment and reuse of *greywater* using small and on-site facilities. Greywater is defined as household wastewater that excludes the flow originating from toilet flushing. Specifically, it includes the streams generated by baths and showers, hand basins, laundry, kitchen, and dishwashing. In general, greywater can be divided into *dark* and *light* greywater. Light greywater excludes kitchen and dishwashing wastewater (although in some places, laundry water is also considered dark greywater). Reuse of greywater allows water to be used at least twice (initial use and local recycling), and in some cases, water can be recovered for a third use (agriculture irrigation), allowing a savings of 10%–20% of urban water consumption. In Israel, if only 30% of households started recycling greywater (i.e., 30% penetration rate), the savings could add up to 25–50 MCM per year in just 20 years (Friedler, 2008). This yearly volume corresponds to the yield of a medium-size desalination plant.

Since greywater does not contain toilet wastewater or (in most cases) kitchen wastewater, its organic load is lower and its microbial quality higher than that of the total domestic wastewater. Therefore, it can be treated using relatively simple means suitable for on-site treatment and reuse systems.

There are many benefits of on-site recycling of greywater. Reduction of water demand in the urban sector following the local reuse of greywater reduces the need to develop new water resources, such as seawater desalination and groundwater abstraction by ever deeper wells. Furthermore, this reduction reduces the volume of domestic wastewater that has to be transported and treated. In places that do not have sewage infrastructure, on-site separation, treatment, and reuse might serve as an effective and inexpensive solution for significantly reducing environmental pollution and sanitary risk. Equally important, decentralized reuse systems could yield both private and public monetary savings with proper planning.

Even if global sustainability is not a personal goal, greywater reuse can play an important role in domestic independence and economics. This approach advocates mimicking the natural ecosystem by planning living space and the local environment to maximize the utilization of resources without exhausting them. Alongside composting, using solar energy for heating, local food cultivation, reductions in utility costs, and reliance on outside sources of food, energy, and water can decrease costs and reliance on utilities.

Despite these advantages, the use of greywater is not devoid of risks and challenges. Greywater contains salts and various organic compounds that could harm plants and over time change the soil properties. In addition, it often contains considerable concentrations of fecal coliforms indicating a potential for the presence of pathogens from the digestive tract. Other opportunistic pathogens such as those related to food handling may also be found in greywater. The unique characteristics of greywater call for different treatment systems than those required for centralized urban wastewater treatment plants. For example, greywater treatment systems have to be compact and able to overcome sharp fluctuations in flow rate and quality of the incoming greywater. Dual reticulation in homes gives rise to the risk of cross-connections between the treated greywater pipes and pipes conveying potable water. Means to prevent cross-connections should therefore be installed. In addition to sanitary and environmental risks, opponents claim that the extensive use of greywater will make it difficult to transport wastewater in the sewer system, place an additional burden on wastewater treatment plants, and reduce the amount of treated wastewater effluent available for agricultural irrigation.

These and other concerns underlie the current Israeli policy, which significantly limits the use of greywater and prohibits its use in private homes. Despite being prohibited, it is estimated that over 15,000 single family households in Israel reuse greywater, mainly for garden irrigation. This number of households is equal in size to a small town of about 50,000 residents. The status quo is for greywater reuse in Israel to be done without regulation or supervision, at the discretion of each household. As a result, many treatment units are built and installed unprofessionally, or reused greywater goes untreated altogether. Regulating the use of greywater in Israel through legislation, as is done in many countries around the world, may help contribute to its safe use. However, this risks the legislation becoming purely *de jure* and not actually practiced. If it does go into action, there is a question of whether such regulation should *allow* the use of greywater or go so far as to *require* individuals to recycle greywater. The law would then have to be fine-tuned to specify who is affected and under what specific circumstances.

A combination of public interest and academic inquiry regarding greywater and its challenges has yielded hundreds of academic research papers and an abundance of information on the Internet (as suggested, not always reliable), which have grown considerably in recent years (Figure I.2). There are at least 493 academic papers written on the subject, identified in a search for *greywater* in the Scopus database between 1997 and the end of 2012. In the last five years, about 43 articles were published per year on average, amounting to approximately half of the total articles published so far. Querying the term *greywater* on the Google search engine returns about 3,900,000 results. However, there are very few books that provide

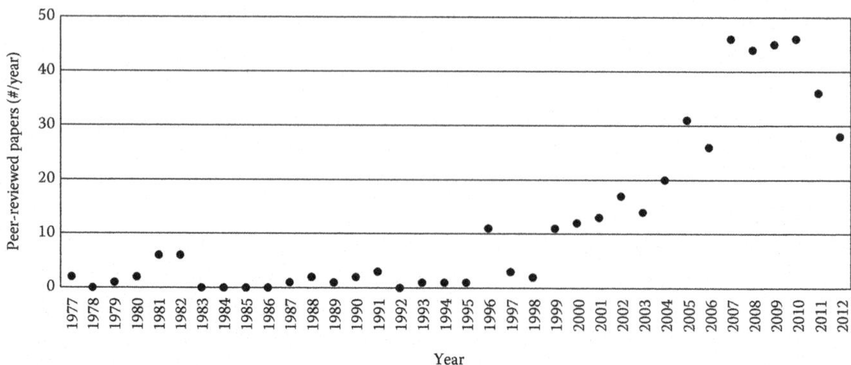


**FIGURE I.2** Number of publications on the subject of greywater from 1977 to 2012, as appearing in the Elsevier's Scopus database.

comprehensive information about greywater. Therefore, the aim of this book is to bind together unbiased information on greywater for students, scientists, professionals, decision makers, and the general public. We hope that this book will provide a broad scientific basis and will serve as a policy-making tool in the global discussion over greywater policy.

In Chapter 1, the chemical, physical, and microbial properties of greywater are described, as well as the flow rate distribution of different greywater streams. In Chapter 2, the common methods for treating potential greywater pollutants and removing them are detailed, and case studies representing some of these methods are presented. In Chapters 3 and 4, the risks involved in the use of greywater are presented, and the regulative possibilities that could help reduce these risks are discussed. The Israeli case is highlighted in Chapter 5. In addition to reviewing the existing situation, this chapter also contains the positions of public and governmental entities and stakeholders on the issue. In Chapter 6, the prevailing perceptions of the public are discussed in detail, as well as its willingness to adopt various uses of greywater. The economic aspect is discussed in Chapter 7, the final chapter, in which economic analyses are presented with regard to the impact of greywater reuse on the individual consumer and on the national water economy.

# 1 Greywater Characteristics

## 1.1 INTRODUCTION

Most domestic water consumption is for washing and cleaning. The domestic use of water creates a flow that contains dissolved, suspended, and solid waste, defined as wastewater.

Household wastewater can be divided into *blackwater* containing wastewater generated by the toilet (feces and urine) and greywater containing all other flows: bathing, washing, laundry, and kitchen water. In addition, there are also those who distinguish between light greywater including bathing and rinsing wastewater (such as that from a shower, bath, and handbasin) and dark greywater consisting of kitchen wastewater and sometimes even washing machine wastewater. The use of water in general and the generation of greywater in particular vary between locations depending on factors such as water availability, consumption habits, and economic status (Figure 1.1).

In general, greywater is less polluted than the total domestic wastewater because it does not contain toilet flush wastewater or, for the most part, kitchen wastewater. Fecal contamination and the amount of solids and fats in the water are significantly decreased by removal of these flows from the greywater stream. In addition, the concentration of organic matter, particularly the biodegradable part, is lower in greywater relative to the total domestic wastewater. Despite the potential advantages over wastewater, the concentrations of pollutants in greywater are not always lower than the pollutant concentrations of the total domestic wastewater. The reason for this lies in the large variance between the relative volumes and flows of each greywater source. For example, it is easy to imagine that the characteristics of wastewater leaving a washing machine full of dirty clothing is very different from that of hand washing before a meal. Other pollutants, such as concentrations of detergents and sometimes even boron, are usually higher in greywater than in general sewage because they are not diluted in the large water volume generated by toilet flushing.

Several factors influence the nature of greywater:

1. *Flows contained in greywater (light or dark)*: Greywater can contain water from washing machines, bathrooms, handbasins, kitchens, and dishwashers. It can combine flows or keep them separate.
2. *Source of greywater*: Greywater can be collected from various sources such as individual house, high-rise building, public showers (e.g., in sports centers), or office buildings.
3. *Cultural variables and characteristics of the occupants*: For example, consumption habits of household chemicals like laundry detergent, clothing softener, and personal care products, the age of home occupants, and the

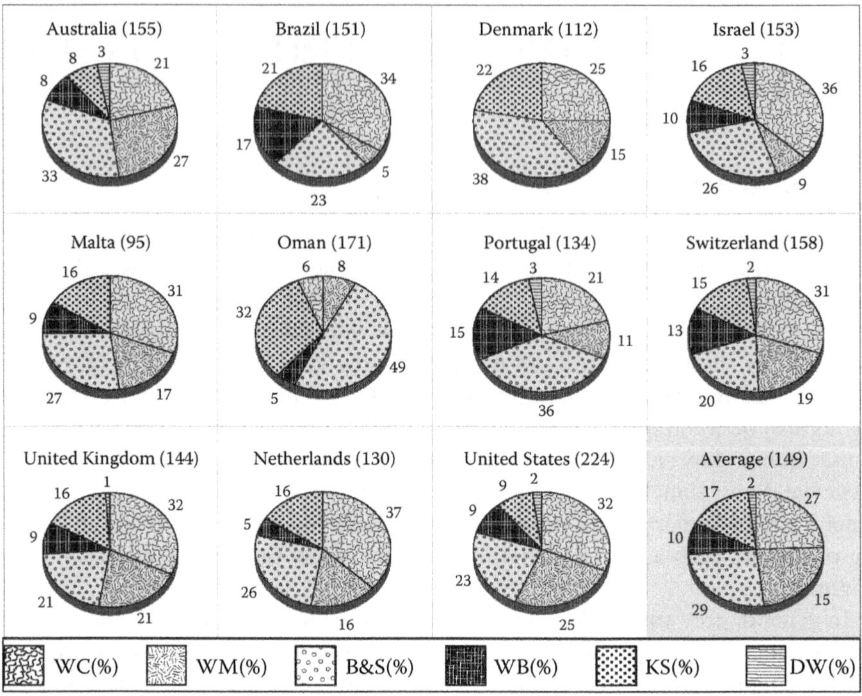


**FIGURE 1.1** The distribution of domestic water consumption by uses in different countries (%). WC, toilet flushing; WM, washing machine; B&S, bath and shower; WB, washing basin; KS, kitchen sink; DW, dishwasher. Numbers in brackets indicate average daily water consumption (L/(person·*d*). (Data were compiled from Loh, M. and Coghlan, P., Domestic water use study: In Perth, Western Australia, 1998–2001, Water Corporation, Stirling, Perth, Western Australia, Australia, 2003; Ghisi, E. and Ferreira, D.F., *Build. Environ.*, 42(7), 2512, 2007; Donner, E. et al., *Sci. Total Environ.*, 408(12), 2444, 2010; Friedler, E., *Int. J. Environ. Stud.*, 65(1), 57, 2008; Butler, D. et al., *Water Sci. Technol.*, 31(7), 13, 1995; Memon, F.A. and Butler, D., Domestic water consumption trends and techniques for demand forecasts, in: Butler, D. and Memon, F.A., eds., *Water Demand Management*, IWA, London, U.K., 2006; Prathapar, S.A. et al., *Desalination*, 186(1–3), 177, 2005; Vieira, P. et al., *Water Sci. Technol. Water Supply*, 7(5–6), 193, 2007; Helvetas, Schweizer Gesellschaft für Internationale Zusammenarbeit, in: *Water Consumption in Switzerland (in German: Wasserverbrauch in der Schweiz)*, 3pp, 2005; Roesner, L. et al., Long term effects of landscape irrigation using household graywater—Literature review and synthesis, Prepared for WERF, published with SDA, 2006.)

number of occupants will influence the volume of water and its content of pollutants. In addition, water use habits will influence the daily distribution of greywater flow rates (Jefferson et al., 2004; Ramon et al., 2004; Abu Ghunmi et al., 2008; Donner et al., 2010).

4. *Climatic and geographic variables*: These influence water consumption and the daily and seasonal consumption distribution.

5. *Supply pipes and the greywater collection piping*: The water piping may release metals such as zinc and copper into the greywater (Eriksson et al., 2002; Meinzinger and Oldenburg, 2009).
6. *Quality of the source water*: For example, groundwater differs from desalinated water in concentrations of various ions, alkalinity, and hardness, and hence, greywater quality is likely to be influenced by the quality of the source water.

Characterizing these influences allows for a preliminary evaluation of the nature of greywater from any given source. This is important for designing the appropriate treatment and recycling scheme for a greywater system and for evaluating the risks involved in using it. For example, characterization of flow rates and loads allows the size of the required treatment facility to be determined, as well as its type and its storage volume. Characterization of the level of microbial pollution allows for the risk associated with reusing greywater to be assessed.

In the first part of this chapter, the physical, chemical, and microbial characteristics of greywater will be presented. In the second part, the sources contributing to greywater will be described in terms of flow rate and quality. Finally, diurnal patterns of the quantity and quality of greywater, resulting from changes in the quantities and loads contributed by these sources, will be explained.

## 1.2 GREYWATER CHARACTERISTICS

Greywater inherently contains traces of the materials that were used within the household premises such as soaps, salts, cosmetic ingredients (e.g., face creams and makeup), food, spices, oils, and minerals. Therefore, in examining the characteristics of greywater, it is appropriate to search for common household products and other such relevant materials (Eriksson et al., 2002).

The variables that characterize greywater can be divided into physical, chemical, and microbial categories. As specified earlier, greywater quality varies between sources. It can even vary within one source over time, a phenomenon that is manifested in the wide range of values for most of the water quality variables. Table 1.1 lists the quality of light greywater as reported by various sources in the literature, while Table 1.2 lists the quality of dark greywater flows.

### 1.2.1 Physical Characteristics

The main physical characteristics that affect the quality of greywater and its treatment are temperature, color, odor, turbidity, suspended solids, and salinity.

#### 1.2.1.1 Temperature

Greywater temperature is influenced by the surrounding temperature and that of the water source. In many cases, greywater will have a higher temperature than the ambient temperature since it is sourced from warm bathing, washing, laundry, and rinsing water. When greywater is collected in a storage or balancing container,

**TABLE 1.1**
**Comparison of Chemical and Physical Characteristics of Light Greywater from Various Places around the World**

| | Jordan (Abu Ghunmi et al., 2008) | | | United Kingdom (Jefferson et al., 2004) | | | United Kingdom (Winward et al., 2008) | | | South England (Birks and Hills, 2007) | | Israel (Friedler et al., 2006) | | Oman (Jamrah et al., 2008) | 135 Studies from More than 20 Countries (Meinzinger and Oldenburg, 2009) | | |
|---|---|---|---|---|---|---|---|---|---|---|---|---|---|---|---|---|---|
| Source | AVG | SD | *n* | AVG | SD | *n* | AVG | SD | *n* | AVG | SD | AVG | SD | AVG | Med | Min | Max |
| pH | 8 | 0.2 | 96 | 7 | 0.3 | 102 | | | | 7 | 0.2 | | | 8 | | | |
| DO | | | | | | | | | | | | | | 4 | | | |
| Alk. | 412 | 106 | 13 | | | | | | | | | | | 18 | | | |
| $HCO_3^-$ | 542 | 21 | 13 | | | | | | | 327 | 23 | | | | | | |
| EC | 1060 | 185 | 96 | | | | | | | | | | | 2800 | | | |
| Tur. | 122 | 78 | 93 | 101 | 109 | 102 | 20 | 14 | 84 | 27 | 21 | 65 | 68 | 279 | | | |
| TS | 876 | 201 | 78 | | | | | | | | | | | 1121 | | | |
| TDS | | | | | | | | | | | | | | 884 | | | |
| VTS | | | 24 | | | | | | | | | | | 492 | | | |
| TSS | 122 | 78 | 26 | 100 | 145 | 102 | 29 | 32 | 82 | 37 | 29 | 92 | 115 | 236 | 228 | | |
| VSS | | | | | | | | | | | | 64 | 76 | | | | |
| COD | 551 | 202 | 96 | 451 | 289 | 102 | 87 | 38 | 51 | 96 | 53 | 211 | 141 | 426 | 535 | 350 | 783 |
| Dissolved COD | | | | | | | | | | | | 108 | 47 | | | | |
| BOD | | | | 146 | 54.3 | 102 | 20 | 11 | 80 | 46 | 27 | 69 | 33 | 408 | 329 | 205 | 449 |
| Dissolved BOD | 149 | 46 | 20 | | | | | | | 31 | | 36 | 20 | | | | |
| TOC | | | | 73 | 79.3 | 102 | | | | | | | | 93 | | | |
| $NH_4^+$–N | 8 | 6 | 22 | | | | | | | | | | | | | | |
| $NO_3^-$–N | | | | | | | | | | | | | | 17 | | | |
| TN | 10 | 14 | 20 | 8.7 | 4.7 | 102 | | | | | | | | | 13 | 7 | 22 |

*(Continued)*

**TABLE 1.1 (*Continued*)**
**Comparison of Chemical and Physical Characteristics of Light Greywater from Various Places around the World**

| | Jordan (Abu Ghunmi et al., 2008) | | | United Kingdom (Jefferson et al., 2004) | | | United Kingdom (Winward et al., 2008) | | | South England (Birks and Hills, 2007) | | Israel (Friedler et al., 2006) | | Oman (Jamrah et al., 2008) | 135 Studies from More than 20 Countries (Meinzinger and Oldenburg, 2009) | | |
|---|---|---|---|---|---|---|---|---|---|---|---|---|---|---|---|---|---|
| Source | AVG | SD | *n* | AVG | SD | *n* | AVG | SD | *n* | AVG | SD | AVG | SD | AVG | Med | Min | Max |
| TKN | | | | | | | | | | 4.6 | 2.8 | | | | | | |
| $PO_4$–P | | | | 0.4 | 0.2 | 102 | | | | | | | | | | | |
| TP | 7 | 7 | 22 | | | | | | | 0.9 | 0.8 | | | | 4.6 | 0.4 | 8 |
| MBAS | | | | | | | | | | | | | | 56 | | | |
| $Cl^-$ | 141 | 22 | 13 | | | | | | | | | | | | | | |
| $Na^+$ | 143 | 24 | 13 | | | | | | | | | | | | | | |
| $Mg^{2+}$ | 29 | 9 | 13 | | | | | | | | | | | | | | |
| $Ca^{2+}$ | 58 | 17 | 13 | | | | | | | | | | | | | | |
| $K^+$ | 10 | 2 | 13 | | | | | | | | | | | | 8.8 | | |
| S | | | | | | | | | | | | | | | 72 | | |
| SAR | 3 | | | | | | | | | | | | | | | | |

*Sources:* Data compiled from Abu Ghunmi, L. et al., *Water Sci. Technol.*, 58(7), 1385, 2008; Jefferson, B. et al., *Water Sci. Technol.*, 50(2), 157, 2004; Winward, G.P. et al., *Ecol. Eng.*, 32(2), 187, 2008; Birks, R. and Hills, S., *Environ. Monit. Assess.*, 129(1–3), 61, 2007; Friedler, E. et al., *Environ. Technol.*, 27(6), 653, 2006; Jamrah, A. et al., *Int. J. Environ. Studies*, 65(1), 71, 2008; Meinzinger, F. and Oldenburg, M., *Water Sci. Technol.*, 59(9), 1785, 2009.

*Notes:* AVG, average; SD, standard deviation; *n*, no. of cases; med, median; min, minimum value; max, maximum value; DO, dissolved oxygen; Alk., alkalinity; EC, electrical conductivity; Tur., turbidity; TS, total solids; TDS, total dissolved solids; VTS, volatile solids; TSS, total suspended solids; VSS, volatile suspended solids; COD, chemical oxygen demand; BOD, biochemical oxygen demand; TN, total nitrogen; TKN, total Kjeldahl nitrogen; TP, total phosphorous; MBAS, methylene blue absorbing substances (anionic surfactants). All units are in mg/L, except for pH, EC (μS/cm), turbidity (NTU), and sodium adsorption ratio (SAR), alkalinity and $HCO_3^-$ (mg $CaCO_3$/L).

**TABLE 1.2**
**Comparison of Chemical and Physical Characteristics of Dark Greywater from Various Places around the World**

| Source | Urban Area, Jordan (Abu Ghunmi et al., 2008) | Urban Area, Jordan (Abu Ghunmi et al., 2008) | Rural Area, Jordan (Abu Ghunmi et al., 2008) | Rural Area, Jordan (Abu Ghunmi et al., 2008) | | Six-Person Farm, Rural Area, Israel (Jefferson et al., 2004) | | Five-Person Household in Midreshet Ben-Gurion, Israel (Winward et al., 2008) | | Urban Area, Israel (Birks and Hills, 2007) |
|---|---|---|---|---|---|---|---|---|---|---|
| | AVG | AVG | AVG | AVG | SD | AVG | SD | AVG | SE | AVG |
| pH | | 7.8 | 6.4 | 6.7 | | 6.7 | 0.1 | | | 7.1 |
| Alk. | | 225 | | 140 | | | | | | |
| $HCO_3^-$ | | | 426 | | | | | | | |
| EC | | 1910 | 1890 | | | 1400 | 0 | 1.2 | 0.1 | 1478 |
| Tur. | | 49 | | | 68 | | | | | |
| TS | 1291 | 1061 | 1919 | | | | | | | 1245 |
| TSS | 253 | 168 | 1074 | 264 | 115 | 138 | 21 | 158 | 30 | 329 |
| VSS | | | | | 76 | | | | | 202 |
| COD | 870 | 78 | 2568 | 1460 | 141 | 686 | 255 | 839 | 47 | 823 |
| Dissolved COD | | | | | 47 | | | | | 473 |
| BOD | 314 | 41 | 1056 | 764 | 33 | 270 | 60 | 466 | 66 | 477 |
| Dissolved BOD | | | | | 20 | | | | | 238 |
| $NH_4^+$–N | | | 75 | 1 | | | | 0.3 | 0.1 | 1.6 |
| $NO_3^-$ | | | | | | | | 3 | 1.3 | |
| $NO_2^-$ | | | | | | | | 0.3 | 0.2 | |

*(Continued)*

**TABLE 1.2 (*Continued*)**
**Comparison of Chemical and Physical Characteristics of Dark Greywater from Various Places around the World**

| Source | Urban Area, Jordan (Abu Ghunmi et al., 2008) | Urban Area, Jordan (Abu Ghunmi et al., 2008) | Rural Area, Jordan (Abu Ghunmi et al., 2008) | Rural Area, Jordan (Abu Ghunmi et al., 2008) | | Six-Person Farm, Rural Area, Israel (Jefferson et al., 2004) | | Five-Person Household in Midreshet Ben-Gurion, Israel (Winward et al., 2008) | | Urban Area, Israel (Birks and Hills, 2007) |
|---|---|---|---|---|---|---|---|---|---|---|
| | AVG | AVG | AVG | AVG | SD | AVG | SD | AVG | SE | AVG |
| TN | 2 | 9 | 128 | 25 | | 14 | 2 | 34.3 | 2.6 | |
| TP | 3 | 9 | 20 | 13.8 | | 17.7 | 5.1 | 22.8 | 1.8 | |
| MBAS | | | | | | 40 | 4 | 7.9 | 1.7 | 37 |
| $Cl^-$ | 192 | 162 | 227 | | | | | | | 281 |
| B | | | | | | 0.6 | 0.2 | 1.6 | 0.1 | 0.4 |
| $Na^+$ | 145 | 120 | 196 | 136 | | | | | | 199 |
| $Mg^{2+}$ | 20 | 18 | 16 | 34 | | | | | | |
| $Ca^{2+}$ | 46 | 36 | 41 | 50 | | | | | | |
| $K^+$ | 8 | | 24 | | | | | | | |
| SAR | 4 | 4 | 7 | 4 | | 4.8 | | | | |

*Sources:* Data compiled from Abu Ghunmi, L. et al., *Water Sci. Technol.*, 58(7), 1385, 2008; Gross, A. et al., *Water Sci. Technol.*, 52(8), 161, 2005; Gross, A. et al., *Chemosphere*, 66(5), 916, 2007; Friedler, E., *Environ. Technol.*, 25(9), 997, 2004.

*Notes:* AVG, average; SD, standard deviation; *n*, no. of cases; Alk., alkalinity; EC, electrical conductivity; Tur., turbidity; TS, total solids; TSS, total suspended solids; VSS, volatile suspended solids; COD, chemical oxygen demand; BOD, biochemical oxygen demand; TN, total nitrogen; TKN, total Kjeldahl nitrogen; TP, total phosphorous; MBAS, methylene blue absorbing substances; SAR, sodium adsorption ratio. All units are in mg/L except for pH, EC (μS/cm), turbidity (NTU), SAR, alkalinity and $HCO_3^-$ (mg $CaCO_3$/L).

temperature variability will be smaller, and the water temperature will be similar to the ambient temperature (or only slightly higher). It should be noted that in some cold climate areas, it may be feasible to *harvest* the heat of the greywater via heat exchangers.

High temperatures, above 30°C–40°C, which is characteristic of greywater, may lead to the development of bacteria and encourage the accumulation of residues (limescale) in collecting containers and piping. However, it may also accelerate biological treatment processes and make them more efficient.

#### 1.2.1.2 Color

Greywater is named because of its color, which in many cases is a shade of grey. The source of the color is mostly coloring substances that are added to products such as soaps and detergents. Color is usually considered an aesthetic challenge, so its removal during treatment is recommended. The color of the water can be measured in several ways (such as those listed in APHA, 2005). It should be noted that the method has to be adapted to the nature and source of the color (from humic substances, other color materials, metals, etc.). However, these techniques are of limited efficiency, and their relationship to the quality of the solution is limited. Usually, color does not cause significant problems in treating and reusing greywater.

#### 1.2.1.3 Odor

The source of the odor in raw greywater is usually household chemicals, such as detergents and other cleaning agents. However, when raw greywater is stored for an extended period of time in a tank or equalization basin, the concentration of dissolved oxygen decreases within hours, and anaerobic decomposition processes begin to take place. These include the reduction of sulfate into sulfide-containing compounds, a process characterized by the release of bad smells (reminiscent of rotten eggs) (Eriksson et al., 2002). The formation of odors is one reason that some treatment systems do not store the greywater, but instead, or convey it almost immediately. Quantitative odor measurements are complex and impractical when it comes to greywater recycling on a local scale.

#### 1.2.1.4 Solids and Turbidity

The main index for describing solids in wastewater in general, and in greywater specifically, is *total suspended solids* (TSS). Suspended solids are defined as solids suspended in the water with a diameter larger than 1 μm (which cannot pass through a fiber glass filter). The common measurement method is moving a known volume of liquid through a filter and weighing the solids left on the filter after drying (APHA, 2005). The customary unit of measurement is milligrams per liter (mg/L). These solids are also a major source of turbidity, which is measured using a turbidity meter and expressed in nephelometric turbidity units (NTU). This measurement expresses the scattering and absorption of light as opposed to the amount of light that can pass through the water. Causes of water turbidity are suspended solids and colloid particles such as clay particles, organic materials, algae, and microorganisms. Turbidity causes aesthetic damage, can lower disinfection efficiency, and can affect the reliability of analytic tests. In recent years, turbidity also became a health index due to the *shelter* that colloid particles and TSS provide to heavy metals and the attached bacteria. This attachment is usually strong due to the high surface area of the particles and their load. In this way, pathogens *sheltered* by particles are less influenced by disinfection. In addition, organic colloids may produce toxic by-products in the process of water disinfection (by chlorination).

The sources of TSS in greywater are food residue, soil, dust, hair and skin particles, and fibers from the washing machine (Eriksson et al., 2002). Thus, greywater originating from washing machines and kitchens contains the highest concentration of TSS relative to other sources (Eriksson et al., 2002). Generally, greywater contains lower concentrations of TSS than household wastewater because the concentration of TSS in blackwater is very high (Birks and Hills, 2007). TSS concentrations in greywater change during the day and depend on the source of the greywater (e.g., washing machine, kitchen, or shower) (Almeida et al., 1999).

TSS is often further divided into volatile suspended solids (VSS) and nonvolatile solids (or fixed suspended solids [FSS] or ash). VSS is usually considered an index of suspended organic matter and bacterial biomass. VSS is measured by subtracting the weight of the ash (FSS) that remains after burning in a furnace at a temperature of approximately 550°C from the net TSS weight obtained after drying the sample at a temperature of 105°C (APHA, 2005).

In some cases, a correlation between the turbidity index and TSS in greywater allows a fast and simple turbidity measurement to be used instead of the time-consuming TSS analysis. The ratio between TSS and turbidity in greywater is somewhat lower than the one in urban wastewater (Jefferson et al., 2004). The low ratio is explained by the colloid nature of greywater and the distribution of particle size.

The size of most of the solids in greywater is 10–100 μm. The $d_{50}$ (the median diameter) of shower, bathroom, and handbasin water is 33.2 (±4.9), 29.6 (±2.7), and 27.1 (±0.11) μm, respectively. Another peak that appears in particle size distribution is between 200 and 2000 μ. This peak probably represents an accumulation of soap (i.e., surfactants), skin residue, and hair. An analysis of molar mass profiles of greywater showed that greywater contains humic and fulvic substances (Jefferson et al., 2004). The combination of colloid particles and surfactants can cause the stabilization of solids since surfactants attach to solids and prevent their cohesion. This phenomenon can interfere with processes of solid separation such as sedimentation (Eriksson et al., 2002). In addition, the materials that cause turbidity may interfere with the disinfection process because they can veil bacteria (Narkis et al., 1995; Ho et al., 1998).

### 1.2.2 Chemical Characteristics

#### 1.2.2.1 Salinity and Electrical Conductivity

The salinity of a solution is defined as the sum of ion concentrations in water, usually expressed in units of mg/L. Salinity can be quantified directly by measuring the concentration of the ions present in water and summing their concentrations. There are several methods of measurement, the most common of which are inductively coupled plasma (ICP), atomic absorption (AA), ion chromatograph (IC), and standard protocols (APHA, 2005). Another technique is the gravimetric method in which the weight of the total dissolved solids (TDS) is calculated (APHA, 2005). Here, a known quantity of filtrate is passed through a filter in order to separate out the TSS. The weight of the salt remaining in the vessel after drying is then measured.

The most common method for assessing salinity in wastewater is by finding the electrical conductivity (EC) of the solution. The EC is measured by a conductivity electrode that measures the passing of current through the water, in units of mS/cm. Since EC is determined as an index of salinity, it is defined as a chemical variable. This method is common, quick, reliable, and simple. In general, as salt concentrations increase, the EC of the water also increases. An empirical relationship was found between EC and the concentrations of salts that characterize wastewater, and it can be used to convert the results of EC into salt concentrations in water with high accuracy (Equation 1.1; following Lewis, 1980):

$$TDS = 0.012 - 0.2174 \cdot \left(\frac{EC}{53.087}\right)^{0.5} + 25.3283 \cdot \left(\frac{EC}{53.087}\right)^{1} + 13.7714 \cdot \left(\frac{EC}{53.087}\right)^{1.5} - 6.4788 \cdot \left(\frac{EC}{53.087}\right)^{2} + 2.5842 \cdot \left(\frac{EC}{53.087}\right)^{2.5} \quad (1.1)$$

where

*EC* is the electrical conductivity (mS/cm)

*TDS* is the total dissolved solids, which is a measure of salinity (g/L)

Another simpler rule of thumb is often used to quickly estimate the salinity vis-à-vis EC data; it can be assumed that each 640 mg/L of chlorides are equivalent to 1 mS/cm (although in reality different ions conduct currents differently). As such, EC can be translated into salinity and vice versa. This correlation is usually quite reliable since the highest ion concentration in greywater is often that of sodium chloride (NaCl) ions, and the salinity of greywater falls in the linear part of the correlation curve.

Important sources of salts are sodium, nitrogen, and phosphorus-based soaps, found in detergents and washing powders (Morel and Diener, 2006). High salinity may affect vegetation and soil properties causing salinization of groundwater. The recommended values are influenced by many factors such as type of vegetation, soil type, climate, and irrigation regime (ANZECC, 2000). When treated greywater is used for irrigation, attention should be paid to its general salinity as well as to the levels of sodium, chloride, and boron (which is toxic to plants).

Of all the salts, sodium ($Na^+$), which is very common in domestic use, can cause the greatest damage to the soil. Sodium is used in food processing, in water softening (ion exchange), and in detergents as a structure material. As such, it has high potential for higher concentrations in greywater. Sodium ions may damage the texture of the soil (clay soils especially) when their concentration is in excess, as compared to calcium (Ca) and magnesium (Mg) ions. The ratio between sodium concentration versus magnesium and calcium ions is called sodium adsorption ratio (SAR) and is described by Equation 1.2:

$$SAR = \frac{[Na^+]}{\sqrt{\frac{([Ca^{2+}]+[Mg^{2+}])}{2}}} \tag{1.2}$$

where

*SAR* is the sodium adsorption ratio (meq/L)

$[Na^+]$, $[Ca^{2+}]$, and $[Mg^{2+}]$ is the concentration of sodium, calcium, and magnesium, respectively, in (meq/L)

A high SAR may alter the structure of the soil and reduce its hydraulic conductivity, especially in clay soils (Qian and Mecham, 2005). The recommended SAR value for irrigation with effluents is less than 5 (The Israeli Government Publicity Office gazette, 2010).

As mentioned earlier, another ion of interest in the greywater stream is boron (B). Boron is an important trace element and plays many different roles in the plant including in the metabolism of carbohydrates, seed germination, nucleic acid synthesis, and cell wall structure. However, there is a thin line between boron shortage and excess, which causes toxicity to plants. Boron toxicity is manifested in the yellowing of leaf edges, leaf loss, and even death of the plant (Parks and Edwards, 2005). The source of boron in greywater is boric acid (borax) used for bleaching in different cleaners, mainly washing powders and powders and tablets for dishwashers. It should be noted that the concentration of boron in these sources is relatively low in Israel, in comparison with other countries of the world, due to a government enforcement policy that led to its replacement by other bleaches (see discussion in the following paragraph). The maximal allowed boron concentration according to Israeli regulations of effluent water for unlimited irrigation is 0.4 mg/L (The Israeli Government Publicity Office, 2010).

During the 1990s, a survey conducted by Israel's Ministry of Environmental Protection (2000) revealed that laundry detergents contributed approximately 7%, 42%, and 85% of the total addition of chlorides, sodium, and boron (respectively) to municipal sewage in Israel. As treated wastewater effluent is used for irrigation in Israel, and the possible potential damage that these ions can do to crops and soil is high, a new amended standard was put into action by Israel's Ministry of Environmental Protection and the Standards Institute of Israel (IS 438; SII, 1999) in 1999. This standard requires manufacturers of detergents for washing machines to reduce the sodium, chloride, and boron levels to 4 $g\_Na^+/1$ kg_laundry, 40 $g\_CL^-/1$ kg_product, and 0.5 g_B/kg_product, respectively. In 2006, following the increase in the use of dishwashers and a study performed by Friedler and Reznitsky (2004), another standard was set (IS 1417; SII, 2006) reducing the concentration of boron in detergents for dishwashers as well to 0.5 g_B/1 kg of detergent. Dishwashers are also a source of sodium and chloride, which are used for regeneration of ion exchangers, but no standard was set for these compounds. Ion exchangers are built into dishwashers to soften the water to prevent limescale from forming on dishes. As water in Israel is considered hard to very hard, NaCl consumption of dishwashers is relatively high. With the increase in the proportion of desalinated water in the total water supply, a decrease in the water hardness is expected. With it, the required amount of salt needed to refresh ion exchangers is expected to drop.

#### 1.2.2.2 Metals

The concentration of heavy metals in greywater is usually very low (Jefferson et al., 2004). The potential sources of metals in greywater—particularly of cadmium (Cd), mercury (Hg), nickel (Ni), and lead (Pb)—are piping materials, cutlery, jewelry, coins, household maintenance products, arts-and-crafts materials and products, and even amalgam from dental fillings (AMSA, 2000; Eriksson et al., 2009). It is not easy to quantify the contribution of these sources because it is impossible to know the metal content and its release rate. Even if the sources do not come into direct contact with the greywater, it is possible for metals to adhere to the skin and reach a greywater stream through the shower or handbasin. In addition, abrasion and tearing of various household products can also be a source of metals. Concentrations of 0.012–2.5 μg/L cadmium, 1.3–28 μg/L nickel, 0.61–10 μg/L lead, and 0.022–36 μg/L mercury have been discovered in greywater (Eriksson and Donner, 2009). In recent years, the content of cadmium, mercury, and lead is restricted in various products in Europe, and so it is expected that their greywater concentrations will be reduced over the following year (Eriksson and Donner, 2009). Current official data were not found.

#### 1.2.2.3 pH and Alkalinity

To express the concentration of hydrogen ions (protons) in water, pH is used. It can also be defined as the activity of hydrogen ions in water (Stumm and Morgan, 1996). For convenience, pH is expressed as follows: $pH = -\log [H^+]$.

Low pH values can cause heavy metals to dissolve in water to a level that is poisonous to plants and can result in acidic soils (Boyd, 1995). Prolonged use of acidic water can also corrode piping (ANZECC, 2000). The appropriate pH for unlimited irrigation ranges from 6.5 to 8.5 (Halperin and Aloni, 2003). The pH of most greywater sources is low, ranging from 7 to 8 (see Table 1.1), and the pH of laundry greywater is even more basic ranging from 7.5 to 10 (Table 1.4). This is because laundry powders and liquids are made up of basic materials containing hydroxide $OH^-$ ions, which raise pH.

Alkalinity is basically the sum of alkali ions in solution or in other words a measure of ions in water that are capable of receiving a proton ($H^+$). The values are expressed in milliequivalents/L or mg/L of $CaCO_3$. Alkalinity expresses the buffer capacity of water, or water's ability to resist a change in pH when base or acid materials are introduced. Water with a low buffer capacity (low alkalinity) undergoes fluctuations in pH, while water with a high buffer capacity has steady pH. In biological treatment systems, it is important to maintain a strong buffer capacity. Irrigation with water of low alkalinity may be detrimental to the plants and microorganisms living in the soil and to the health of the soil itself. The main contributor to alkalinity is the water's source. For instance, laundry and dishwashing powders and liquids contain ions that belong to the carbonic and phosphoric systems, which thus contribute to the alkalinity of greywater.

In greywater treatment systems, natural chemical reactions can alter the water's alkalinity. For example, the nitrification process consumes alkalinity and the denitrification process contributes to alkalinity; however, it is not expected that these processes will occur in raw greywater. It is possible to measure alkalinity using Gran titration (Stumm and Morgan, 1996) or with titration to pH 4.5 (APHA, 2005).

#### 1.2.2.4 Organic Matter

Organic matter is the product of an organism, a product of an organism's activity, or the remains of an organism. An organic compound is one that contains hydrogen and carbon atoms that are connected to each other. It is called organic because, when the concept was coined, it was thought that the source of these compounds was limited to living creatures. As it turned out, this was not the case, but the name *organic* has survived despite the fact that synthetic organic compounds have been produced for several decades with sources that are not organisms. Further complication surrounding the term *organic* comes from the use of the term *organic matter* to include organic compounds.

Chemically, organic matter in greywater can be expressed as $C_nH_aO_bN_c$, where $n$ is the number of carbon moles, $a$ is the number of hydrogen moles, $b$ is the number of oxygen moles, and $c$ is the number of nitrogen moles. The concentration of organic matter in greywater is an important factor in assessing its quality. When greywater is released into the ground or a source of water, the properties of the water as well as the ecological fabric of the receiving environment can be affected. The result can be environmental and health risks. For example, organic matter that reaches irrigation water may cause dispersion of clay, thus affecting the hydraulic conductivity of the ground (Rozin, 1997). Furthermore, excess organic material can also modify the water-holding capacity of soil, as well as the ability of particles to move through it. In addition, microbial decomposition processes can reduce the concentration of dissolved oxygen in the water, so high concentrations of organic matter encourage anaerobic decomposition processes, which release toxic gases, such as $H_2S$ and/or methane, into the environment (Eriksson et al., 2002).

Finally, the efficiency of some treatment processes changes as a result of the existence of organic matter in the water. For example, when pathogenic bacteria and viruses are attached to organic particles in the water, their survival rate may increase. These particles serve as a substrate in the water, thus reducing the efficiency of its disinfection (Narkis et al., 1995; Ho et al., 1998). To disinfect water rich in organic matter, more chlorine is needed, and carcinogenic by-products may be formed. Organic matter can be decomposed by physical means (heat), chemical means (a strong oxidizer such as persulfate), or microbial means (e.g., decomposition by bacteria) into carbon dioxide. The primary methods used to estimate the concentration of organic matter measure the concentration of oxygen consumed (i.e., biochemical oxygen demand [$BOD_5$], chemical oxygen demand [COD]) or the carbon emitted as carbon dioxide, produced as the result of full mineralization of organic matter (total organic carbon [TOC]). Alternatively, another means to estimate the concentration of organic matter is to weigh the material lost after burning in high temperature.

#### *1.2.2.4.1 Biochemical Oxygen Demand and Chemical Oxygen Demand*

In the decomposition process of organic matter under aerobic conditions, oxygen acts as the electron acceptor and transforms to water. The organic matter acts as the electrons donor; it loses electrons and is oxidized into carbon dioxide. As such, there is usually a connection between the consumption of oxygen and the concentration of organic matter. COD is defined as the amount of oxygen that would have been consumed for full oxygenation of the organic matter in water. In the test for COD, oxidation is achieved using a strong chemical oxidizer (e.g., dichromate) under acidic conditions and at high temperature. The quantity of oxidizer consumed is measured and then adjusted stoichiometrically to represent the amount of oxidizer that would have been consumed if it had been oxygen and not dichromate (APHA, 2005). Since oxidation is performed in this test under extreme conditions, barely (or non) biodegradable organic matter is also decomposed, and nonorganic reduced species in the water are also oxidized. As compared to COD, $BOD_5$ describes the oxygen consumption resulting from microorganisms decomposing organic matter in the sample being analyzed. $BOD_5$ serves as an indicator of the concentration of biodegradable organic matter in water (greywater in this case), or the ability of organic matter to be biologically decomposed by microorganisms under controlled conditions (usually at 20°C) within a certain time period (usually 5 days, hence the label $BOD_5$) (APHA, 2005).

In greywater generally, and even in light greywater, concentrations of COD and $BOD_5$ vary, sometimes reaching hundreds of mg/L (Tables 1.1 and 1.2, pp. 10–11). The main sources of COD and $BOD_5$ in greywater are surfactants in laundry and washing powders or liquids for laundry, washing, and dishwashing (Eriksson et al., 2002). Additional sources are skin cells, fat, and leftover food. While many western countries have banned the use of surfactants with branched carbon chains that exhibit very low biodegradability (such as alkylbenzenesulfonates [ABS]) and replaced them with easily biodegraded surfactants (such as linear alkylbenzenesulfonates [LAS]), these materials are still used in many countries (Morel and Diener, 2006).

One way to assess the extent of biodegradability of organic matter in water is by computing the $COD/BOD_5$ ratio. The higher this ratio, the smaller the proportion of biodegradable organic matter in relation to the total organic material (biodegradable + nonbiodegradable) and vice versa. Some researchers maintain that the organic matter in greywater is more available for microbial decomposition than that in ordinary household wastewater since the latter contains only gradually degradable ingredients such as feces and toilet paper (Lindstrom, 2000). In contrast, others argue that greywater is less biodegradable due to shortage of easily biodegraded organic matter. For example, average $COD/BOD_5$ ratios were reported at about 2.9 (±1.3) in greywater from the bathroom, 2.8 (±1.0) from the shower, and 3.6 (±1.6) from the handbasin (Jefferson et al., 2004). These ratios are higher than the characteristic ratio of $COD/BOD_5$ in ordinary household wastewater, which ranges from 1.6 to 2.8 (Al-Jayyousi, 2003; Jefferson et al., 2004; Gethke et al., 2007).

$BOD_5$ plays a central and important role in the analysis of wastewater; however, several problems arise in using this index (Metcalf and Eddy, 2003). Complications include the long duration of analysis (5 days), potential interruptions to the process such as microorganism deficiency or the presence of toxic/inhibitory substances, and the limited interpretability of the results. It is acceptable to assume that all of the oxygen consumed in domestic wastewater over 5 days is the result of carbonaceous oxygen demand (CBOD). However, there are a number of oxygen-demanding processes that occur in water that are not microbial respiration. The main one in this context is nitrification, where ammonia is oxygenized by nitrifying bacteria. This oxygen demand is called nitrogenous oxygen demand (NBOD).

The assumption that oxygen consumption is carbonaceous is based on the fact that a high concentration of organic matter in wastewater requires significant dilution (using the standard analysis method) of the concentration of bacteria in general and of nitrifying bacteria in particular. Since one may assume that most nitrifying bacteria are autotrophic (i.e., utilizing an inorganic carbon source), and their growth rate in general is significantly lower than the reproduction rate of heterotrophic bacteria (utilizing organic carbon as their energy and carbon source), the nitrification rate during the 5 days of the test is significantly lower than the CBOD. In greywater, the need for dilution is smaller, especially after treatment, and NBOD is significant. It may reach 18% or more of the total oxygen consumption on the fifth day of the test (Bondrenko et al., 2006). This observation is significant for determining the $BOD_5$ value that should be allowed in greywater reuse. For example, it is possible that in setting the standard for greywater use, the threshold value of $BOD_5$ can be raised or that CBOD testing (i.e., to use nitrification inhibitor) can be required rather than the generic $BOD_5$.

There are several techniques for measuring $BOD_5$, the standard being to use $BOD_5$ bottles (APHA, 2005). This method is to (1) aerate the sample to reach saturation of dissolved oxygen, (2) measure its concentration in the sample at the beginning of the test, (3) incubate for 5 days at 20°C in a full and closed vessel, and then (4) measure the concentration of dissolved oxygen again at the end of the experiment. The difference between the two measurements reflects the oxygen consumption during the 5 days of the experiment. Recently, the use of manometric heads has been increasing and some laboratories are using respirometers. It should be noted that there is not always a relationship between the results of tests using various methods, making it difficult to compare samples that undergo analysis by different methods. In particular, it is important to note that the manometric methods are not sensitive to $BOD_5$ concentrations (lower than 10–20 mg/L) or to especially high concentrations (more than 300–400 mg/L) (Bondrenko et al., 2006).

#### *1.2.2.4.2 Total Organic Carbon*

In the TOC analysis, organic matter is fully oxidized to $CO_2$ and then passed through an infrared detector (in most cases) that identifies quantitatively the concentration of $CO_2$ emitted. To obtain the organic carbon component through this method, the various forms of the carbonaceous system containing inorganic carbon (i.e., $CO_2$, $HCO_3^-$, and $CO_3^{2-}$) have to be removed as a preliminary step, which is done automatically by most commercial testing equipment. The disadvantages of TOC measurement are the high cost of the device and the fact that the method expresses also nonbiodegradable organic matter (Droste, 1997).

Another method of comparing various concentrations of organic matter is the spectrometric method, based on the absorption of light at a wavelength of 254 nm (in the UV range). It was found that for many types of water, including greywater, there is a direct relation between the absorption of light and the concentration of organic matter. Aromatic compounds have more double bonds than aliphatic ones, and double bonds absorb photons in the UV range. Therefore, it is possible to test light absorption by greywater to evaluate the content of organic matter. However, it should be noted that this index does not apply to the total organic matter but mainly to aromatic organic substances, and the test has to be calibrated according to the type of greywater. This method is primarily used for research and is not defined as a standard method of testing water. Its advantage lies in its simplicity, while its disadvantage is the dependency on the type of the greywater tested, as well as on the presence of inorganic substances that absorb within the UV range (which will artificially elevate the calculated organic matter concentration).

#### 1.2.2.5 Nutrients: Nitrogen and Phosphorus

Nutrients are elements that make up the living cell. In addition to carbon (C), hydrogen (H), and oxygen (O), living cells need other ions in order to grow such as nitrogen (N), phosphorus (P), potassium (K), calcium (Ca), chloride (Cl), sulfur (S), magnesium (Mg), and sodium (Na). Normal cell growth is not possible if these elements are lacking. Despite this reliance on diverse elements, nitrogen and phosphorus are most commonly being referred to by the term *nutrients*. This is because they are required in a much higher amount than the others and therefore may limit the growth of plants, algae, and cells in general. When greywater is used for irrigation, the presence of a certain level of nitrogen and phosphorus is advantageous, since they can serve as alternative to fertilizers.

Conversely, when present in surplus amounts, nitrogen and phosphorus can pollute water bodies and soils. Excess nutrients may cause eutrophication. In other words, they can change the trophic level of water bodies and soils, from oligotrophic (containing low levels of nutrients) to mesotrophic and finally to eutrophic (having excess nutrients) (Droste, 1997). In a common situation, a state of excess nutrients in water bodies causes the algal blooms, disrupts the balance between the various species of algae, increases turbidity, decreases dissolved oxygen concentration, and subsequently injures aquatic organisms. In addition, excess nitrogen and phosphorus can cause biological clogging of water conveyance pipelines and overstimulation of plant growth (ANZECC, 2000).

##### *1.2.2.5.1 Nitrogen*

Greywater contains about a tenth of the nitrogen concentration compared to that of household sewage (Birks and Hills, 2007). It is likely that the bulk of nitrogen in greywater (excluding kitchen wastewater) originates from urine, because its concentration in urine is so high that even a small quantity is sufficient to contribute measurable amounts to the total nitrogen count in greywater. The presence of the steroid estrogen (17b-estradiol) lends evidence to this position. It is released in urine, in samples of greywater (Eriksson et al., 2009), and by the presence of coprostanol—a product of cholesterol decomposition and evidence of fecal contamination

(Ottoson and Stenstrom, 2003). Another source of nitrogen is found in washing (in general) and in washing the human body (in particular), which releases nitrogen-containing skin particles. The most significant amount of nitrogen in greywater is from kitchen wastewater, originating from food leftovers, such as meat, which contains protein. Protein contains approximately 16.25% nitrogen (Eriksson et al., 2002). It should be remembered that in many countries, recycling kitchen wastewater is prohibited, and therefore this nitrogen source is not mentioned. Contrary to domestic wastewater, in greywater, there is generally no problem of excess nitrogen, and even the opposite problem may occur: low nitrogen levels may limit microbial processes and delay the decomposition of organic matter in the biological treatment systems common for treating greywater.

Greywater from showers contains relatively low levels of nitrogen and phosphorus (Table 1.7) (Eriksson et al., 2002, 2009; Jefferson et al., 2004; Ramon et al., 2004). Other researchers indicate that when treating light greywater (from showers and handbasins), no lack of nitrogen was observed (Friedler et al., 2006). There are several methods to test nitrogen species and its total concentration in water and wastewater (APHA, 2005). It is customary to express nitrogen concentrations in different forms (such as $NH_3$, $NO_3$) in mg N/L nitrogen (N). The regulations concerning the effluent reuse in Israel (Inbar, 2007) allow a maximum concentration of 10 mg/L ammoniacal nitrogen and 25 mg/L of total nitrogen for unrestricted use.

The nitrogen cycle is very complex, and its description is beyond the scope of this book (Figure 1.2). As a general rule, most of the organic nitrogen breaks down to ammonia ($NH_3/NH_4^+$) within a very short time. Depending on the treatment or its fate in the soil, assuming that the greywater is used for irrigation, ammonia is oxidized in the presence of oxygen by ammonia-oxidizing bacteria. It is turned into nitrite ($NO_2^-$) and with the help of nitrite-oxidizing bacteria into nitrate ($NO_3^-$). Under anoxic conditions, nitrate is recirculated by bacteria into gaseous nitrogen ($N_2$) in the denitrification process.

#### *1.2.2.5.2 Phosphorus*

The little phosphorus found in greywater mainly originates from washing powders and liquids (Eriksson et al., 2002; Jefferson et al., 2004) (Tables 1.5 and 1.6), and some of it comes from urine (see discussion on nitrogen earlier). In recent decades in Europe, manufacturers were forbidden to add phosphorus to detergents; however, in countries where no such prohibition exists on the use of detergents containing phosphorus, higher concentrations can be found. Since the popularity of dishwashers is increasing, it is expected that the amounts of phosphorus in wastewater will increase in the future (Friedler and Reznitsky, 2004; Meinzinger and Oldenburg, 2009). This increase is expected to be expressed in places where the water stream generated from the kitchen is included in greywater.

Unlike the nitrogen cycle, which is affected primarily by biological processes, the phosphorus cycle is chiefly influenced by chemical processes and therefore is easier to predict. A simple schematic description of the phosphorus cycle is shown in Figure 1.3. Phosphorus species appear in water as orthophosphate in the phosphorous 3-proton system (i.e., $H_3PO_4$, $H_2PO_4^-$, $HPO_4^{-2}$, $PO_4^{-3}$). Phosphorus is also found in organic matter, which is released during decomposition, and the dissolved

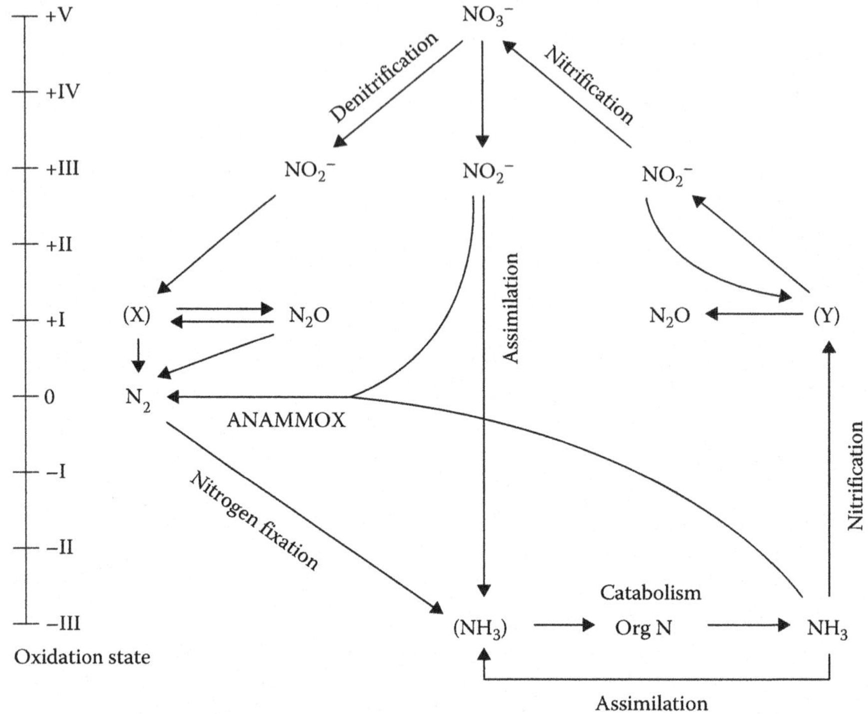


**FIGURE 1.2** The nitrogen cycle.

phosphorus is discharged into the water. The final elimination of phosphorus is accomplished by its fixation in the solid phase (using biological processes and/or chemical deposition) and disposal of the solid phase.

There are several methods for measuring phosphorus species and total phosphorus (APHA, 2005), and it is common to express the concentrations of phosphorus in mg/L phosphorus (P). According to Israeli regulations for effluent quality, the concentration of total phosphorus should not exceed 5 mg/L in effluent that is used for unrestricted irrigation (The Israeli Government Publicity Office gazette, 2010).

#### 1.2.2.6 Ratio between Nitrogen, Phosphorus, and Organic Matter

One of the factors that affect the biodegradation ability of organic matter in wastewater is the balance between nutrients and organic substances. The C/N/P ratio that is considered optimal for biological treatment is 100:20:5. As mentioned earlier, the quality of greywater varies widely between different sources and even within the same source at different times, and therefore its biodegradation is not obvious. For example, in measurements of the C/N/P ratio in bath, shower, and handbasin greywater, the ratios were 100:6:0.2, 100:8:0.1, and 100:5:0.2, respectively (Jefferson et al., 2004). These ratios indicate that the greywater tested was nitrogen and phosphorus poor and not optimal for biological treatment.

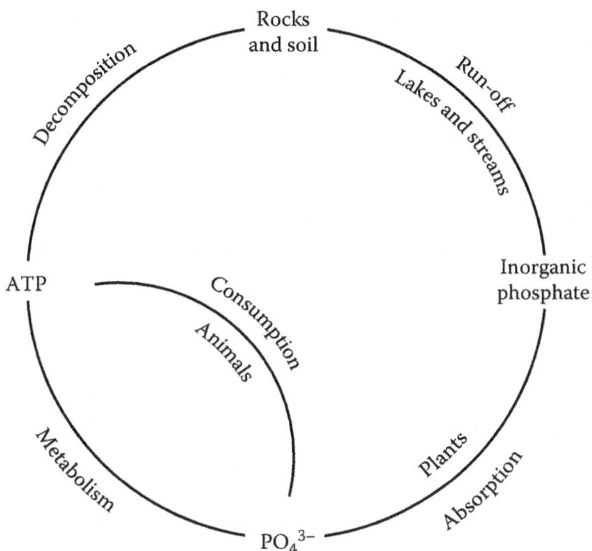


**FIGURE 1.3** The phosphorus cycle.

However, when working with light greywater, higher ratios of C/N/P (100:11:2) were found, and in this case, the researchers concluded that there was no shortage of nutrients in their greywater (Friedler et al., 2006).

#### 1.2.2.7 Xenobiotic Organic Compounds

A xenobiotic organic compound (XOC) is material foreign to the biological system, specifically artificial substances that did not exist in nature before man created them. In greywater, XOC compounds originate primarily in chemicals used in the home. Some of these compounds are often of low degradability and can pose an environmental hazard when they reach soil or water bodies. Studies have found approximately 900 xenobiotic organic substances in greywater, originating from substances used in showers and laundry. A study of greywater from showers in Denmark found 200 xenobiotic substances, including surfactants, foaming agents, flavor enhancers, preservatives, antioxidants, and softeners (Eriksson et al., 2002). The authors of this study recommended that only the most common compounds that generate environmental risk be monitored, due to the numerous materials found in low concentration. It was also found that the consumer habits influencing the release of these substances differed between countries and regions, but the order of magnitude is the same. For example, the Danish consumer uses about 2.3 kg/year of shampoo and conditioner compared to a consumption of 0.9–1.1 kg by the Swedish consumer (Eriksson et al., 2002).

These researchers divided xenobiotic materials into 14 groups according to the role it played in a product: amphoteric surfactants, anionic surfactants, cationic surfactants, nonionic surfactants, bleachers, dyes, emulsifiers, enzymes,

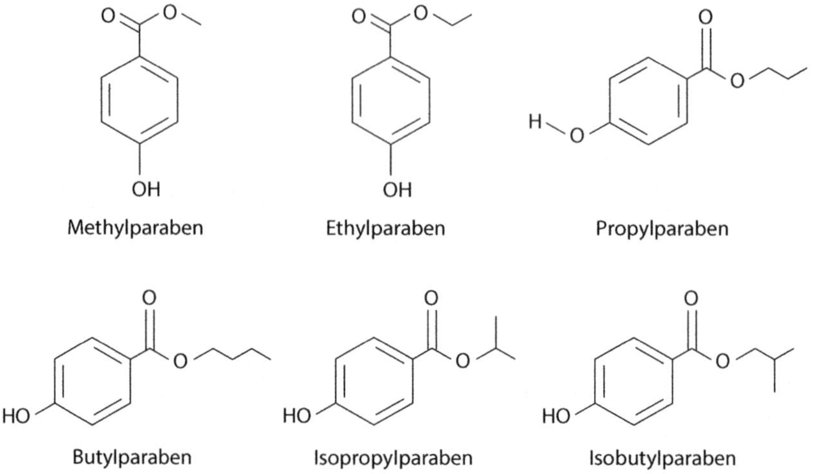

Methylparaben Ethylparaben Propylparaben

Butylparaben Isopropylparaben Isobutylparaben

**FIGURE 1.4** The molecular structure of several members of the paraben group.

flavor enhancers, preservatives, softeners, solvents, UV radiation filters (UV protection providers), and other substances with varying properties. Nine hundred compounds were rated through a risk assessment on a scale from 1 to 8 according to the extent of their damage to the environment. Sufficient information was found concerning degradability, toxicity, and accumulation in the environment for only 211 compounds. Sixty-six substances were ranked in the top three levels (as having the greatest impact). Of these, 34 were surfactants (amphoteric, anionic, cationic, and nonionic), 6 were preservatives, and 7 were softeners. It can be assumed that when information for the other 700 compounds is collected, the number of substances that have a significant impact on the environment will increase (Eriksson et al., 2002).

Among the preservatives, the presence of parabens in greywater was extensively studied. Parabens are used in personal hygiene products, shampoos, toothpastes, and deodorants and as preservatives in the cosmetic industry (Eriksson et al., 2009). Parabens differ from each other primarily in the length of their carbonaceous chain. For example, parabens with short chains, such as methyl-, ethyl-, and propylparabens, are usually found in water-based materials such as shampoo and liquid soap. Long-chain parabens (butyl- and isobutyl-parabens) are found in oil-based materials, such as various creams (Figure 1.4). Generally, parabens undergo hydrolysis in a basic environment but are not affected in an acid environment and are biodegradable in the range of 5–6 h residence time in a fixed-bed reactor (rotating biological contactor [RBC]). Xenobiotic compounds and parabens can be measured only by using advanced analytical methods such as HPLC.

#### *1.2.2.7.1 Surfactants*

The main source of surfactants in domestic wastewater is detergents. The surfactant molecule is organic (typically synthetic) and is composed of two parts: one hydrophobic and one hydrophilic. The hydrophobic group, known also as the

| | | |
|---|---|---|
| Ionic | Sulfonate | $-SO_3^-$ |
| | Sulfite | $-OSO_3^{2-}$ |
| | Carboxylate | $-CO_2^-$ |
| Nonionic | Quaternary ammonium | $-R_3N^+$ |
| | Polyoxyethylene | $-O-CH_2-CH_2-O-CH_2-CH_2-....-O-CH_2-CH-OH$ |
| | Sucrose | $-O-C_6H_7O(OH)_3-O-C_6H_7O(OH)_4$ |
| | Polypeptide | $-NH-CHR-CO-NH-CHR'-CO-....-NH-CHR''-CO_2H$ |

**FIGURE 1.5** The molecular structure of surfactants.

*hydrophobic tail*, typically contains a hydrocarbon skeleton (R) that contains between 10 and 20 carbon atoms; its structure can be defined within the entire range from aliphatic to aromatic. The hydrophilic part contains a chemical group that is electrically charged or polarized (i.e., capable of creating hydrogen bonds) (Dental et al., 1994). Hydrophilic groups that are common in surfactants are shown in Figure 1.5.

In aquatic solutions, surfactants tend to accumulate in the interphase of air/solution or solid/solution, thus reducing the solution's surface tension. This property is advantageous when using these substances in cleaning processes. Another important property of surfactants is the self-grouping of surfactant molecules in aquatic solutions to create micelles, or small clusters (of several dozen nanometers in size) of diverse geometry. Micelles can include ellipsoids, elongated cylinders, or spheres, and they are built so that the hydrophobic groups are in the inner side of the structure and the hydrophilic groups face the solvent (the water). This structure also lowers the system's free energy. This phenomenon called the critical micelle concentration (CMC) occurs only above a certain concentration (Figure 1.6). Below this concentration, the molecules exist in the solution as dissolved monomers.

The vast number of uses for surfactants is attributable to their physical–chemical properties. The fact that they have hydrophilic and hydrophobic groups allows them to attach to polar and nonpolar substances at the same time, imparting them with the following features:

- *The wettability effect*: Reducing the surface tension of water causes it to disperse over the entire area equally, thus improving the cleaning process (Figure 1.7).
- *The emulsion effect*: Micelles created in the aquatic solution *capture* non-soluble organic matter in water (such as oil) in the inner part of the cluster, thus making it dissolvable in a nonorganic solvent (such as water). This effect is important in processes of textile cleaning and rinsing.
- *The dispersion effect*: Due to reduced surface tension, water penetrates into clusters of soil and dirt. This causes their disaggregation and the suspension of particles in the solution, which facilitates dirt removal. The ability to create foam and control its amount is another property that improves cleaning, and it is made possible by the reduction in the water's surface tension.

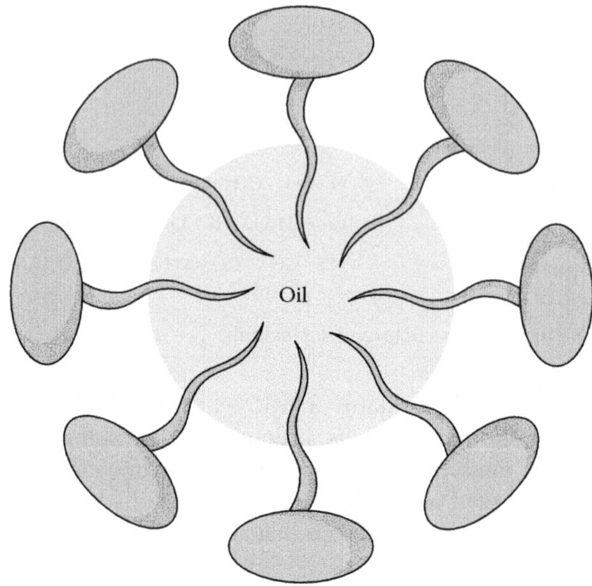


**FIGURE 1.6** A surfactant molecule with a hydrophilic head and a hydrophobic tail and a spherical micelle created when the surfactant concentration exceeds the critical concentration.

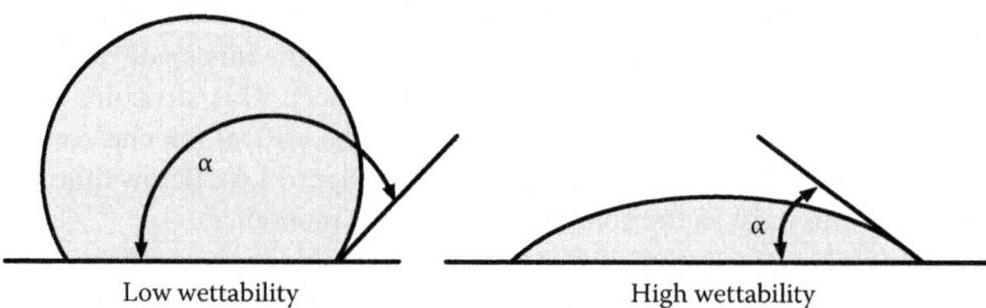


**FIGURE 1.7** The wettability effect; the contact angle of the liquid with the solid serves as a wettability index. When $\alpha < 90°$, the liquid wets the solid; when $\alpha > 90°$, the liquid does not wet the solid.

*Surfactants are classified into groups according to their electrical properties (electric charge):*

*Anionic surfactants* are negatively charged in aquatic solutions due to the presence of one of the following hydrophilic groups: sulfonate, sulfate, carboxylate, or phosphoric acid. They consist of a chain of variable length—depending on the surfactant type—which constitutes the hydrophobic component (Figure 1.8). These surfactants are mostly foam forming at high levels, but are sensitive to hard water (i.e., the presence of calcium and magnesium ions in the aqueous solution). Anionic surfactants

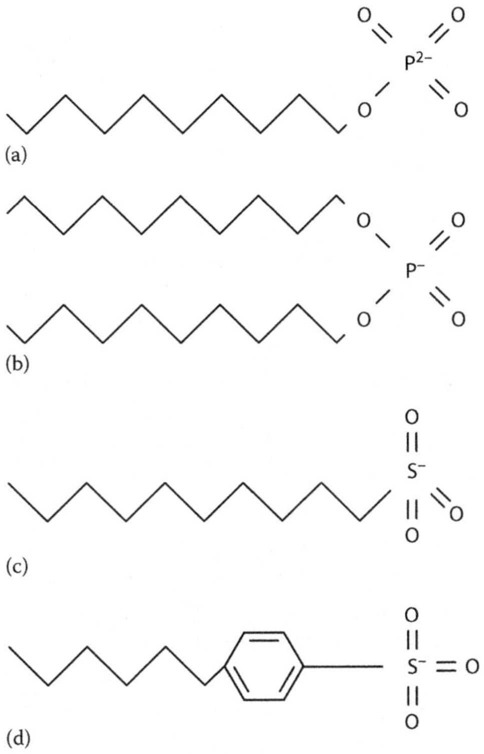

**FIGURE 1.8** Anionic surfactant—a structure containing (a, b) two alkyl phosphates, (c) alkyl sulfonate, and (d) ABS.

are more effective than other surfactants when it comes to removing soil and dirt particles. As they are made up of straight molecules, anionic surfactants biodegrade relatively quickly in the environment.

*Nonionic surfactants* are electrically neutral and contain hydrophilic groups that do not undergo ionization in aquatic solutions (Figure 1.9). To create hydrophilic

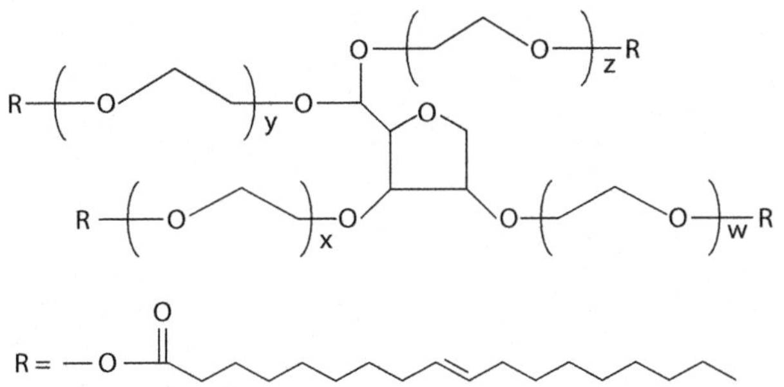

**FIGURE 1.9** Nonionic surfactant—the structure of *Tween 85* (sorbitan trioleate poly(ethylene oxide)).

properties in nonionic molecules, several groups need to be combined. For instance, here are several common polar substances that could together form nonionic molecules: alcoholic hydroxyl (–OH–), single ether (–O–), primary and secondary amides $\left(\underset{\underset{O}{\|}}{-C}-NH_2 \quad \underset{\underset{O}{|}}{-C}-NH- \quad \underset{\underset{O}{\|}}{-C}-N\langle\right)$, and oxide amine $\left(-N^+\rightarrow \overline{\underline{O}}|^{(-)}\right)$. Nonionic surfactants are not usually affected by hard water (because they are electrically neutral), thus reducing the need for additive addition to detergents. The ratio between hydrophilic and hydrophobic groups, ethylene oxide (EO)/hydrophobe, is low, enhancing emulsification ability. In addition, the larger the ratio, the more soluble the molecule is in water and vice versa. Therefore, these substances are more efficient than other surfactants in removing fatty dirt from synthetic fabrics. Most of the nonionic surfactants are considered poor foam formers, but dissolve well in cold water. They have a low CMC value and are therefore efficient in low concentrations. The nonionic surfactants are mostly composed of branched chains with an extremely slow biodegradation and could constitute an environmental risk if they accumulate.

*Cationic surfactants* are positively charged in aquatic solutions and therefore do not react with hard water ions. The central atom in these molecules is usually nitrogen (Figure 1.10). The important cationic surfactants in industry are salts of primary amine ($R{-}NH_3^+$), salts of secondary amine ($R{-}NH_2^+{-}CH_3$), structures of tertiary sulfonium, and structures of quaternary sulfonium. They are mainly used as fabrics softeners, corrosion suppressants, antibacterial substances, and insecticides. Since they do not provide a cleaning effect in neutral pH, they are not used for general cleaning and hence are not common in greywater.

Amphoteric surfactants vary in their electrical charge (positive or negative) according to their pH and possess both acidic and basic properties. Two main groups make up these surfactants, betaines and real amphoterics, which are derivatives of base fatty alkyl imidazolines. The latter group contains two functional key groups in their chemical structure, quaternary nitrogen and a carboxylic group. The betaines

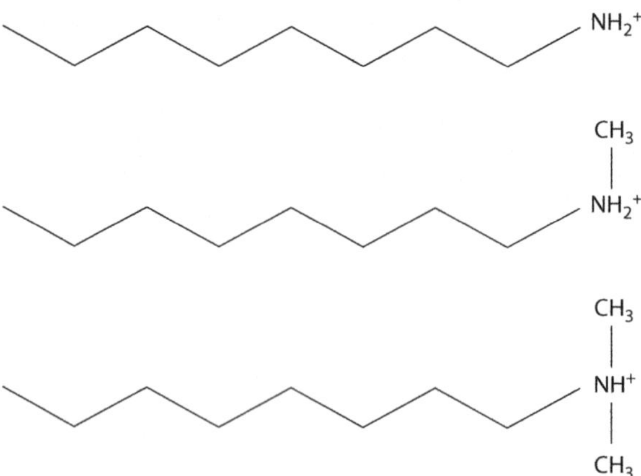


**FIGURE 1.10** Cationic surfactant.

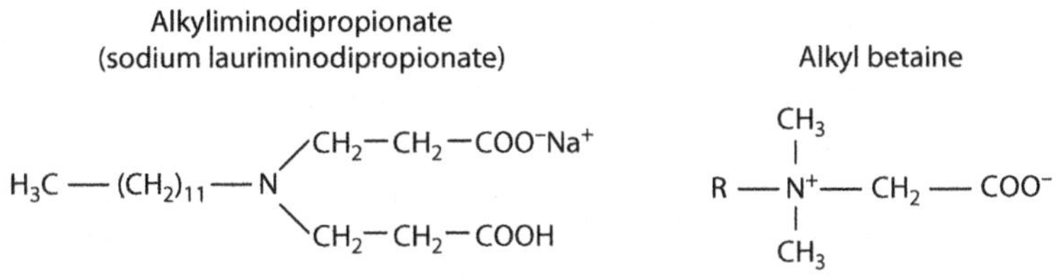

Alkyliminodipropionate
(sodium lauriminodipropionate)

$$H_3C - (CH_2)_{11} - N \begin{matrix} \diagup CH_2 - CH_2 - COO^-Na^+ \\ \diagdown CH_2 - CH_2 - COOH \end{matrix}$$

Alkyl betaine

$$R - \overset{\overset{\large CH_3}{|}}{\underset{\underset{\large CH_3}{|}}{N^+}} - CH_2 - COO^-$$

**FIGURE 1.11** Amphoteric surfactant—derivatives of imidazolines and betaines.

are characterized by quaternary nitrogen atoms and do not display anionic properties in basic solutions (Figure 1.11). The amphoteric surfactants cause less skin irritations as compared to the other surfactants and are used extensively in personal care products like shampoos and conditioners, body lotions, and liquid soaps. They are also used in general household detergents and in industrial detergents.

Today, anionic surfactants occupy the greatest volume in household and industrial use of surfactants (50%). The global market of surfactants is estimated to be over 18 million tons per annum, of which about 40% is used for household cleaning. Although the general definition of surfactants includes many compounds, in about 80% of the required surfactants (globally), only about 10 compounds appear. Of these, the most common and researched surfactant is the anionic surfactant LAS (Fernández et al., 2008).

Most of the surfactants decompose in an aerobic environment, and only a minority decomposes in an anaerobic environment. Sometimes, the degradation products are more hazardous to the environment than the parent compound. For example, octylphenol (OP) and nonylphenol (NP) were found to have an endocrine impact (Scott and Jones, 2000) and are more toxic to marine organisms than their parent compound, a nonionic surfactant called alkylphenol ethoxylate (APE) (Ying, 2006). Among the most degradable surfactants in both aerobic and anaerobic environments are the fatty alcohol sulfate (FAS) surfactants and alkyl polyglycosides (APGs) (Cirelli et al., 2008).

Surfactants originating in household detergents create the highest concentration of organic chemicals in household greywater (Abu-Zreig et al., 2003). Although regulations in many countries allow for only biodegradable ingredients in household detergents, it is possible that strong adsorption of the surfactants to the soil hinders their decomposition (Wiel-Shafran et al., 2006). In soils that were irrigated with untreated greywater, surfactant concentrations of up to 60 mg/kg were found (Gross et al., 2005; Wiel-Shafran et al., 2006).

Most of the research on the effects of surfactants on the environment focuses on their toxicity to aquatic animals and plants, while only scant data exist regarding the effect of surfactants on the hydrophysical properties of soil. However, evidence is accumulating that indicates that surfactants can cause increased environmental pollution, specifically hydrophobicity in soil (an adverse change in soil structure).

There are several methods for evaluating the concentration of surfactants in solution, including chemical colorimetric methods (APHA, 2005), as well as HPLC- and LC-based advanced methods.

### 1.2.3 Microbial Characteristics

The microbial quality of greywater depends on many factors such as water source, temperature, and personal hygiene habits. Bacteria, viruses, worms, and pathogenic protozoa from four main sources can infiltrate greywater: inside users' bodies (mainly fecal pathogens), external body parts (e.g., skin, nose, mouth, ears), food preparation (Coagan et al., 1999), and dirty laundry. It has thus been shown that greywater may contain high levels of bacteria (Birks and Hills, 2007). For example, sometimes, intestinal bacteria, such as salmonella and campylobacter, may penetrate greywater as a result of processing food in the kitchen (Ottoson and Stenstrom, 2003). Nevertheless, it should be stressed that the levels of pathogens in greywater are usually much lower than in blackwater. Table 1.3 displays a list of relatively common pathogens associated with water and secretions.

Traditionally, there is a tendency for water quality research to focus mainly on fecal contamination. As such, most water quality tests refer to fecal contamination indicators, primarily *Escherichia coli*. Although fecal contamination does occur in greywater, the extent is usually lower than in full household wastewater. In greywater, concentrations of pathogens from other sources may actually be higher and thus of greater concern. Even so, despite the presence of some pathogenic bacteria in greywater, many countries allow the use of greywater without treatment for

**TABLE 1.3**
**Concentration Ranges of Bacteria and Viruses Found in Greywater**

| | $Greywater_{Light}$ | $Greywater_{Mixed}$ |
|---|---|---|
| Total coliforms | 1.7–7.4 | 7.2–8.8 |
| Fecal coliforms | 1.0–6.9 | 3.0–8.0 |
| Enterococci/fecal streptococci | 1.0–3.4 | 2.4–4.6 |
| Heterotrophic plate count | 5.6–8.3 | 5.0–7.0 |
| *P. aeruginosa* | 0–3.5 | 2.3–4.3 |
| *S. aureus* sp. | 4.0–5.7 | 4.0–5.7 |
| *Legionella pneumophila* sp. | 0–3.5 | 1.5–2.9 |
| *Clostridium perfringens* sp. | 0.66 | |
| *Salmonella* spp. | ND | |
| *Cryptosporidium* spp. | ND | 0–8.3 |
| *Giardia* spp. | ND | 0–7.9 |
| F-RNA phages | ND | 5.6 |
| Somatic phages | ND | 3 |

*Sources:* Data compiled from Burrows, W.D. et al., *Wat. Sci. Techol.*, 24 (9), 81, 1991; Christova-Boal, D. et al., *Desalination*, 106, 391, 1996; Nolde, E., *Urban Water*, 1, 275, 1999; Casanova, L.M. et al., *J. Am. Water Resources*, 37(5), 1313, 2001; Birks, R. et al., *Water Sci. Technol.*, 50(2), 165, 2004; Birks, R. and Hills, S., *Environ. Monitor. Assess.*, 129(1–3), 61, 2007; Friedler, E. et al., *J. Environ. Manag.*, 81, 360, 2006; Gilboa, Y. and Friedler, E., *Water Res.*, 42(4–5), 1043, 2008; Winward, G.P. et al., *Ecol. Eng.*, 32(2), 187, 2008.

*Note:* ND, not detected.

relatively small flows (up to 1 $m^3$/day, equivalent to greywater flow of a detached house) and they make no reference to microbial contamination.

When using greywater for irrigation or flushing toilets, pathogenic microorganisms can pose risks to human health in different ways. For example, sometimes, they are distributed through aerosols, or an individual makes direct contact with the water. In addition, bacteria and viruses can infiltrate groundwater causing contamination. This is in contrast to protozoa and nematodes, which are relatively large and do not reach groundwater (Eriksson et al., 2002). Such risk increases in the presence of organisms that are resistant to a variety of disinfection treatments and thus can spread in greywater systems, such as *Cryptosporidium* and *Giardia*. The microbial risks involved in the use of greywater are elaborated in Chapter 4 (see Section 4.4).

#### 1.2.3.1 Indicator Bacteria

Greywater contains available and degradable organic matter, which may encourage the growth of intestinal bacteria. These bacteria are sometimes used as a fecal indicator because it has been established that the presence of human and animal intestinal bacteria often indicates the presence of pathogenic bacteria. Thus, tracking indicator bacteria that has grown disproportionately due to available organic matter in greywater could lead to overestimation of the fecal load and subsequent miscalculation of risk to health (Ottoson and Stenstrom, 2003).

Ideal indicator bacteria should meet several criteria (NRMMC and EPHC, 2006):

- Be present in a relatively high concentration as compared to pathogens
- Survive in a wide variety of water types
- Have survival rates similar to, or higher than, that of pathogens
- Not multiply in water
- Originate in the human body and not be common in other environments
- Originate in the digestive system

It is customary to use the bacterium *E. coli* and *fecal enterococci* as indicators of fecal contamination and the presence of pathogenic bacteria (Eriksson et al., 2002; Birks and Hills, 2007). That said, the absence of these bacteria does not necessarily mean that the greywater is clean of pathogenic bacteria, nor does their presence guarantee the presence of pathogens. In fact, it has been found that the use of *E. coli* as a proxy for pathogenic contamination and risk sometimes yields an overassessment of risk. It was also found that an increase of indicator bacteria sometimes occurs with the treatment systems unconnected to the presence of pathogens, which also leads to a miscalculation of risk. As a result, it has been suggested that biomarkers such as coprostanol, which is the degradation product of cholesterol generated by intestinal microflora and decomposes cholesterol, be used as a replacement for indicator bacteria (Ottoson and Stenstrom, 2003). This indicator also does not guarantee accuracy for various reasons (e.g., young children do not secrete coprostanol because the microflora that produces it does not yet exist in their intestine). Another method proposed by these authors to assess fecal contamination is to measure detergents' load, bile content, fatty acids, and cholesterol. However, the ability to translate these measurements into fecal load is still limited. In summary, counting *E. coli* bacteria from fecal origin to assess microbial contamination is still the most widely used and acceptable method in the world and in Israel to calculate the presence of pathogenic bacteria.

A considerable amount of greywater originates from rinsing water, which tends to have lower fecal contamination content (Ottoson and Stenstrom, 2003). As such, it was suggested that the presence of pathogens from nonfecal sources such as *Staphylococcus aureus* and *Pseudomonas aeruginosa* be examined. These bacteria are found on the skin or in a mucus environment in humans (Gross et al., 2007; Gilboa and Friedler, 2008).

## 1.3 SOURCES OF GREYWATER

Most domestic water consumption is for the purpose of cleaning or rinsing (such as bathing, washing dishes, and laundry). Each water flow resulting from rinsing that is discharged into the wastewater collection system has different characteristics. The method of consumption of each stream is different, and hence the level of pollution it produces is also different. The following sections analyze individual greywater streams and assess their contribution. The relative volumetric contribution of each individual stream to the greywater, as reported in the literature, is presented in Figure 1.1. It is worth noting that the quantities of greywater contributed by modern washing machines and even more so by modern dishwashers have fallen sharply in the last decade. This implies that in the future as existing appliances are gradually replaced by new ones, their contribution is expected to further decrease. In addition, the introduction of water-efficient fittings, such as faucet aerators, into extensive use may also decrease the volume of the water.

### 1.3.1 Washing Machines

Washing machines are the most significant contributor of pollutants to greywater (when kitchen water is excluded from the definition of greywater) (Table 1.4).

Washing machines and dishwashers have approximately 4–5 cleaning stages, each of which produces greywater of different quality. Most of the contamination is released in the first two cleaning phases, and the level of pollutants then decreases sharply with each stage. The first stage in operating the washing machine contributes only 18% of the volume, but approximately 64%, 40%, 30%, and 34% of the total COD, ammonia, phosphorus, and anionic surfactants, respectively.

### 1.3.2 Baths and Showers

Water consumption for washing differs between people and cultures. A survey found that the average Israeli water consumption for baths was 53 (±27.5) L/use and for showers 28 (±18.8) L/use (Friedler, 2004). In a report from southeast England (Almeida et al., 1999), it was found that water consumption in baths was 61.4 L/use and 42.3 L/use in showers. In Israel, the contribution of showers to domestic greywater stands at approximately 20%. Baths and showers are the major contributors to fecal coliforms in light greywater, with an average of $4 \cdot 10^6$ cfu/100 mL (Friedler, 2004). The characterization of greywater originating from baths and showers is summarized in Tables 1.5 and 1.6.

**TABLE 1.4**
**Characterization of Greywater from Washing Machine**

| Source | Several Homes, Israel (Abu Ghunmi et al., 2008) | Students Dormitory, Jordan (Jefferson et al., 2004) | | Several Homes, Oman (Winward et al., 2008) | Overall Average from Three Sources | | |
|---|---|---|---|---|---|---|---|
| | AVG | AVG | SD | AVG | AVG | SD | Range |
| pH | 7.5 | 9.6 | 0.72 | 8.5 | 8.5 | 1.1 | 7.5–9.6 |
| Alk. | | | | 33 | 33 | | |
| DO | | | | 3.6 | 4 | | |
| EC | 2457 | 4542 | 1960 | 3500 | 3500 | 1043 | 2457–4542 |
| Tur. | | | | 328 | 328 | | |
| TS | 2021 | 3940 | 2885 | 2384 | 2782 | 1019 | 2021–3940 |
| TDS | | | | 2140 | 2140 | | |
| VTS | 765 | | | 649 | 707 | 82 | 649–765 |
| TSS | 188 | 760 | 384 | 244 | 397 | 315 | 188–760 |
| VSS | 106 | | | | 106 | | |
| COD | 1339 | 2500 | 1865 | 471 | 1437 | 1018 | 471–2500 |
| Dissolved COD | 996 | | | | 996 | | |
| BOD | 462 | 1266 | 232 | 296 | 675 | 519 | 296–361 |
| Dissolved BOD | 381 | | | | 381 | | |
| TOC | 361 | | | 170 | 266 | 135 | 170–361 |
| DOC | 281 | | | | 281 | | |
| Total oil | 181 | | | | 181 | | |
| $NH_4^+$–N | 4.9 | | | | 4.9 | | |
| $NO_3^-$ | | | | 16 | 16 | | |
| TN | | 2.8 | 0.45 | | 3 | | |
| $PO_4$ | 169 | | | | 169 | | |
| TP | | 9 | 0.7 | | 9 | | |
| MBAS | 42 | | | 101 | 72 | 42 | 42–101 |
| $Cl^-$ | 450 | 300 | 25 | | 375 | 106 | 300–450 |
| B | 0.4 | | | | 0.4 | | |
| $Na^+$ | 530 | 220 | 30 | 667 | 472 | 229 | 220–667 |
| $Mg^{2+}$ | | 31 | 8 | 61 | 46 | 21 | 31–61 |
| $Ca^{2+}$ | | 50 | 15 | 19 | 34 | 22 | 19–50 |
| $K^+$ | | 10 | 2 | 23 | 17 | 9.5 | 10–23 |
| Zn | | | | 0.14 | 0.14 | | |
| Al | | | | 0.08 | 0.08 | | |
| Pb | | | | 0.08 | 0.08 | | |
| Cu | | | | 0.01 | 0.01 | | |
| Ni | | | | 0.12 | 0.12 | | |
| SAR | | 6 | | | 6.00 | | |
| FC | $4.00 \cdot 10^6$ | | | | 4E+06 | | |

*Notes:* AVG, average; SD, standard deviation; DO, dissolved oxygen; Alk., alkalinity; EC, electrical conductivity; Tur., turbidity; TS, total solids; TDS, total dissolved solids; VTS, volatile solids; TSS, total suspended solids; VSS, volatile suspended solids; COD, chemical oxygen demand; BOD, biochemical oxygen demand; TOC, total organic carbon; DOC, dissolved organic carbon; TN, total nitrogen; TP, total phosphorous; MBAS, methylene blue absorbing substances (anionic surfactants); SAR, sodium adsorption ratio; FC, fecal coliforms. All units are in mg/L, except for pH, EC (µS/cm), turbidity (NTU), SAR, alkalinity (mg $CaCO_3$/L), FC (CFU/100 mL).

**TABLE 1.5**
**Characterization of Greywater from Showers**

| | Several Homes, Israel (Abu Ghunmi et al., 2008) | | Dormitories, Jordan (Jefferson et al., 2004) | | Public Showers, Germany (Winward et al., 2008) | | Homes, United Kingdom (Birks and Hills, 2007) | | Public Showers, Israel (Friedler et al., 2006) | | Homes, Oman (Jamrah et al., 2008) | Overall Average | |
|---|---|---|---|---|---|---|---|---|---|---|---|---|---|
| Source | AVG | SD | AVG | SD | MED | SD | AVG | SD | AVG | SD | AVG | AVG | Range |
| pH | 7.43 | 0.36 | 7.15 | 0.87 | 7.5 | 0.6 | 7.52 | 0.28 | 7.5 | 0.2 | 7.3 | 7.4 | 7.2–7.5 |
| Alk. | | | | | | | | | | | 13 | 13 | |
| DO | | | | | | | | | | | 3.6 | 3.6 | |
| EC | 1565 | 485 | 830 | 190 | 281 | 87 | | | 1241 | 143 | 2000 | 1183 | 281–2000 |
| Tur. | | | | | 40 | 13 | 85 | 71 | 23 | 8.5 | 346 | 123 | 23–346 |
| TS | 1090 | 440 | 750 | 94 | | | | | | | 520 | 787 | 520–1090 |
| TDS | | | | | | | | | 599 | 43 | 279 | 439 | 279–599 |
| VTS | 533 | 281 | | | | | | | | | 346 | 440 | 346–533 |
| TSS | 303 | 205 | 150 | 109 | 22 | 33 | 89 | 113 | 30 | 11 | 242 | 139 | 22–303 |
| VSS | 102 | 84 | | | | | | | | | | 102 | |
| COD | 645 | 289 | 537 | 165 | 95 | 33 | 420 | 245 | 170 | 49 | 375 | 374 | 95–645 |
| Dissolved COD | 319 | 218 | | | | | | | 106 | 42 | | 213 | 106–319 |
| BOD | 424 | 219 | 120 | 40 | 38 | 18 | 146 | 55 | 78 | 26 | 380 | 198 | 38–424 |

(*Continued*)

**TABLE 1.5 (*Continued*)**
**Characterization of Greywater from Showers**

| | Several Homes, Israel (Abu Ghunmi et al., 2008) | | Dormitories, Jordan (Jefferson et al., 2004) | | Public Showers, Germany (Winward et al., 2008) | | Homes, United Kingdom (Birks and Hills, 2007) | | Public Showers, Israel (Friedler et al., 2006) | | Homes, Oman (Jamrah et al., 2008) | Overall Average | |
|---|---|---|---|---|---|---|---|---|---|---|---|---|---|
| **Source** | **AVG** | **SD** | **AVG** | **SD** | **MED** | **SD** | **AVG** | **SD** | **AVG** | **SD** | **AVG** | **AVG** | **Range** |
| Dissolved BOD | 237 | 125 | | | | | | | | | | 237 | |
| COD/BOD | 1.5 | | 4.5 | | 2.5 | 0.5 | | | | | | 2.8 | 1.5–4.5 |
| TOC | 120 | 70 | | | | | 65 | 45 | | | 66 | 84 | 65–120 |
| DOC | 59 | 32 | | | | | | | | | | 59 | |
| Total oil | 164 | 150 | | | | | | | | | | 164 | |
| $NH_4^+$–N | 1.2 | 0.83 | | | | | | | 2.7 | | | 2.0 | 1.2–2.7 |
| $NO_3^-$ | | | | | | | | | 0.67 | | 23.6 | 12 | 0.7–24 |
| TN | | | 2.4 | 1.5 | 20 | 21 | 8.7 | 4.8 | | | | 10 | 2.4–20 |
| $PO_4$ | 10 | 13.7 | | | | | 0.3 | 0.1 | 0.09 | | | 3.5 | 0.09–10 |
| $SO_4$ | | | | | | | | | 58 | | | 58 | |
| TP | | | 1.2 | 1.1 | 1.5 | 1.5 | | | | | | 1.4 | 1.2–1.5 |
| MBAS | 61 | 47 | | | | | | | | | 15 | 38 | 15–61 |
| $Cl^-$ | 284 | 167 | 170 | 20 | | | | | | | | 227 | 170–284 |
| B | 0.35 | 0.12 | | | | | | | | | | 0.35 | |
| $Na^+$ | 151 | 83 | 130 | 25 | | | | | 106 | | 185 | 143 | 106–185 |
| $Mg^{2+}$ | | | 18 | 5 | | | | | 47 | 56 | 41 | 20 | 18–56 |

(*Continued*)

**TABLE 1.5 (*Continued*)**
**Characterization of Greywater from Showers**

| | Several Homes, Israel (Abu Ghunmi et al., 2008) | | Dormitories, Jordan (Jefferson et al., 2004) | | Public Showers, Germany (Winward et al., 2008) | | Homes, United Kingdom (Birks and Hills, 2007) | | Public Showers, Israel (Friedler et al., 2006) | | Homes, Oman (Jamrah et al., 2008) | Overall Average | |
|---|---|---|---|---|---|---|---|---|---|---|---|---|---|
| **Source** | **AVG** | **SD** | **AVG** | **SD** | **MED** | **SD** | **AVG** | **SD** | **AVG** | **SD** | **AVG** | **AVG** | **Range** |
| $Ca^{2+}$ | | | 45 | 10 | | | | | 80 | 16 | 47 | 32 | 16–80 |
| $K^{+}$ | | | 7 | 2 | | | | | 10 | 43 | 20 | 20 | 7–43 |
| Zn | | | | | | | | | | 2.4 | 2.4 | | |
| Al | | | | | | | | | | 0.014 | 0.014 | | |
| Pb | | | | | | | | | | 0.1 | 0.1 | | |
| Cu | | | | | | | | | | 0.01 | 0.01 | | |
| Ni | | | | | | | | | | 0.035 | 0.035 | | |
| SAR | | | 4 | | | | | | | | 4 | | |
| FC | $4 \cdot 10^{6}$ | $8.5 \cdot 10^{6}$ | | | 60 | | $1.5 \cdot 10^{3}$ | $4.9 \cdot 10^{3}$ | | | $1.3 \cdot 10^{6}$ | $2.3 \cdot 10^{6}$ | $6 \cdot 10^{1}$–$4 \cdot 10^{6}$ |
| TC | | | | | $6 \cdot 10^{7}$ | | $6.8 \cdot 10^{3}$ | $9.7 \cdot 10^{3}$ | | | $3.0 \cdot 10^{7}$ | $4.2 \cdot 10^{7}$ | $7 \cdot 10^{3}$–$6 \cdot 10^{7}$ |
| IE | | | | | $5 \cdot 10^{2}$ | | | | | | $5.0 \cdot 10^{2}$ | | |
| FS | | | | | | | $2.1 \cdot 10^{3}$ | $4.4 \cdot 10^{3}$ | | | $2.1 \cdot 10^{3}$ | | |

*Notes:* AVG, average; SD, standard deviation; MED, median; DO, dissolved oxygen; Alk., alkalinity; EC, electrical conductivity; Tur., turbidity; TS, total solids; TDS, total dissolved solids; VTS, volatile solids; TSS, total suspended solids; VSS, volatile suspended solids; COD, chemical oxygen demand; BOD, biochemical oxygen demand; TOC, total organic carbon; DOC, dissolved organic carbon; TN, total nitrogen; TP, total phosphorous; MBAS, methylene blue absorbing substances (anionic surfactants); SAR, sodium adsorption ratio; FC, fecal coliforms. TC, total coliforms; IE, intestinal enterococci; FS, fecal streptococci. All units are in mg/L, except for pH, EC (μS/cm), turbidity (NTU), SAR, alkalinity (mg $CaCO_3$/L), FC, TC, IE and FS (CFU/100mL).

**TABLE 1.6**
**Characterization of Greywater from the Bathroom**

| | Several Homes, North Israel (Abu Ghunmi et al., 2008) | | Homes, United Kingdom (Jefferson et al., 2004) | | Several Homes, South Israel (Winward et al., 2008) | | Overall Average | | |
|---|---|---|---|---|---|---|---|---|---|
| **Source** | **AVG** | **SD** | **AVG** | **SD** | **AVG** | **SD** | **AVG** | **SD** | **Range** |
| pH | 7.14 | 0.04 | 7.57 | 0.29 | 7.3 | 0.3 | 7.3 | 0.22 | 7.1–7.6 |
| EC | 1200 | 409 | | | 1130 | 50 | 1165 | 49 | 1130–1200 |
| Tur | | | 64.8 | 70.5 | | | 65 | | |
| TS | 777 | 303 | | | | | 777 | | |
| VTS | 318 | 244 | | | | | 318 | | |
| TSS | 78 | 105 | 58 | 46 | 153 | 83 | 96 | 50 | 58–153 |
| VSS | 76 | 98 | | | | | 76 | | |
| COD | 230 | 195 | 367 | 246 | 435 | 130 | 344 | 104 | 230–435 |
| Dissolved COD | 165 | 105 | | | | | 165 | | |
| BOD | 173 | 218 | 129 | 57 | 44 | 5 | 115 | 66 | 44–173 |
| Dissolved BOD | 75 | 65.6 | | | | | 75 | | |
| TOC | 91 | 89 | 60 | 43 | | | 75 | 22 | 60–91 |
| DOC | 47 | 28 | | | | | 47 | | |
| Total oil | 77 | 114 | | | 7.2 | 1.1 | 42 | 49 | 7.2–77 |
| $NH_4^+$–N | 0.9 | 1.5 | | | 0.65 | 0.2 | 0.77 | 0.17 | 0.65–0.89 |
| TN | | | 6.6 | 3.4 | 7.2 | 1.8 | 6.9 | 0.42 | 6.6–7.2 |
| $PO_4$ | 4.6 | 5.34 | | | | | 4.6 | | |
| TP | | | | | 2.8 | 1.3 | 2.8 | | |
| MBAS | 15 | 15 | | | 4.1 | 0.6 | 9.6 | 7.7 | 4.1–15 |
| $Cl^-$ | 166 | 128 | | | | | 166 | | |
| B | 0.41 | 0.09 | | | 0.31 | 0.06 | 0.36 | 0.07 | 0.31–0.41 |
| $Na^+$ | 112 | 44 | | | | | 112 | | |
| FC | $4 \cdot 10^6$ | $6.9 \cdot 10^6$ | 82.7 | 120 | | | $2.0 \cdot 10^6$ | $2.8 \cdot 10^6$ | $83–4.0 \cdot 10^6$ |
| TC | | | $6.35 \cdot 10^3$ | $9.71 \cdot 10^3$ | | | $6.4 \cdot 10^3$ | | |
| FS | | | 40.1 | 48.6 | | | 40.1 | | |

*Notes:* AVG, average; SD, standard deviation; DO, dissolved oxygen; Alk., alkalinity; EC, electrical conductivity; Tur., turbidity; TS, total solids; TDS, total dissolved solids; VTS, volatile solids; TSS, total suspended solids; VSS, volatile suspended solids; COD, chemical oxygen demand; BOD, biochemical oxygen demand; TOC, total organic carbon; DOC, dissolved organic carbon; TN, total nitrogen; TP, total phosphorous; MBAS, methylene blue absorbing substances (anionic surfactants); SAR, sodium adsorption ratio; FC, fecal coliforms; TC, total coliforms; FS, fecal streptoccoci. All units are in mg/L, except for pH, EC (μS/cm), turbidity (NTU), alkalinity (mg $CaCO_3$/L), FC, TC and FS (CFU/100mL).

### 1.3.3 Washbasins

The water consumption of handbasins (washbasins) is widely varied depending on the nature of its use. One survey found that the average water consumption of the handbasin in Israel is 3.66 (±5.96) L/use (Friedler and Butler, 1996). Another study reported that the average water consumption of the handbasin is 1.9 (±1.7) L/use (Friedler, 2004). Washbasins contribute about 15% of greywater's total volume and mostly has low concentrations of pollutants (Table 1.7).

### 1.3.4 Kitchen Sinks and Dishwashers

The kitchen sink produces about 26% of domestic greywater and it is reported that a single use consumes, on average, about 12 L of water (Almeida et al., 1999; Friedler, 2004). Kitchen sink water contains a high load of pollutants (Table 1.8) as evidenced by the finding that the kitchen sink stream constitutes about 58% of suspended volatile solids, 42% of the general COD, 48% of the $BOD_5$, 43% of total fats, and 40% of anionic surfactants in greywater. Despite this high concentration of pollutants, it was found that kitchen wastewater is only a secondary contributor to fecal coliforms and boron (Friedler, 2004). However, washing uncooked meat in the kitchen sink can contribute to the existence of additional bacteria such as salmonella. Testing of poultry in England found salmonella and campylobacter bacteria on chickens and around work surfaces in the kitchen (Coagan and Humphrey, 1999). The risk of this microbial pollution can be reduced by excluding the kitchen sink water from the greywater flow (Ottoson and Stenstrom, 2003). Soil irrigated with greywater that includes (or is limited to) water from the kitchen contained the highest microbial contamination (Casanova et al., 2001). It is likely that this is due to the growth of bacteria in the soil as a result of a high organic load.

Dishwashers were found to be the most significant contributors of boron to greywater (Table 1.9), with an average close to 4 mg/L (Friedler, 2004). However, as mentioned earlier, an Israeli standard that recently came into force (IS 1417; SII, 2006) limiting the level of boron in dishwashers in Israel has most likely diminished the problem, but no research has been conducted to examine the issue. As aforementioned, a dishwasher has 4–5 cleaning stages, and it was found that the most polluting stage is the second phase with 33% of the volume and 50%, 70%, 90%, 27%, 64%, and 84% of the COD, ammonia, phosphorus, anionic surfactants, chlorides, and boron load, respectively (Friedler, 2004).

### 1.3.5 Combining Flows

Water from multiple sources can usually be found in reused greywater. Due to the different properties of each source, selecting the supply for reuse has an influence on both the quantity and quality of the greywater. Therefore, a combination of sources should be selected to produce the optimal result. For example, since in the urban sector the amount of greywater required is only about 65% of the total amount

**TABLE 1.7**
**Characterization of Greywater from Hand Washing and Rinsing Basins**

| | Several Homes, Israel (Abu Ghunmi et al., 2008) | | Individuals, United Kingdom (Jefferson et al., 2004) | | Homes, Oman (Winward et al., 2008) | Overall Average | | |
|---|---|---|---|---|---|---|---|---|
| **Source** | **AVG** | **SD** | **AVG** | **SD** | **AVG** | **AVG** | **SD** | **Range** |
| pH | 7 | 0.3 | 7.32 | 0.27 | 7.2 | 7.2 | 0.16 | 7.0–7.3 |
| Alk. | | | | | 14 | 14 | | |
| DO | | | | | 4 | 4 | | |
| EC | 1200 | 401 | | | 1400 | 1300 | 141 | 1200–1400 |
| Tur. | | | 164 | 171 | 211 | 188 | 33 | 164–211 |
| TS | 835 | 263 | | | 679 | 757 | 110 | 679–835 |
| TDS | | | | | 361 | 361 | | |
| VTS | 316 | 194 | | | 397 | 357 | 57 | 316–397 |
| TSS | 259 | 130 | 153 | 226 | 318 | 243 | 84 | 153–318 |
| VSS | 86 | 52 | | | | 86 | | |
| COD | 386 | 230 | 587 | 379 | 110 | 361 | 239 | 110–587 |
| Dissolved COD | 270 | 173 | | | | 270 | | |
| BOD | 205 | 43 | 155 | 49 | 100 | 153 | 53 | 100–205 |
| Dissolved BOD | 93 | 57 | | | | 93 | | |
| TOC | 119 | 44 | 99 | 142 | 63 | 94 | 28 | 63–119 |
| DOC | 74 | 26 | | | | 74 | | |
| Total oil | 135 | 177 | | | | 135 | | |
| $NH_4^+$–N | 0.39 | 0.29 | | | | 0.4 | | |
| $NO_3^-$ | | | | | 10.2 | 10 | | |
| TN | | | 10 | 4.8 | | 10 | | |
| $PO_4$ | 15 | 14 | 0.4 | 0.3 | | 7.7 | 10 | 0.4–15 |
| MBAS | 3.3 | 31 | | | 42 | 23 | 27 | 3.3–42 |
| $Cl^-$ | 237 | 118 | | | | 237 | | |
| B | 0.44 | 0.2 | | | | 0.44 | | |
| $Na^+$ | 131 | 57 | | | 149 | 140 | 13 | 131–149 |
| $Mg^{2+}$ | | | | | 21 | 21 | | |
| $Ca^{2+}$ | | | | | 20 | 20 | | |
| $K^+$ | | | | | 5.5 | 5.5 | | |
| Zn | | | | | 0.04 | 0.04 | | |
| Al | | | | | 0.01 | 0.01 | | |
| Pb | | | | | 0.06 | 0.06 | | |

(*Continued*)

**TABLE 1.7 (*Continued*)**
**Characterization of Greywater from Hand Washing and Rinsing Basins**

| | Several Homes, Israel (Abu Ghunmi et al., 2008) | | Individuals, United Kingdom (Jefferson et al., 2004) | | Homes, Oman (Winward et al., 2008) | Overall Average | | |
|---|---|---|---|---|---|---|---|---|
| **Source** | **AVG** | **SD** | **AVG** | **SD** | **AVG** | **AVG** | **SD** | **Range** |
| Cu | | | | | | | | |
| Ni | | | | | 0.04 | 0.04 | | |
| FC | $3.5 \cdot 10^3$ | $7.4 \cdot 10^3$ | 10 | $8.8 \cdot 10^3$ | | $1.8 \cdot 10^3$ | $2.5 \cdot 10^3$ | $10–3.5 \cdot 10^3$ |
| TC | | | $9.4 \cdot 10^3$ | $1.0 \cdot 10^4$ | | $9.4 \cdot 10^3$ | | |
| FS | | | $1.7 \cdot 10^3$ | $5.5 \cdot 10^3$ | | $1.7 \cdot 10^3$ | | |

*Notes:* AVG, average; SD, standard deviation; DO, dissolved oxygen; Alk., alkalinity; EC, electrical conductivity; Tur., turbidity; TS, total solids; TDS, total dissolved solids; VTS, volatile solids; TSS, total suspended solids; VSS, volatile suspended solids; COD, chemical oxygen demand; BOD, biochemical oxygen demand; TOC, total organic carbon; DOC, dissolved organic carbon; TN, total nitrogen; MBAS, methylene blue absorbing substances (anionic surfactants); FC, fecal coliforms; TC, total coliforms; FS, fecal streptoccoci. All units are in mg/L, except for pH, EC (μS/cm), turbidity (NTU), alkalinity (mg $CaCO_3$/L), FC, TC and FS (CFU/100mL).

produced, the kitchen flow can be eliminated, thus reducing the costs of treatment without decreasing the volume of greywater below that required.

A study was undertaken on the impact of removing a particular source on the quality and quantity of greywater and revealed that excluding the shower and bath from the greywater flow resulted in two undesired consequences: a reduction in the volume of water by 20% and an increase in the concentration of chemical pollutants. Removing the stream generated by the kitchen sink resulted in a 25% decrease in greywater volume, while the concentrations of TSS, COD, $BOD_5$, total volatile solids, total fat, and anionic surfactants decreased. On the other hand, the concentrations of boron, sodium, nutrients, and fecal coli increased. When greywater from the kitchen sink, washing machine, and dishwasher was removed, the water volume dropped by 45%; however, the greywater quality improved significantly in most of the variables (Friedler, 2004). It should be noted that for some uses, the resulting volume of greywater would not be sufficient (e.g., a combined use of flushing toilets and irrigating a large garden of a detached house).

### 1.3.6 Evaluating the Amount of Greywater and Potential for Water Saving

Domestic water consumption in industrialized countries ranges between 100 and over 200 L/(person-day) (LPD) as mentioned. Of this, approximately 60%–70% becomes greywater, and most of the remainder is used for flushing toilets.

## TABLE 1.8
## Characterization of Greywater from the Kitchen Sink

| Source | Homes, N. Israel (Abu Ghunmi et al., 2008) | | Students Dormitory, Jordan (Jefferson et al., 2004) | | Homes, Oman (Winward et al., 2008) | Homes, S. Israel (Birks and Hills, 2007) | | Overall Average | | |
|---|---|---|---|---|---|---|---|---|---|---|
| | AVG | SD | AVG | SD | AVG | AVG | SE | AVG | SD | Range |
| pH | 6.48 | 0.6 | 6.83 | 0.65 | 6.7 | 5.7 | 0.3 | 6.4 | 0.5 | 5.7–6.8 |
| Alk. | | | | | 11 | | | 11 | | |
| DO | | | | | 1.4 | | | 1.4 | | |
| EC | 1040 | 294 | 1244 | 310 | 4200 | 1220 | 120 | 1926 | 1519 | 1040–4200 |
| Tur. | | | | | 140 | | | 140 | | |
| TS | 1272 | 1020 | 4101 | 419 | 1037 | | | 2137 | 1705 | 1037–4101 |
| TDS | | | | | 903 | | | 903 | | |
| VTS | 661 | 593 | | | 808 | | | 735 | 104 | 661–808 |
| TSS | 625 | 518 | 1180 | 300 | 134 | 1250 | 860 | 797 | 523 | 134–1250 |
| VSS | 459 | 370 | | | | | | 459 | | |
| COD | 1340 | 1076 | 8071 | 3535 | 486 | 2180 | 690 | 3019 | 3438 | 486–8071 |
| Dissolved COD | 679 | 549 | | | | | | 679 | | |
| BOD | 890 | 480 | 1850 | 890 | 562 | 1042 | 320 | 1086 | 547 | 562–1850 |
| Dissolved BOD | 377 | 194 | | | | | | 377 | | |
| TOC | 582 | 214 | | | 76 | | | 329 | 358 | 76–582 |
| DOC | 316 | 141 | | | | | | 316 | | |
| Total oil | 323 | 218 | | | | 195 | 67 | 259 | 91 | 195–323 |
| $NH_4^+$–N | 0.6 | 0.81 | | | | 0.99 | 0.6 | 0.80 | 0.28 | 0.60–0.99 |
| $NO_3^-$ | | | | | 8 | | | 8 | | |
| TN | | | 26 | 34 | | 30 | 10 | 28 | 2.9 | 26–30 |
| $PO_4$ | 22 | 27 | | | | | | 22 | | |
| TP | | | 4.3 | 3.9 | | 3.7 | 0.8 | 4 | 0.42 | 3.7–4.3 |
| MBAS | 59 | 41 | | | 26.5 | 8.2 | 1 | 31 | 26 | 8.2–59 |
| $Cl^-$ | 223 | 152 | 140 | 25 | | | | 182 | 59 | 140–223 |
| B | 0.02 | 0.025 | | | | 0.54 | 0.17 | 0.28 | 0.37 | 0.02–0.54 |
| $Na^+$ | 89 | 43 | 160 | 20 | | | | 125 | 50 | |
| $Mg^{2+}$ | | | 21 | 7 | | | | 21 | | |
| $Ca^{2+}$ | | | 47 | 10 | | | | 47 | | |
| $K^+$ | | | 7 | 2 | | | | 7 | | |
| SAR | | | 5 | | | | | 5 | | |
| FC | $1.2 \cdot 10^6$ | $2.4 \cdot 10^6$ | | | | | | $1.2 \cdot 10^6$ | | |

*Notes:* AVG, average; SD, standard deviation; DO, dissolved oxygen; Alk., alkalinity; EC, electrical conductivity; Tur., turbidity; TS, total solids; TDS, total dissolved solids; VTS, volatile solids; TSS, total suspended solids; VSS, volatile suspended solids; COD, chemical oxygen demand; BOD, biochemical oxygen demand; TOC, total organic carbon; DOC, dissolved organic carbon; TN, total nitrogen; TP, total phosphorous; MBAS, methylene blue absorbing substances (anionic surfactants); SAR, sodium adsorption ratio; FC, fecal coliforms. All units are in mg/L, except for pH, EC (μS/cm), turbidity (NTU), SAR, alkalinity (mg $CaCO_3$/L), FC (CFU/100mL).

**TABLE 1.9**
**Characterization of Greywater from the Dishwasher**

| | Israel (Abu Ghunmi et al., 2008) | United States (Jefferson et al., 2004) | Overall Average | |
|---|---|---|---|---|
| **Source** | **AVG** | **AVG** | **AVG** | **SD** |
| pH | 8.2 | | 8.2 | |
| EC | 2721 | | 2721 | |
| TS | 2819 | 1500 | 2160 | 933 |
| VTS | 1045 | 870 | 958 | 124 |
| TSS | 525 | 440 | 483 | 60 |
| VSS | 424 | 370 | 397 | 38 |
| COD | 1296 | | 1296 | |
| Dissolved COD | 547 | | 547 | |
| BOD | 699 | 1040 | 870 | 241 |
| Dissolved BOD | 262 | 650 | 456 | 274 |
| TOC | 234 | 600 | 417 | 259 |
| DOC | 150 | 390 | 270 | 170 |
| Total oil | 328 | | 328 | |
| TN | | 40 | 40 | |
| $NH_4^+$–N | 5.4 | 4.5 | 5 | 1 |
| $NO_3$ | | 0.3 | 0 | |
| TP | | 68 | 68 | |
| $PO_4$ | 537 | 32 | 285 | 357 |
| MBAS | 11 | | 11 | |
| $Cl^-$ | 716 | | 716 | |
| B | 3.8 | | 4 | |
| $Na^+$ | 641 | | 641 | |
| FC | $6.00 \cdot 10^4$ | | 6.0E+04 | |

*Notes:* AVG, average; SD, standard deviation; DO, dissolved oxygen; EC, electrical conductivity; Tur., turbidity; TS, total solids; TDS, total dissolved solids; VTS, volatile solids; TSS, total suspended solids; VSS, volatile suspended solids; COD, chemical oxygen demand; BOD, biochemical oxygen demand; TOC, total organic carbon; DOC, dissolved organic carbon; TN, total nitrogen; TP, total phosphorous; MBAS, methylene blue absorbing substances (anionic surfactants); FC, fecal coliforms. All units are in mg/L, except for pH, EC (μS/cm), turbidity (NTU), FC (CFU/100mL).

Using greywater to flush toilets can save between 40 and 60 LPD, which could cover 10%–20% of urban water consumption (Friedler, 2008; Kim et al., 2009). In a report from Jordan, researchers estimated that reusing greywater for toilet flushing, laundry, and garden irrigation could lead to savings of between 38% and 54% of the water consumption in Jordan (Abu Ghunmi et al., 2008).

According to literature, rural and urban areas differ in greywater yield, especially in developing countries. For example, a comparison in Jordan found that water consumption in the students' dormitory in Amman, the capital of Jordan, was three times higher than water consumption in rural areas. In this study, the greywater yield in Amman stood at 70% of the consumption as compared with 64% in the rural areas (Abu Ghunmi, 2008). As expected, water consumption was significantly higher in public areas including recreational establishments (since there it does not cost directly) than it is in private homes. For example, water consumption in hotels is approximately 500 LPD, and in restaurants and pubs, it is about 60 LPD (Gethke et al., 2007).

## 1.4 DIURNAL PATTERN OF GREYWATER FLOWS

The diurnal pattern of greywater discharge is important for determining the storage volume. Calculating appropriate storage space for greywater is critical because the temporal pattern of greywater reuse in both rural and urban areas does not line us with greywater production. This is true both for flushing toilets and washing laundry (Abu Ghunmi et al., 2008). Figure 1.12 displays the diurnal pattern of individual streams that make up the total greywater volume.

The peak greywater flow was recorded between 07:00 and 09:30 and again from 17:30 to 20:00—before people leave for work and after they return home (Butler et al., 1995; Eriksson et al., 2009). Greywater diurnal flow pattern in student dormitories was nearly identical: the peaks occurred between 07:00 and 09:00 and between 20:00 and 00:00 (Abu Ghunmi et al., 2008). In dormitories for married students in southern England, the peak flow was also observed in the morning at 08:00 and in the evening at 22:00, and the lowest flows were observed at 05:00 and 15:00 (Birks and Hills, 2007). In a residential building, no flow was recorded for several hours, especially during the late morning and early afternoon. This diurnal pattern was found to be regular and very similar each day. A similar diurnal pattern held on

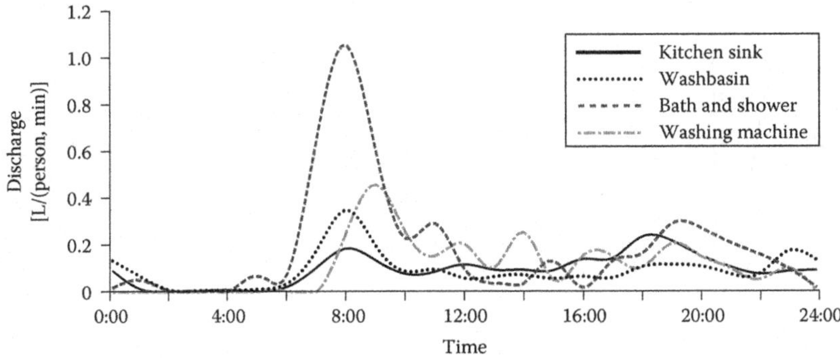


**FIGURE 1.12** The diurnal pattern of flows from individual greywater-generating appliances. (Based on Butler, D. et al., *Water Sci. Technol.*, 31(7), 13, 1995.)

weekends, but the peaks and troughs appeared later (Eriksson et al., 2009). It should be noted that differences between peaks and low flows are more significant in small collection systems (single houses) than in relatively large collection systems such as high-rise buildings.

## 1.5 DIURNAL CHANGES IN GREYWATER CHARACTERISTICS

### 1.5.1 Diurnal Changes in Organic Matter Content

Figure 1.13a displays the variability of $BOD_5$ in greywater throughout the day, and Figure 1.13b presents the relative contribution of each individual greywater stream to the $BOD_5$ load. According to the chart, the concentration increases from 06:00 and the peak concentration is obtained at noon. In the afternoon, the concentration decreases until it reaches a minimum during the night. In the early morning hours, showering is the main contributor of organic matter. Late in the morning and in the

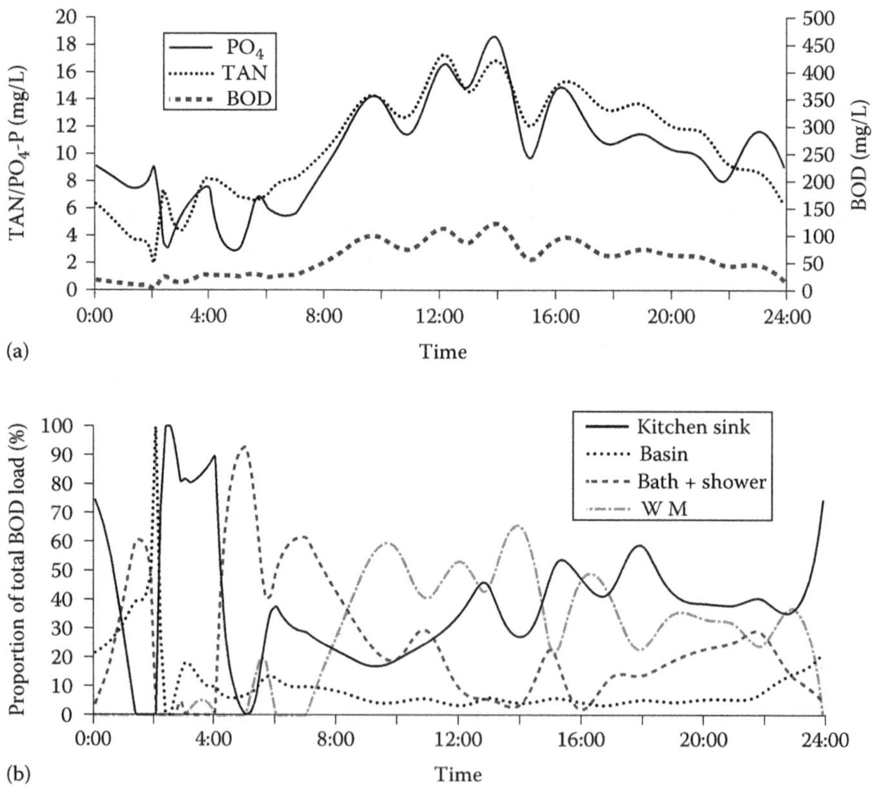


**FIGURE 1.13** Diurnal variation of greywater characteristics: (a) of $BOD_5$, ammonia, and phosphorus concentrations, (b) relative contribution of individual streams to the $BOD_5$ load. *(Continued)*

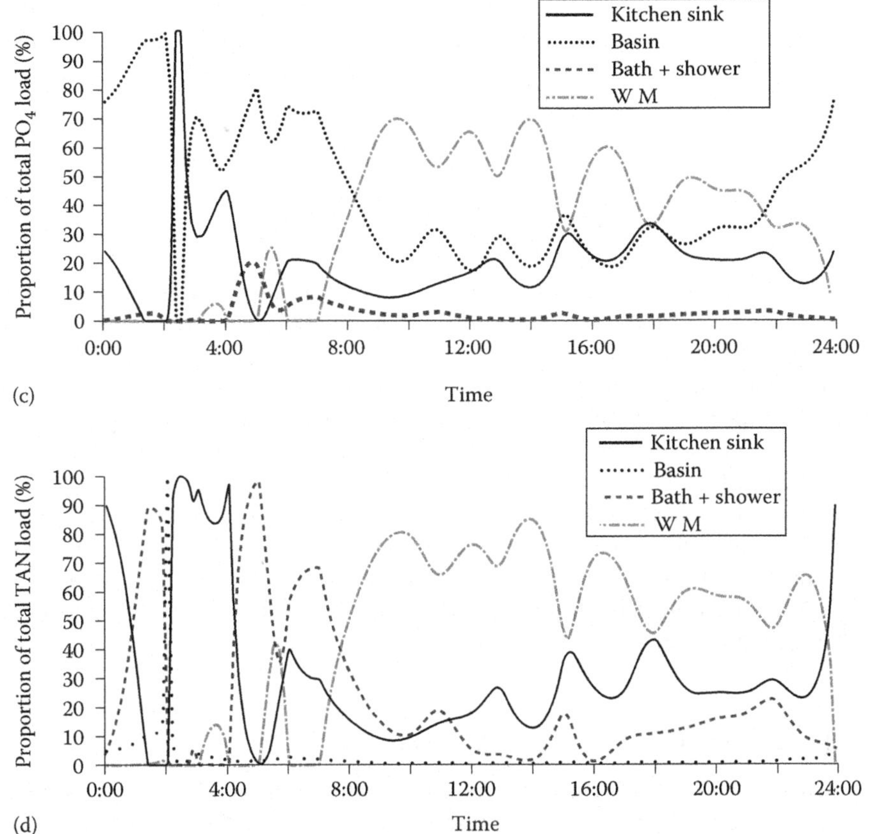


**FIGURE 1.13 (Continued)** Diurnal variation of greywater characteristics: (c) relative contribution of individual streams to the phosphorus load, and (d) relative contribution of individual streams to the ammonia load. (Based on Butler, D. et al., *Water Sci. Technol.*, 31(7), 13, 1995.)

afternoon, the washing machine contributes organic matter, and in the evening, the kitchen sink is the main contributor. At night, the main contribution to greywater comes from the handbasin, so the concentration of organic matter is low as this use is low in organic matter (Butler at al., 1995).

A study conducted in Denmark found that the average concentration of organic matter (as expressed in COD measures) was 142 mg/L, and this concentration ranged between <25 and >650 mg/L throughout the day (Eriksson et al., 2009). The peak concentration was measured in low flow periods. During peak flows in the morning, the COD was under 150 mg/L.

In greywater originating from student dormitories, the COD load daily peak was observed in the evening (16:30 to 0:30) and in the morning (9:00 to 11:00). The high load was the result of a peak flow as opposed to a change in the concentrations.

In fact, there was no daily fluctuation in COD concentrations. This was due to the greywater flows in the student dormitories coming mostly from showers and laundry (Abu Ghunmi et al., 2008). Hourly changes in substance concentrations are typical of small recycling systems of greywater (Eriksson et al., 2009).

### 1.5.2 Diurnal Changes in Content of Nutrients: Phosphorus and Ammonia

Figure 1.13a also shows the diurnal pattern of phosphorus and ammonia concentrations, which follow the same pattern over time as the concentration of $BOD_5$: an increase in the morning, a peak at noon, and a decrease during the afternoon and evening. The increase in phosphorus concentration is steeper as compared to the increase of ammonia (Butler et al., 1995). This difference is due to the fact that detergents contain more phosphorus than ammonia. It should be noted that since the publication of this research, the amount of phosphorus in detergents has decreased. As a result, this pattern of daily variability is likely to have changed. The relative contribution of each use compared to the total phosphorus and ammonia in greywater during the day can also be seen (Figure 1.13c and d, respectively). It appears that the washing machine is the most significant contributor in regard to phosphorus and ammonia in greywater.

### 1.5.3 Diurnal Changes in Salt Contribution

An increase of 50–100 μS/cm was detected in the morning, noon, and late afternoon. The increase corresponds to an increase in COD values. This finding suggests the presence of ions (cations and anions) such as sodium, chloride, and long-carbon-chain fatty acids found in personal care substances (Eriksson et al., 2009).

### 1.5.4 Diurnal and Seasonal Changes in Concentration of Microorganisms

No significant differences between weekdays and weekends were found in the concentrations of indicator microorganisms nor in the total plate count (TPC; at 37°C incubation). Likewise, no differences were found in TPC and total coliforms between the morning and the afternoon hours. However, the TPC at 22°C incubation was higher on weekdays than on weekends. In addition, the level of *E. coli* and fecal enterococci was higher in the afternoon (Birks and Hills, 2007), but the researchers hypothesize that the difference is due to the afternoon samples staying longer in the collection container. Seasonal changes were found in the concentration of fecal coliforms. In spring (April–June) and autumn (November–December), there were lower concentrations of fecal coliforms than in other seasons (Casanova et al., 2001). The seasonal and hourly changes can be attributed to varying environmental conditions during the different seasons and at different hours, the most important of which is temperature.

## 1.6 SYNTHETIC GREYWATER

As demonstrated in this chapter, the composition and flow of greywater vary significantly throughout the day. In addition, greywater in different regions has different levels of contamination. When a greywater treatment unit is being developed, greywater of fixed levels of pollution and flow should be used (at least in the initial stages), which means that an accurate greywater schedule cannot be applied. In addition, *real* greywater poses a challenge in regard to transportation and storage. A possible solution is to use synthetic greywater *produced* according to a fixed formula, made of substances of fixed concentration. In addition to mirroring *real* greywater levels over time, the synthetic greywater can be changed and adapted according to the needs of the study or test.

In Australia (Diaper et al., 2008) and in England (Brown and Palmer, 2002), standards were offered for testing greywater treatment systems. In these studies, it is suggested that synthetic greywater be used for testing systems. Setting a standard for a method of testing, and a formula for synthetic greywater, would allow a comparison of different greywater treatment technologies. Systems that meet the test standard are expected to produce treated greywater that is safe to use (Brown and Palmer, 2002). A formula for greywater should contain the ingredients typically found in *real* greywater such as a variety of personal hygiene products, materials used in the home, and bacteria. The mixture of these substances should yield the concentrations of pH, COD, $BOD_5$, TSS, and surfactants usually found in actual greywater. Usually, these substances are poor in nutrients, but in reality, greywater contains a small amount of nutrients. It is thus helpful to add nutrients as well. Table 1.10 presents formulae for synthetic greywater, as found in the literature and in government standards.

## 1.7 SUMMARY

This chapter reviewed the characteristics of greywater and the contribution of various streams (kitchen, bathroom, laundry, and handbasins) to the overall pollutant level and discharge. Additionally, the possibility of combining these streams was analyzed.

It was determined that when treating greywater for reuse, a significant improvement in its quality is often needed such as a reduction in the pollutant load and/or the removal of pathogens. As such, even the least contaminated flows (e.g., from handbasins) of greywater must be treated before reuse. Using untreated greywater may jeopardize public health (see Chapter 4) and may affect the environment (see discussion in Chapter 4 on irrigation damage).

In Chapter 2, we review various options for treating greywater according to the type of greywater, reuse possibilities, and reuse region.

**TABLE 1.10**
**Various Formulae of Synthetic Greywater**

| Reference | Friedler et al. (2008) (MG per L) | Diaper et al. (2008) (G per 100 L) | NSF Bath (Amount per 100 L) | NSF Laundry (Amount per 100 L) | BSI Laundry and Bath (Amount per 100 L) |
|---|---|---|---|---|---|
| Ammonium chloride | 75 | | | | |
| Soluble starch | 55 | | | | |
| Potassium sulfate | 4.5 | | | | |
| Sodium sulfate $Na_2SO_4$ | | 3.5 | | 4 g | |
| $Na_2PO_4$ | | | | 4 g | |
| Sodium dihydrogen phosphate | 11.4 | 3.9 | | | |
| Sodium bicarbonate $NaHCO_3$ | | 2.5 | | 2 g | |
| Boric acid | | 0.14 | | | |
| Lactic acid | | 2.8 | 3 g | | |
| Synthetic soap | | | | | |
| Body wash with moisturizer | | | 30 g | | |
| Conditioner | | | 21 g | | |
| Shampoo | 0.022 | 72 | 19 g | | 86 mL |
| Liquid hand soap | | | 23 g | | |
| Bath cleaner | | | 10 g | | |
| Liquid laundry fabric softener | | | | 21 mL | |
| Liquid laundry detergent | | | | 40 mL | |
| Laundry | | 15 | | | At recommended concentrations for hard water |
| Kaolin | 25 | | | | |
| Clay | | 5 | | | |
| Test dust | | | 10 g | 10 g | |
| Sunscreen/ moisturizer | | 1/1.5 | | | |
| Toothpaste | | 3.25 | 3 g | | |
| Deodorant | | 1 | 2 g | | |
| Vegetable oil | | 0.7 | | | 1 mL |
| Secondary effluent | | 2 L | 2 L | 2 L | To give final concentration of $10^5$–$10^6$ cfu of total coliforms |

# 2 Greywater Treatment

## 2.1 CHALLENGES IN TREATING GREYWATER

Recycling of greywater is often performed in decentralized systems and is distinct from the prevailing paradigm in most western countries where treatment is centralized. Specifically in Israel, the centralization of all water-related systems and wastewater treatment is almost complete. As noted in the previous chapter, the safe use of greywater is important because if unaddressed, the presence of pollutants can pose health and environmental hazards. However, greywater treatment risk from a decentralized system is not the same as in centralized treatment of full domestic wastewater. First, there are constraints related to the construction of decentralized systems including space, cost, maintenance, and monitoring. Also, unlike in a centralized treatment structure, the volume and quality of water flowing into systems serving a single family (or a small number of families) will likely experience significant variations. For example, when there are guests, the amount of water used will be proportionately greater than usual, while there may be dry periods when the residents are away on vacation. Fluctuations in water quality may also result from a one-time intensive use of soap, acid, or hypochloride solution (such as during holiday time) and from the entry of small amounts of urine and feces, paint, solvents, and more. Hence, the characteristics of greywater present challenges that require targeted solutions different from those of domestic wastewater treatment by central systems.

The first part of this chapter presents the basic principles behind greywater treatment. The second part is a review of technologies commonly used to treat greywater, and examples of case studies are presented. In this section, various technologies are also analyzed for their suitability to the specific characteristics of greywater. In the third portion, different methods for disinfection are presented as well as their individual advantages and disadvantages. Finally, the last section of this chapter compares various greywater treatment technologies as a whole.

## 2.2 TREATMENT PRINCIPLES

### 2.2.1 Physical Treatment

#### 2.2.1.1 Sedimentation

Sedimentation is generally referred to gravitational deposition of organic and inorganic particles in solution at the bottom of a sedimentation basin. The basin can also be used as a separator for oil and grease if the water outlet is located above the sediment and below the upper water level where the oil and grease accumulate. Removal of suspended solids (SSs) from raw effluents by gravitational sedimentation is one of the most common wastewater pretreatment methods. Often, sedimentation is also used to separate sludge and supernatant after secondary treatment.

Using the classic laws of sedimentation, it is possible to describe the sedimentation of discrete particles that are not flocculants. These laws determine the final velocity of the particle by equalizing the gravitational force and the drag forces acting on the particle. The drag coefficient varies depending on the nature of the flow around the particle, whether laminar or turbulent.

The common procedure in planning sedimentation basins is to choose a particle with finite velocity ($V_{SC}$) and design the basin so that all particles with finite speed equal to or greater than ($V_{SC}$) would be completely removed. This is considered the *critical particle*, which enters the sedimentation basin at the water level and sinks to the bottom exactly at the far end (near the outlet) of the basin.

The deposition velocity of the critical particle can be described as being dependent on the effluent flux and on the horizontal area of the basin:

$$V_{SC} = \frac{Q}{A} \tag{2.1}$$

where

$Q$ is the discharge ($m^3/s$)
$A$ is the area (horizontal) of the sedimentation basin ($m^2$)
$V_{SC}$ is the sedimentation critical velocity of the particle (m/s)

The $Q/A$ ratio is also called *hydraulic load* or *overflow rate*, and its unit is $m^3/(m^2 \cdot h)$.

Assuming that the particles in raw water are uniformly dispersed throughout the entire water profile (regardless of their size and density) beginning from entry to the sedimentation basin, an analysis of the particle route shown in Figure 2.1 indicates that particles with a sedimentation velocity ($V_p$) lower than $V_{SC}$ will be removed according to the ratio

$$R.R. = \frac{V_p}{V_{SC}} \tag{2.2}$$

where $R.R.$ is the removal ratio.

Inserting this equation into Equation 2.1 yields an expression of the removal ratio as a function of the sedimentation velocity of the particle and the hydraulic load:

$$R.R. = \frac{V_p}{Q/A} \tag{2.3}$$

The equation shows that according to the general planning for sedimentation of a discrete particle (in an ideal deposition basin), the sedimentation effectiveness does not depend on the basin's depth.

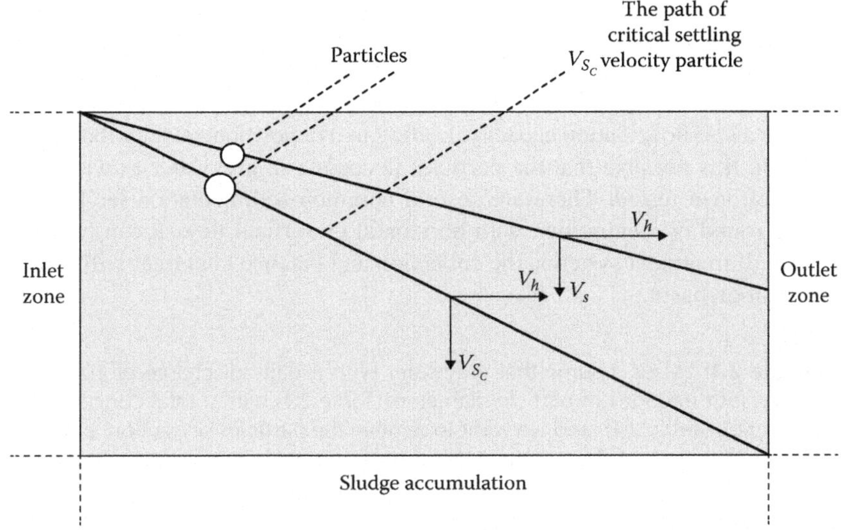


**FIGURE 2.1** Deposition scheme of a discrete ideal particle. (Based on Metcalf and Eddy, *Wastewater Engineering: Treatment and Reuse*, Tchobanoglous, G., Burton, F.L., and Stensel, H.D., eds., McGraw-Hill, New York, 2003.)

The concentration of SSs at the exit from the sedimentation basin is estimated by

$$C_e = C_i \cdot (1 - R.R.) \tag{2.4}$$

where

- $C_e$ is the solid concentration at the exit
- $C_i$ is the solid concentration at the entrance
- $R.R.$ is the removal ratio

and the amount of daily sludge to be removed ($M$) is

$$M = C_i \cdot (R.R.) \cdot Q_d \tag{2.5}$$

where $Q_d$ is the daily feed discharge. The integration of the equations suggests that the removal ratio is a function of the particles' sedimentation velocity and hydraulic load:

$$R.R. = f\left[V_S : \frac{Q}{A_H}\right] \tag{2.6}$$

Most wastewater suspensions have a wide range of particle sizes. To determine the effectiveness of a given deposition, the entire sedimentation velocity range existing in the system has to be considered. The range can be determined using the sedimentation column test. In addition, it should be noted that in urban wastewater most particles have a self-flocculation capacity, leading to a deposition velocity that increases with depth. It is possible that the particles flocculate in greywater as well, but this has not yet been tested. There are several common sedimentation facility shapes including round or rectangular, with horizontal or vertical flow. It can be assumed that in small treatment systems, the collection and balance container will also serve as a deposition basin.

**Example 2.1**: Let us assume that greywater with a daily discharge of 10 $m^3$/day contains four fractions of particle diameters (Table 2.1) with a total concentration of 100 mg/L (Table 2.1), and we want to remove the particles of fraction 2 at a rate of 100%. The area of the sedimentation basin, removal ratio, effluent concentration, and sludge concentration can be calculated as follows:

$$R.R. = \frac{V_S}{\left(\frac{Q}{A_H}\right)} \Rightarrow A_H = \frac{Q \cdot R.R}{V_{SC}} \tag{2.7}$$

$$A_H = \frac{(10/24/3600)\cdot 1}{5.10^{-2}/100} = 0.23\left(m^2\right) \tag{2.8}$$

### 2.2.1.2 Filtration

Filtration is defined as the removal of particulate or colloidal matter from liquid. It is an ancient and widespread means of removing particulate matter from water. The use of sand and gravel filters for water purification has been documented since 2000 BC (Crittenden et al., 2005).

**TABLE 2.1**
**Sedimentation Calculation Example: Fractions of Particles with a Total Concentration of 100 mg/L**

| | | | | Given | |
|---|---|---|---|---|---|
| **Fraction** | **Relative (%)** | **Cumulative (%)** | **$V_S$ (cm/s)** | **Removal Ratio Datum (%) $R.R. = V_S/V_{SC}$** | **Concentration in Effluent (% Relative Mass Fraction in Effluent)** |
| 1 | 10 | 10 | $10^{-1}$ | 100 | 0 (0%) |
| 2 | 20 | 30 | $5\cdot10^{-2}$ | 100 | 0 (0%) |
| 3 | 50 | 80 | $10^{-2}$ | 20 | 40 (69%) |
| 4 | 20 | 100 | $5\cdot10^{-3}$ | 10 | 18 (31%) |
| | | | | Total in effluent | 58 (100%) |
| | | | | Total in sludge | 0.42 kg |

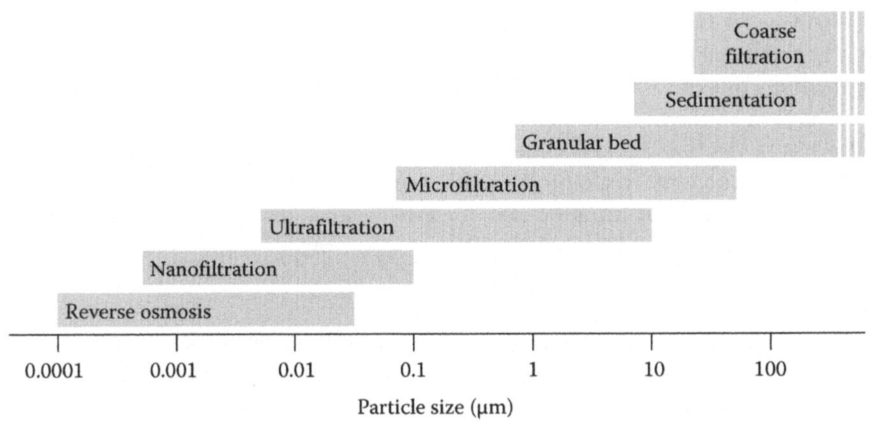


**FIGURE 2.2** Size of the particle removed by different filtration technologies.

A variety of filtration levels can be achieved, using different techniques (Figure 2.2). These range from coarse filtration to remove hair and large particles up to membrane filtration including nanofiltration and reverse osmosis (RO). Coarse filtration and/or sedimentation is often the initial treatment phase, and in some systems (see filtration and routing systems), they constitute the only treatment.

#### *2.2.1.2.1 Sand Filtration*

A sand filter is constructed of a container filled with sand grains of a certain size. Above the sand layer is a gravel layer, with a network of thin tubes discharging water at low pressure to disperse it evenly across the sand. The water seeps through the gravel and sand layer down to another layer of gravel, located at the bottom of the filter, and is then discharged for use. The upper part of the filter (up to about 50 mm) occasionally becomes clogged, and backwash has to be performed or the sand replaced with clean sand.

The design of the sand filter is based on Darcy's law, which describes the flow of fluid through porous media:

$$Q = K \cdot i \cdot A = V \cdot A \tag{2.9}$$

where

$Q$ is the discharge ($m^3/s$)
$K$ is the hydraulic conductivity (m/s)
$A$ is the sectional area in the flow direction ($m^2$)
$i$ is the hydraulic gradient calculated according to

$$i = \frac{h + l}{l} \tag{2.9.1}$$

where

$h$ is the average water head above the filter (m)
$l$ is the sand depth (m)

When water flows through the sand, the filtration rate is represented by $V$. It is usually measured by a permeation test, $V = K \times i$, which depends on the head ($h$) and the medium thickness ($i$). Unlike the filtration rate, hydraulic conductivity $K$ does not change with the head. It depends not on the medium thickness, but only on the properties of the liquid and the medium. The system designer should determine the hydraulic conductivity based on the sand used in the filter. A rule of thumb for planning gravitational sand filtration unit for greywater is to assume a filtration rate of 5–10 $m^3/m^2/h$ when the depth of the filtration layer ranges from 50 to 150 cm.

##### *2.2.1.2.2 Membrane Treatment*

Membrane filtration is an extension of the general definition for filtration of dissolved materials (Metcalf and Eddy, 2003). There are many types of membranes designed for various water treatments, the most common of which are microfiltration (MF), ultrafiltration (UF), and RO membranes. These three membranes operate by applying pressure to one side of a semipermeable membrane, forcing the water to move to the other side while the particles and dissolved matter are unable to pass through. The filtered particle size depends on the pore size of the membrane (Figure 2.2).

An MF membrane has a pore size of 0.5 μm. In the filtration process, a matrix of large particles settles across the membrane and serves as a porous filter that captures the smaller particles. Filtration through MF membranes removes suspended and colloid particles. It leads to complete removal of bacteria and protozoa and to a reduction of the virus content by 2–3 orders of magnitude. The characteristic pore diameter of UF membranes is 0.02 μm, and these are intended to filter all viruses. Nevertheless, due to the lack of uniformity in the membranes, there is no certainty that all pathogens will be removed, so the use of another means of disinfection may be considered according to need (Kennedy et al., 2008). The smaller the pores, the greater the water pressure needed to pass the water through and the greater the membrane's sensitivity to clogging. Therefore, delicate membranes such as RO membranes require a greater investment in operation and maintenance. As such, they are not used in greywater treatment and therefore will not be discussed further. To prevent membrane clogging, the water has to undergo other treatments in advance such as sand filtration. To this end, MF and UF membranes can be used as pretreatment before RO.

Table 2.2 presents the common uses of membrane technology for wastewater treatment. In recent years, the use of membranes in wastewater treatment has become more common, but the use of membranes in greywater treatment is minimal because of its high cost.

### 2.2.2 Chemical Treatment

#### 2.2.2.1 Coagulation and Flocculation

Coagulation is a process during which one or more chemical coagulants (e.g., alum, ferric chloride, ferric sulfate) are introduced into water to prepare suspended, colloidal, and dissolved materials for the flocculation process. It can also create the conditions that allow the removal of dissolved and particulate material from water

**TABLE 2.2**
**Common Uses of Membrane Technologies for Wastewater Treatment**

| Treatment | Use |
|---|---|
| *MF and UF* | |
| Aerobic MBR | The membrane is used to separate effluent from active biomass in the activated sludge process.<br>The membrane unit can be submerged in the biological reactor or externally. |
| Anaerobic MBR | The membrane is used to separate effluent from the active biomass in the anaerobic reactor. |
| Membrane aeration bioreactor | Different membranes are used to transfer oxygen to the biomass attached on the external wall of the membrane. |
| Extractive MBRs | The membranes are used to extract decomposable organic molecules from inorganic ingredients in wastewater, such as acids, bases, and salts, which pass on to biological treatment. |
| Membrane pretreatment for effective disinfection | The membranes are used to remove residual SSs from secondary effluent or from deep or field filtration effluent to achieve effective disinfection through chlorine or UV radiation. |
| Membrane pretreatment for nanofiltration and RO | Microfilters are used to remove SSs and residual colloids as a stage before further processing. |
| *Nanofiltration* | |
| Reuse of effluents | The membranes are used for the treatment of treated effluent (mostly after MF) for indirect reuse as drinking water, such as by introduction into groundwater. Nanofiltration also serves as a kind of disinfection. |
| Sewage softening | The membrane is used to reduce the multivalent ion concentration that contributes to the hardness of water for specific uses. |
| *RO* | |
| Reuse of effluents | The membranes are used for the treatment of treated effluents (mostly after MF) for reuse as drinking water indirectly, such as insertion into groundwater. Also, RO serves as a kind of disinfection. |

*Source:* Metcalf and Eddy, *Wastewater Engineering: Treatment and Reuse*, Tchobanoglous, G., Burton, F.L., and Stensel, H.D., eds., McGraw-Hill, New York, 2003.

(Crittenden et al., 2005). The small size of colloidal particles and their overall negative electric charge result in repulsion forces between them that are greater than the attraction forces. Under these stable conditions, Brownian motion keeps the colloidal particles suspended, and they cannot be removed by precipitation. However, chemical coagulant upsets the stability of these colloidal particles so they can grow following collisions between them (Metcalf and Eddy, 2003) and can thus be removed by precipitation.

In flocculation, destabilized particles, such as those resulting from the coagulation process, form clusters (flocs) that are relatively large and can thus be removed from the water by sedimentation and/or filtration (Crittenden et al., 2005).

The use of physicochemical processes may allow for better treatment of greywater in small on-site treatment systems characterized by changing volumes and qualities. While simple physical technologies can successfully cope with varying volumes, they remove the total organic matter less effectively (Pidou et al., 2008) and may clog quickly (Friedler and Alfiya, 2010). Examination of the effectiveness of physicochemical processes (coagulation and magnetic ion exchange resin) found that the processes remove organic matter satisfactorily from light greywater, but when raw greywater contains a high concentration of organic matter, the results do not meet the strictest standards (Pidou et al., 2008). Other researchers found that coagulation can be used as a preliminary efficient step before sand filtration and RO (Friedler et al., 2008; Friedler and Alfiya, 2010). Finally, one literature review (Pidou et al., 2007) reported that in three systems that used coagulation as part of greywater treatment, the treated effluent was of high quality (i.e., in two of these systems, the concentration of biological oxygen demand [$BOD_5$] and total SSs [TSSs] in greywater effluent was 10 mg/L or less).

## CASE STUDY: PHYSICOCHEMICAL TREATMENT SYSTEM

Basic treatment systems typically consist of physical treatment plus supplementary disinfection. An example of such a system is presented here by March et al. (2004).

The authors describe the treatment system of a hotel in Spain that has 81 suites, and greywater is collected from the baths and handbasins.

The treatment system includes filtration using a nylon sack–type filter with pores of 0.3 mm and a total surface area of 1 $m^2$, followed by sedimentation and disinfection using sodium hypochlorite. The treated greywater was collected in a tank of 4.5 $m^3$ and pumped into six upper storage tanks with a total capacity of 4 $m^3$, with connection to potable water for backup if necessary. The tank system made it possible to control the storage volume and consequently the residence time in the storage tanks.

The treated greywater was used for flushing toilets. The water from the storage tanks flowed through gravitation straight to the flushing tanks. The average water consumption was 146 L/person/day, and the average water consumption for toilet flushing was 36 L/person/day. The average residence time in the storage tanks was 38 h.

### Disadvantages

Because treatment did not reduce the organic load significantly, chlorine use was very high (36 mg/L in 12 h). As a very high dose of chlorine is required, the cost of treatment is increased and undesirable compounds may result as by-products.

Despite the simplicity of the system, it required frequent maintenance. It was necessary to empty the solids that accumulated in the filter after 15–16 $m^3$ of greywater had passed through the system; otherwise, the filtration rate dropped and did not meet the demand for treated greywater. Once 150 $m^3$ had flown through the storage tank, it had to be cleaned, and the same was true for every 50 $m^3$ of greywater in the settling tank. In addition, in unoccupied rooms it was necessary to empty the flushing tank once a day (to prevent the emission of bad odors).

It is estimated that it took 2.5 min of maintenance for each cubic meter of treated greywater.

**Advantages**

The system ran for 7 months a year while the hotel had an occupancy rate of 85%. In the tourism off-season when the hotel was not active, no maintenance of the system was required. This gives the system an advantage over biological treatment systems that would require maintenance even when the hotel is not active.

The cost of the system is relatively low: the capital cost including construction was 17,000 €, and its operation and maintenance cost was 0.75 €/$m^3$. In total, the system led to a savings of 1.09 €/$m^3$.

The guests responded positively toward the use of greywater.

The water quality produced by this system in comparison with other case studies is shown in Table 2.9.

### 2.2.3 Biological Treatment

Biological treatment is designed to remove dissolved and suspended organic matter from wastewater, but under the right configurations, it could also treat nitrogen, phosphorus, and even to some extent salts (Green et al., 1998; Shelef et al., 2010; Travis et al., 2012). The difference between treatment efficiency is reflected in the following factors: flow regime (i.e., continuous vs. batch mode, completely mixed vs. plug flow), type of flow (i.e., saturated or unsaturated), operating conditions (i.e., aerobic or anaerobic), and type of microbial growth (i.e., fixed or suspended). There are different types of biological treatment, and their efficiency is often described by the mass balance principle (Figure 2.3, Equation 2.10).

Biological treatment processes can be described by depicting the mass balance of the pollutant (e.g., organic matter): the pollutant influx, pollutant output flux, pollutant removal, and pollutant production, as in the following equation:

$$V\frac{dC}{dt} = Q \cdot C_{in} + P - Q \cdot C_{out} - R \tag{2.10}$$

where

$V$ is the volume of liquid
$Q$ is the flow rate
$C_{in}$ is the inlet concentration of tested pollutant
$C_{out}$ is the outlet concentration of tested pollutant
$C$ is the concentration of tested pollutant at the treatment system
$R$ is the removal rate from the system
$P$ is the production rate into the system

When the treatment system is in a steady state relative to the pollutant (usually after the bacteria population in the system has acclimated), it can be said that the rate of change in the concentration of the tested (organic and biodegradable) pollutant equals 0 ($dC/dt=0$). Figure 2.4 describes the system acclimation process in a biological system.

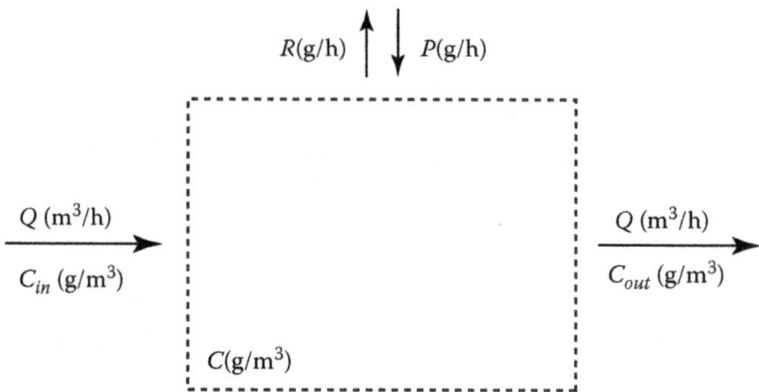


**FIGURE 2.3** Schematic illustration of mass balance in a biological treatment system.

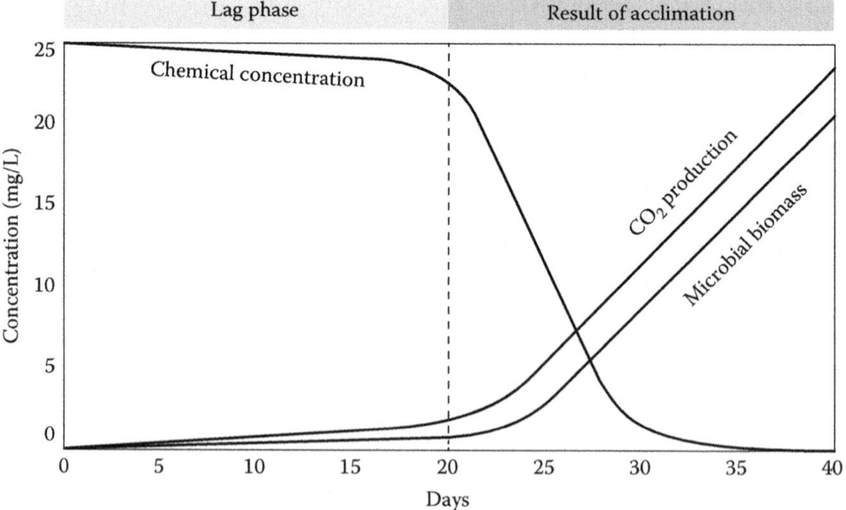


**FIGURE 2.4** Acclimation process in a biological treatment system.

Biological treatment produces optimal conditions for the natural activity of microorganisms that decompose available organic matter in water. Biodegradation of organic matter by bacteria is used for two main purposes:

1. The creation of biomass (anabolism) that increases the biomass concentration while removing nutrients such as phosphorus and nitrogen. Biomass can be described with the following formula: $C_5H_7NO_2$. When normal phosphorus content is introduced into the cells, the equation becomes $C_5H_7NO_2P_{0.08}$ (Droste, 1997) and is parallel to a COD/N/P ratio of 65:5:6:1, based on weight. Grady et al. (1999) suggested another ratio of 0.087 mg nitrogen and 0.017 mg phosphorus per mg of produced biomass, representing a COD/N/P ratio of 83:5:1, based on weight. Typically, the COD/N/P ratio of biomass in the literature ranges between 50:10:1 and 100:10:1.
2. Cell maintenance and degradation (catabolism) in treatment produce water and carbon dioxide ($CO_2$) under aerobic conditions or methane ($CH_4$) and carbon dioxide under anaerobic conditions.

Full biodegradation of the organic matter consists of several stages and is called mineralization. For example, mineralization of glucose can be described as follows:

| Glucose | Microorganisms | New cells (biomass) |
|---|---|---|
| $C_6H_{12}O_6 + O_2 + NH_3$ | $\longrightarrow$ | $C_5H_7O_2 + CO_2 + H_2O$ |

Incomplete biodegradation of organic matter can occur in several cases: when enzyme activity is inhibited, when there is a shortage of nutrients (mainly nitrogen and phosphorus), or when there is a shortage of suitable final electron acceptors (i.e., oxygen in aerobic decomposition). Moreover, the availability of organic matter for biodegradation can be reduced due to low solubility of a substance (such as oil) in water or when the matter is attached to soil particles (Pepper et al., 2006). Additional influences on the biological process are physical factors that depend on the type of reactor, the way it is operated and its medium, hydraulic load and residence time, organic load, aeration (i.e., actively dissolving oxygen in liquid), as well as environmental factors such as temperature, pH, and the presence of toxic/inhibiting substances in the water.

Many chemical processes can be characterized on a kinetic basis, such as kinetics of zero, first, or second order (Figure 2.5).

The common kinetics used for characterizing processes of microbial growth and substrate decomposition, such as biologic organic matter, is the Monod kinetics (Metcalf and Eddy, 2003). The Monod kinetics is described by the following equations and in Figure 2.6:

$$\frac{dS}{dt} = -q_{\max} \cdot \frac{S \cdot X}{K_S + S} \tag{2.11}$$

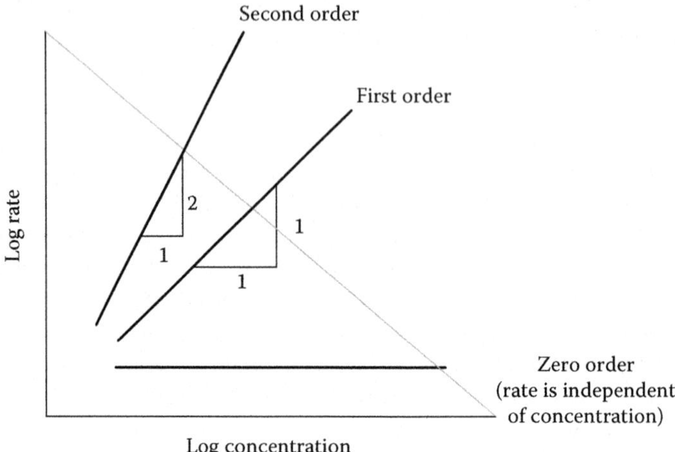


**FIGURE 2.5** Kinetics of chemical processes—reaction orders.

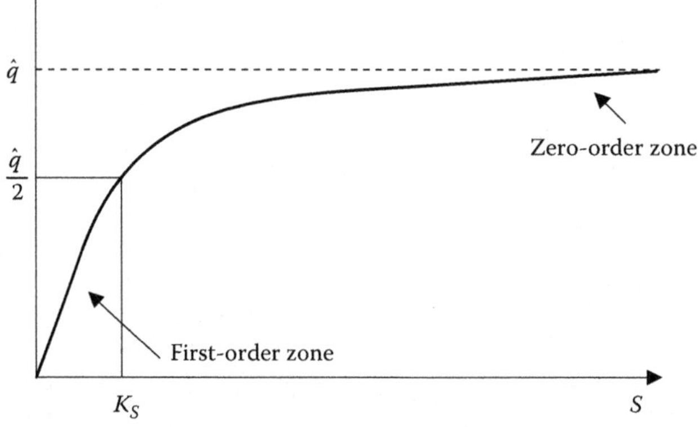


**FIGURE 2.6** The Monod model for biological removal of substrate.

$$q = -\frac{1}{X} \cdot \frac{dS}{dt} \tag{2.12}$$

$$q = q_{\max} \cdot \frac{S}{K_S + S} \tag{2.13}$$

where

- $dS/dt$ is the substrate removal rate (mg/L/day)
- $q_{\max}$ is the (constant) maximal specific removal rate (day$^{-1}$)
- $K_s$ is the half saturation coefficient (mg/L), which is the substrate concentration for which the removal rate equals half the maximal rate

$S$ is the substrate concentration (mg/L)
$X$ is the biomass concentration (mg/L)
$t$ is the time (day)

An analysis of this equation indicates that when the substrate concentration ($S$) is significantly lower than the half saturation constant ($K_s$), the equation becomes a first-order equation (Equation 2.14). For example, the addition of a substrate will proportionally increase its rate of removal. On the other hand, when the substrate concentration is significantly larger than the half saturation constant, the equation obtained is of zero order. In other words, beyond the maximal point, the removal rate is not affected by the addition of higher substrate concentration (Equation 2.15):

$$S \ll K_S; \frac{dS}{dt} = -\frac{q_{\max} \cdot X}{K_S} \cdot S \tag{2.14}$$

$$S \gg K_S; \; q = q_{\max} = \text{const.} \tag{2.15}$$

The substrate removal rate can also be described in relation to the detention time, according to the following equation:

$$q = \frac{S_o - S_e}{X \cdot \theta} \tag{2.16}$$

where
$q$ is the substrate removal rate—described as the substrate consumption rate per unit biomass (g BOD/$g_{VSS}$/day)
$S_e$ is the substrate concentration in the process tank (in the case of a well-mixed reactor—the concentration at the exit from the reactor) (mg/L)
$X$ is the biomass concentration in the reactor (mg/L)
$\theta$ is the hydraulic detention time in the reactor (day)

Similarly, the rate of microbial growth in a wastewater or greywater treatment system can also be described as

$$\mu = \frac{\mu_m \cdot S_o}{K \cdot S_o} = \frac{1}{X_a} \cdot \frac{dX_a}{dt} \tag{2.17}$$

where
$\mu$ is the specific biomass growth rate (1/day)
$\mu_m$ is the maximal specific biomass growth rate (1/day)
$K_s$ is the half saturation coefficient—the concentration of substrate for which growth rate equals half of the maximal value (mg/L)
$S_o$ is the substrate concentration at the entrance to the aeration basin (e.g., general $BOD_5$) (mg/L)
$X_a$ is the concentration of active biomass (mg/L)

#### 2.2.3.1 Suspended Growth

In suspended growth processes, the microorganisms responsible for treatment (i.e., removal of organic matter and nutrients) are maintained in the body of water through mixing. Most treatment facilities that use suspended growth are activated at positive oxygen concentrations (i.e., aerobic treatment); however, there are situations that call for anaerobic reactors (e.g., for high concentrations of organic matter in industrial wastewater or in the treatment of organic sludge). The most common treatment process using suspended growth is activated sludge (Metcalf and Eddy, 2003).

#### 2.2.3.2 Treatment Processes with Substrate-Attached Growth

In treatment processes with substrate-attached growth, the microorganisms responsible for breaking down the nutrients or organic matter are attached to inert material. The organic matter and nutrients are removed from the water when they pass through the substrate-attached microorganisms (called biofilm). The substrate material can be stone, gravel, sand, and a variety of plastics or other synthetic materials. Substrate-attached processes can occur in aerobic and anaerobic environments, and the substrate can either be fully immersed in liquid or not immersed at all (i.e., unsaturated flow). Examples of substrate-attached treatment technologies are constructed wetlands, rotating biological contactor (RBC), and trickling filter.

While in suspended substrate processes, efficiency of organic matter removal is determined by the concentration of suspended biomass, in substrate-attached processes, the determining factor is the substrate surface on which the biofilm can grow. Therefore, in innovative systems of fixed substrate, synthetic bedding (usually made of plastic) is used in many cases with a high specific surface area. Another advantage of a synthetic substrate is its low density relative to stone or gravel substrate. Its disadvantage is a relatively high cost.

#### 2.2.3.3 Oxidation–Reduction Reactions

The degradation process of organic matter occurs when the oxidizer (the final electron acceptor) receives electrons from the organic compound (the electron donor). As a result of the chemical reaction, there is an energy gain that varies according to the oxidizer (e.g., $O_2$ or $NO_3$). The energy gain from the oxidation–reduction (redox) reaction of the organic matter with molecular oxygen is much higher than would have been obtained in the reaction with another oxidant, such as nitrate, sulfate, or carbon dioxide (Pepper et al., 2006). For this reason, the rate of organic matter degradation in aerobic conditions is the highest, and thus this is the most common way to treat wastewater.

#### 2.2.3.4 Aerobic Treatment

In aerobic greywater treatment processes, some of the carbon involved becomes carbon dioxide and is released into the air, and the rest becomes cellular matter. The more aerated the water, the more efficient the decomposition and the higher the decomposition rate. Oxygen demand for anabolism and catabolism can be calculated from the redox equation of known organic matter in wastewater. Figure 2.7 shows a general formula for the degradation of organic matter found in wastewater.

$$C_{18}H_{19}O_9N + 8.75O_2 + 0.75NH_3 \longrightarrow 1.75C_5H_7O_2N + 9.25CO_2 + 4.5H_2O$$

Carbon source + Electron acceptor + Nitrogen source ⟶ Biomass (anabolism) + Products (catabolism)

**FIGURE 2.7** Example of the biological aerobic degradation process.

The preceding formula lacks reference to phosphorus, which according to the formula $C_5H_7NO_2P_{0.08}$ can be an element in biomass. In greywater, it is expected that the substrate formula (source of carbon) will be different, and consequently, a different biomass will develop. The C/N/P ratio of greywater ranges between 100:5:0.2 and 100:11:2 (Kovlio, 2004; Jefferson et al., 2004, respectively), and the usual ratio in urban sewage is 100:23:4 (Droste, 1997) (see Chapter 1). That is, in greywater the concentrations of nitrogen and phosphorus are lower than in full domestic wastewater because their main source is human excrement (pumped from toilets into sewage).

Once the oxygen requirement for oxidizing organic matter in wastewater has been established, the energy consumption for this process can be calculated according to the means used for aeration in the treatment process (area aeration, diffusers, etc.). An advantage of aerobic decomposition is that it does not produce toxic gases and/or gases with a bad odor. In addition, aerobic conditions allow ammoniacal nitrogen compounds to be turned into nitrate through the nitrification process (see Section 2.2.4). Many examples of aerobic greywater treatment can be found in Section 2.3.

#### 2.2.3.5 Anaerobic Treatment

Anaerobic treatment is an inclusive name given to any biodegradation process of organic matter whose final electron acceptor is not molecular oxygen. In anaerobic conditions, microorganisms can utilize the oxygen found in the organic compound itself or in oxidized inorganic compounds such as sulfate or carbon dioxide. In general, the anaerobic decomposition can be described by the following mass balance:

$$C_6H_{12}O_6 \rightarrow 3CO_2 + 3CH_4 \tag{2.18}$$

While anaerobic decomposition is slower than aerobic decomposition, it has two main advantages. First, the biomass (yield of new cells) is lower relative to the yield in aerobic process, so the amount of sludge generated in the process is smaller. Second, the process is often designed for the production of methane ($CH_4$) by methanogens (methane-forming microorganisms) that can be utilized for energy. In addition, there are certain organic compounds, such as in pesticides and halogen compounds, that do not decompose in aerobic processes. They require an anaerobic process for their decomposition (Mara and Horan, 2003). Besides the relatively slow rate of anaerobic treatment, other drawbacks of this process are the release of odorous gases, greenhouse gases, and other toxic gases such as sulfide, ammonia, and methane (although as the aforementioned, methane gas can be collected under certain conditions and used as an energy source, turning it into an advantage). In addition, the anaerobic process is more sensitive than the aerobic process to environmental and operational conditions such as pH, nutrient composition, and mixing (Mara and Horan, 2003).

Traditionally, anaerobic digestion was used primarily for the treatment of sludge and wastewater that had a high concentration of organic matter. Recently, however, this technology has increasingly been used for thinner sewage (Metcalf and Eddy, 2003). Anaerobic conditions are common in deposition tanks used for pretreatment in numerous greywater systems. It was found that greywater is highly biodegradable under anaerobic conditions and the use of an upflow anaerobic sludge blanket (UASB) facility has been proposed to create biogas in the pretreatment phase of greywater (Elmitwalli and Otterpohl, 2007). Sometimes the transition between aerobic treatment and anaerobic treatment is used for removing nitrogen as detailed in Section 2.2.4.

### 2.2.4 Nitrogen Removal

The nitrogen cycle is a very complex process and is sensitive to environmental conditions. It combines chemical and microbial processes and will only be partially addressed in this book (Figure 1.2). The source of nitrogen in greywater is mostly various organic materials such as food scraps, urine (often found in shower water), and the low concentration of nitrate in the water itself. Organic nitrogen usually becomes ammonia in a short time through biodegradation processes, and therefore the ammonia concentration will typically be highest before treatment. Its concentration and nature in effluent depend on the type of treatment. When the effluent is discharged to the environment (e.g., natural water bodies), it is necessary to remove nitrogen from the water because of its toxic potential and due to the risk of eutrophication. Even when effluents are used for irrigation, the plants require only a certain amount, and the surplus may reach groundwater and other water sources.

Nitrogen removal from wastewater in general, and from greywater in particular, can be accomplished in many ways such as membrane filtration, ion exchanger, ammonia evaporation, assimilation by bacteria and/or plants, the anaerobic ammonia oxidation (ANAMMOX) process (turning ammonia into atmospheric nitrogen without the presence of oxygen by using bacteria), and most commonly the combination of nitrification in aerobic conditions and denitrification in anoxic conditions (without dissolved oxygen, but in the presence of nitrate and nitrite). These will be briefly discussed in Sections 2.2.4.1 and 2.2.4.2.

#### 2.2.4.1 Nitrification

Nitrification is described as biological transformation of ammonia to nitrate by microorganisms, namely, bacteria and archaea (which will not be discussed). In the first phase, ammonia is oxidized into nitrite ($NO_2$–N), and in the second phase, nitrite is oxidized to nitrate ($NO_3$–N) as described in the following equations:

$$2NH_4^+ + 3O_2 \rightarrow 2NO_2^- + 2H_2O + 4H^+ \qquad (2.19)$$

$$2NO_2^- + O_2 \rightarrow 2NO_3^- \qquad (2.20)$$

A number of different bacterial groups are capable of oxidizing ammonia into nitrite and nitrite into nitrate. These bacteria are generally autotrophic aerobic (i.e., inorganic carbon fixers and producers of energy from inorganic compounds such as ammonia). The most common bacteria that oxidize ammonia into nitrite include those from the groups *Nitrosomonas*, *Nitrosospira*, *Nitrosococcus*, *Nitrosovibrio*, and *Nitrosolobus*. The oxidation of nitrite into nitrate is performed by other bacterial groups, the best known of which are *Nitrobacter*, *Nitrospina*, *Nitrococcus*, and *Nitrospira*.

The biomass of heterotrophic bacteria (organic carbon fixers), which is responsible for the removal of available organic matter (i.e., $BOD_5$), increases at a much faster rate than nitrification bacteria. As such, systems that are designed to remove nitrogen should have longer hydraulic and sludge retention times than those designed for the removal of $BOD_5$ only.

#### 2.2.4.2 Denitrification

In the denitrification process, nitrate is reduced to nitrogen gas $N_2$ in a multistep biological process as part of the respiratory electron transfers, as described in the following equation:

$$NO_3^- \rightarrow NO_2^- \rightarrow NO \rightarrow N_2O \rightarrow N_2 \tag{2.21}$$

In this process, the enzymes nitrate reductase and nitrite reductase, which exist in certain heterotrophic bacteria, allow them to use the oxygen associated with the nitrate and nitrite molecules as the final electron acceptor in the oxidation of a variety of organic and inorganic electron donors. The groups of bacteria that perform denitrification are mostly facultative aerobic—that is, in anoxic conditions they will use nitrogen compounds, and in aerobic conditions they will use gaseous oxygen as the final electron acceptor because of the higher energy gain of the process. Therefore, in the presence of oxygen, no denitrification will usually occur in the dissimilation process.

An example of a mass balance equation of denitrification based on methanol as a source of carbon is presented as

$$NO_3^- + 0.833CH_3CH \rightarrow 0.5N_2 + 0.833CO_2 + 1.167H_2O + OH^- + E \tag{2.22}$$

#### 2.2.4.3 Anaerobic Ammonia Oxidation

In ANAMMOX process, certain microorganisms can produce gaseous nitrogen ($N_2$) from ammonia and nitrite to obtain energy under anoxic conditions:

$$NH_4^+ + NO_2^- \rightarrow N_2 + 2H_2O \tag{2.23}$$

In recent years, the ANAMMOX reaction has been used in wastewater treatment facilities to remove nitrogen, but due to the sensitivity of the process, its lower pace, and the relatively low N concentration, it is not used in decentralized greywater treatment facilities.

#### 2.2.4.4 Assimilation

Assimilation is nitrogen fixation into the organism with the purpose of cell reproduction. The pace of nitrogen assimilation to cells and its amount depend on environmental conditions such as temperature, pH value, C/N ratio, nitrogen availability, and the concentration of the assimilating organisms. It should be noted that in the case of assimilation, nitrogen remains in the treatment facility as long as the assimilating organisms are not removed, for example, when sludge is pumped or in the case of constructed wetlands after harvesting the plants.

#### 2.2.4.5 Ammonia Evaporation

Nitrogen can also be removed physically from greywater by evaporating gaseous ammonia. This process is not significant, and it is believed that no greywater treatment facilities use it; however, the process can occur naturally to some extent. Ammonia is found in its liquid state in equilibrium according to the following equation:

$$NH_4^+ + OH^- \rightarrow NH_3 \uparrow + H_2O \tag{2.24}$$

This formula implies that the concentration of gaseous ammonia ($NH_3$) in water increases with the elevation in the pH value, and in standard conditions, its molar ratio equals the ammonia ion ($NH_4^+$) at a pH value of 9.25. The concentration of gaseous ammonia decreases logarithmically with the decrease in pH.

In biological greywater systems, the pH value typically ranges from 6.5 to 8.5. For example, the typical greywater ammonia ($NH_4^+ + NH_3$) concentration is 30 mg/L, the concentration of gaseous ammonia at pH value of 8.5 is 4.6 mg/L (15% of the overall concentration of ammonia), and the concentration of the ammonium ion is 25.4 mg/L. Thus, it is clear that this mechanism of nitrogen removal is marginal.

#### 2.2.4.6 Ion Exchanger and Membrane Filtration

Nitrogen can be removed through membrane filtration such as that found in RO systems or by ion exchange that binds charged nitrogen species (e.g., ammonia, nitrite, nitrate) to selective specific medium. Neither method is common in greywater treatment.

### 2.2.5 Phosphorous Removal

The phosphorus cycle is less complex than the nitrogen cycle because it is mainly chemical in nature and more predictable (Figure 1.3). However, phosphorous is difficult to remove from water with any treatment method. Usually, its removal in wastewater treatment facilities takes place by fixing it to sediment, sludge, or vegetation and removing the latter from the water. Chemically, phosphorus is removed by introducing salts of multivalent metal ions that combine with dissolved forms of phosphorus to produce insoluble precipitate. The most common multivalent metal ions used for this purpose are calcium ($Ca^{2+}$), aluminum ($Al^{3+}$), and iron ($Fe^{3+}$) (Metcalf and Eddy, 2003).

Usually, calcium is added in the form of slaked lime ($Ca(OH)_2$). When this is added to the water, it reacts first with the bicarbonate alkalinity and produces depositions of calcium carbonate ($CaCO_3$). Only when the pH value rises above 10 do the remaining calcium ions react with phosphorus to create the hydroxylapatite that precipitates, according to the following equation:

$$10Ca^2 + 6PO_4^{3-} + 2OH^- \rightarrow Ca_{10}\left(PO_4\right)_6\left(OH\right)_2 \tag{2.25}$$

Due to the reaction of the lime with the alkalinity, the amount of lime consumed is independent of the concentration of phosphorus but depends on the alkalinity. In this method, it is necessary to lower the pH value after treatment.

Phosphorus deposition by aluminum and iron works according to the following equations:

$$Al^{3+} + H_nPO_4^{3-n} \rightarrow AlPO_4 + nH^+ \tag{2.26}$$

$$Fe^{3+} + H_nPO_4^{3-n} \rightarrow FePO_4 + nH^+ \tag{2.27}$$

Phosphorus can also be biologically removed by treatment with the *luxury uptake* method. This pattern is attributed to bacteria of the acinetobacter type. These bacteria use sugars as a source of carbon and energy, but only in the presence of oxygen. In anaerobic conditions, they are able to decompose volatile fatty acids like acetate to create poly-2-hydroxy butyrate (PHB) using the decomposition of polyphosphate as an energy source. As a result, the concentration of phosphorus in water increases in anaerobic conditions. When most available organic matter is consumed in the anaerobic phase, the acinetobacter moves to aerobic conditions in an environment with a low concentration of available organic matter. To compensate for this, decomposition of the PHB compound into acetate takes place, and as a result, the bacteria get both a carbon source and a source of energy by creating ATP. The ATP production leads to a higher rate of polyphosphate consumption from the body of water into the bacteria cells than does its release into the water in the anaerobic phase (Horan, 1990). Bacteria are then removed from the water.

When water is treated through constructed wetlands, the vegetation planted in the water also consumes phosphorus. The vegetation can be removed, thus removing a certain amount of phosphorus from the water. This method is not feasible as a method of phosphorous removal because plants are not usually a significant reservoir for phosphorus and also because of the large amount of work involved in removing the vegetation from the water in an ongoing and efficient manner (Vymazal et al., 1998).

### 2.2.6 Importance of Removing Nitrogen and Phosphorus from Greywater

In general, there is no significant removal of phosphorus and nitrogen in decentralized greywater treatment facilities. Since greywater is commonly used for irrigation, nitrogen and phosphorus in the effluent can be used as a fertilizer substitute to some extent. Even if the water is used for flushing toilets, the relatively high concentrations of nitrogen and phosphorus are insignificant, because the effluent is routed at the end of the process to a central wastewater treatment facility. High concentrations of phosphorus and nitrogen may create a problem if and when the water is used for washing or is released into nature.

## 2.3 TREATMENT TECHNOLOGIES

Most greywater treatment systems include a combination of different treatment methods. Physical treatment, such as sedimentation or filtration, is almost always a necessary step before and/or after biological or chemical treatment. In addition, there are systems that combine aerobic and anaerobic phases in the biological treatment. The following review first presents common options for using greywater without treatment and then a variety of treatment technologies.

### 2.3.1 Use of Untreated Greywater

#### 2.3.1.1 Direct Irrigation with a Garden Hose or Bucket

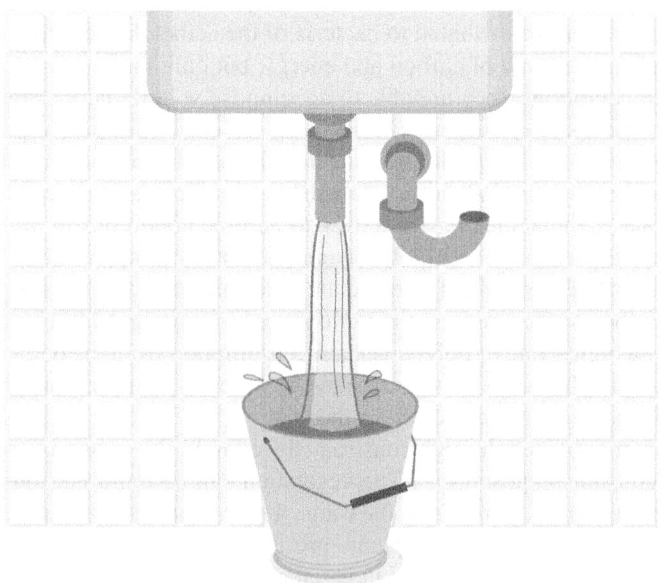

Greywater can be used for direct irrigation by simply connecting the washing machine or shower outlet to a flexible hose that can be moved to different areas in the garden or by placing a bucket under the sink and using the outgoing water. Simple solutions like these seemingly save the time and money involved in installing treatment systems. However, such uses may lead to unintentional health and environmental risks. These methods of greywater use usually involve significant human exposure to the water, either directly or indirectly via contact with irrigated areas. In addition, direct discharge of water onto the ground, especially in one place, may bring with it a bad smell and mosquitoes if seepage of water is insufficient. If the water is not used immediately (i.e., if the greywater is standing in a tank before being used), the sanitary risk is increased because certain pathogenic bacteria may survive and even grow in the tank.

Another sanitary risk is the seepage of pollutants into groundwater, especially in areas and seasons where groundwater level is high. In addition to sanitary risk, there is an environmental risk: detergents, oils, and organic load might change soil properties and cause damage, especially in the long run (see discussion of greywater irrigation in Chapter 3). This way of using greywater is approved (but not recommended) in some countries that allow greywater to be used for irrigating home gardens, probably due to the assumption that the quantity of greywater used is limited, as well as the duration of its usage.

#### 2.3.1.2 Mulch Basins

In this greywater treatment method, a trench at a depth of 30 cm is dug around the irrigated plant and filled with dry organic material such as dry leaves, crushed trimmings, and straw. Greywater enters through a perforated pipe into the trench, so no water is exposed on the surface. The organic matter absorbs the greywater, and, according to proponents of the system, this accelerates aerobic decomposition of organic matter by organisms living in the soil. The sanitation hazard is lower with this method than with direct irrigation on the surface because the water is always below the surface and does not come into contact with humans or animals.

Using the ground itself as a *treatment facility* for raw greywater is not recommended because its efficiency is unknown and there is a risk of contamination of groundwater, particularly in areas where the groundwater level is high. Moreover, soil might be negatively affected by this practice. This concern arises mainly when greywater is used over a long period, much as with the use of untreated wastewater (Figure 2.8).

#### 2.3.1.3 Filtering and Diversion Systems

Filtering and diversion systems typically include a coarse filter and sometimes a fat separator, as well as piping leading to underground irrigation. In most Australian states, the use of these systems is conditional on the system receiving a stamp of approval (from the authorities) and sometimes on approval of its installation. The emphasis is on the use of greywater for irrigation only, according to a garden's needs, so that flooding does not result in an increase in the water on the ground or leaving the plot. There are simple systems in which the user is responsible for directing water down a drain, presenting a risk of overwatering during rain. There are also more sophisticated systems that can automatically divert the irrigation to different lots, or to the drain, based on soil moisture sensors.

As with mulch basins, the sanitation and health risk with filtering and diversion systems is low (except in areas where the groundwater level is high) thanks to the subsurface irrigation. The environmental risk is comparable to that of mulch basins,

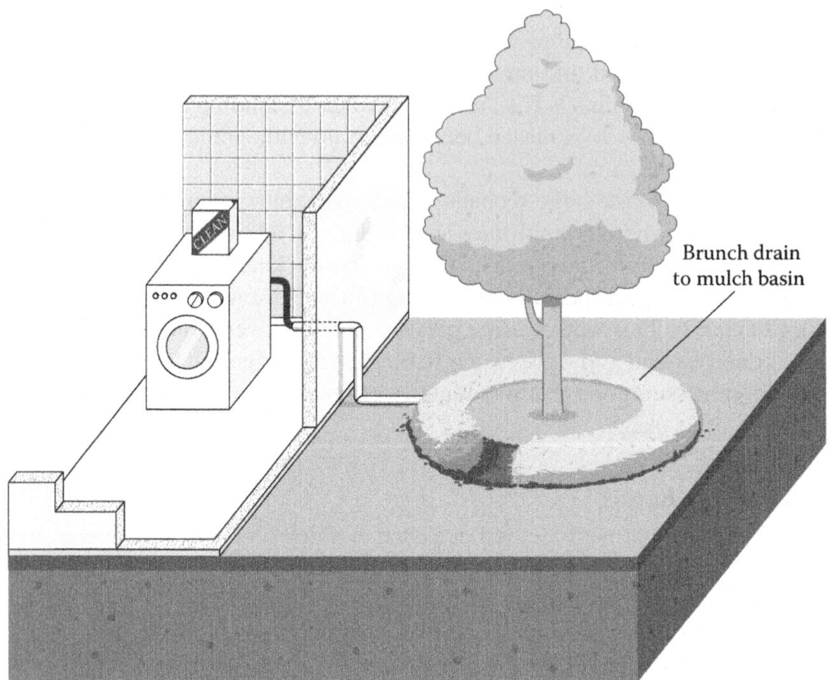


**FIGURE 2.8** Schematic description of a mulch basin. (From http://www.warrendesign.com/altsepticsystem.htm.)

but there is better control over the amounts of water used for irrigation, thus decreasing the risk of overwatering.

In both of these methods, the capacity of the soil to hold water and treat it is exploited. Therefore, to avoid excessive watering, the irrigation system must be carefully designed according to the nature of the soil and the needs of the plants (Figure 2.9).

### 2.3.2 Constructed Wetlands

Constructed wetlands are engineering systems for wastewater treatment that mimic natural habitats. In the past, natural wetlands were used as sites to absorb sewage because of their ability to tolerate high organic loads. Over the years, in order to protect the natural basins and to improve wastewater treatment efficiency, the construction of artificial wetlands took off. Constructed wetlands were found to be suitable for recovering wastewater, in particular in remote areas. Recently, their use has increased even in urban areas, especially in the suburbs. The treatment they provide is relatively inexpensive, their operation and maintenance are simple, and their treatment effectiveness is high. The principle of the operation of constructed wetlands is based on a slow water flow, in relatively shallow basins, during which the biodegradable material is metabolized by the microorganisms that develop naturally in an aqueous

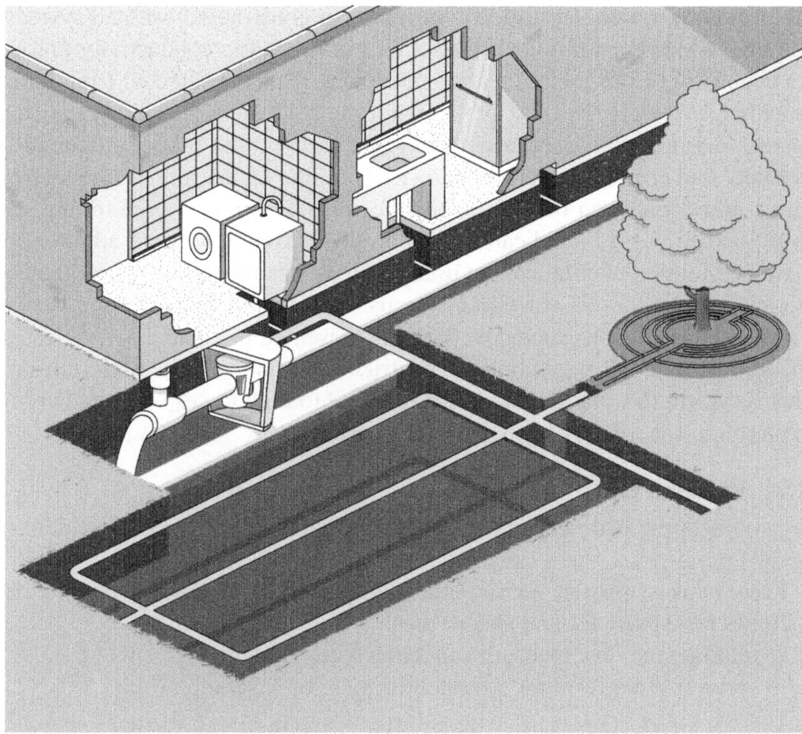

**FIGURE 2.9** Schematic description of a filtering and routing system. (From http://www.awws.us/category/products.)

environment. To encourage the development of a microbial population, a substrate material of large surface area, and/or plant roots, are often used as a substrate for the microorganisms' settlement. Constructed wetlands are complex ecological systems that utilize natural processes for water treatment, taking advantage of physical mechanisms (filtration and sedimentation), chemical mechanisms (adsorption), and biological mechanisms (microbial processes, use of vegetation) (Kadlec and Knight, 1996).

### 2.3.2.1 Role of Vegetation in Constructed Wetlands

One of the basic elements in constructed wetlands systems is vegetation. Some of the plants in constructed wetlands have adapted so that they are able to cope with anoxic conditions in the root area. Some can also tolerate large amounts of organic matter as well as the by-products of the aerobic and anaerobic decomposition of organic materials (EPA, 2000). The role of vegetation in constructed wetlands is still not entirely clear, and different researchers have come to different conclusions. Some stress the importance of vegetation in the water treatment process (Brix, 1997; Vymazal, 2005), for example, the fact that plants are able to absorb about 3–3.5 g of nitrogen per $m^2$ per day (Kadlec and Knight, 1996). Others downplay the importance of this fact, proposing that the benefits of vegetation are mainly aesthetic (EPA, 2000).

The means by which vegetation contributes to constructed wetland systems can include the following: a physical impact on the hydraulic conductivity and on the filtering capacity of the basin; creation of a microclimate protected from radiation, cold weather, and wind; use as a substrate for the settlement of microorganisms that decompose organic matter; release of oxygen, carbon, and other compounds from the root area that encourage the breakdown of organic matter, nitrification, and denitrification and could even be used as disinfectants; absorption of nutrients, metals, and salts; and use as a bioindicator of the system's functioning. In addition, plants that have economic potential can be used, such as ornamental plants, vegetation for bioenergy, or fibers for construction materials (Shelef et al., 2013). A summary of the roles of vegetation in constructed wetlands is presented in Table 2.3.

One disadvantage of vegetation in constructed wetlands is the high transpiration rate that is typical to water plants, which can lead to high losses of treated greywater by evaporation, specifically in small on-site systems.

#### 2.3.2.2 Wetland Bed

The functions of the bed in constructed wetlands are as follows (EPA, 2000):

1. To serve as a substrate for settlement and growth of microbial populations
2. To filter the particles and absorb them
3. To regulate the flow entering and leaving the system
4. To serve as a platform for the establishment of vegetation roots

The bed can be made of gravel, tuff, woodchips, and/or beads with a relatively large surface area. Plastic beads produced especially for constructed wetlands have the advantage of a large surface area, and they are lighter than gravel. Tuff has a greater surface area than gravel, but it tends to crumble over time. The substrate dimensions are highly significant in terms of surface area available for attachment of bacteria, porosity, and hydraulic conductivity. As a rule of thumb, for larger grains like sand and pebbles, porosity is proportional to *hydraulic conductivity* (but this is not the case for fine soils like clays). The size of the pores in these types of soil determines the nature of the flow of the water through the substrate and the nature of air transport (in the case of unsaturated flow). Smaller pores allow fine filtering but quickly become clogged. For example, fine sand has small pores that tend to clog faster and therefore come less recommended. Thus, the recommended substrate diameter is about 30–20 mm. To reduce the risk of clogging, it is advised that a substrate with a larger diameter grain (40–80 mm) be used in the entrance and exit areas of the basin (EPA, 2000).

For the vegetation establishment, it is advised to add planting bedding at a thickness of about 100 mm on the treatment substrate. The planting substrate should be a finer grain (>20 mm), but large enough not to enter the spaces between the substrate particles underneath.

Constructed wetlands can be divided into three main types, as shown in Figure 2.10: (a) free-water surface and (b) subsurface horizontal-flow constructed wetlands, in which the greywater flows horizontally through the wetland, and (c) vertical-flow constructed wetlands (VFCW), in which the water flows vertically through a highly permeable bed

**TABLE 2.3**
**Possible Roles of Vegetation in Constructed Wetlands**

| Roles of Vegetation in Constructed Wetlands | Sources |
|---|---|
| *Roots' structure—physical influences* | |
| Filtration | Vymazal (2011) |
| Reducing the speed of flow encourages deposition and reduces suspension | Vymazal (2011) |
| Preventing clogging in the substrate | Brix (1994) |
| Improving hydraulic conductivity | Petticrew and Kalff (1992), Brix (1997) |
| Macrophytes do not contribute to hydraulic conductivity and even cause clogging | Brix (1997), Stottmeister et al. (2003) |
| No influence on removal of SSs | Vymazal (2011) |
| *Roots as substrate for microorganisms* | |
| Surface supply for clinging of microorganisms | Brix (1997), Vymazal (2011) |
| *Gas and secretion release by the roots* | |
| Oxygen release—creating another aerobic niche. | Hammer and Bastian (1989), Armstrong and Armstrong (1990), Luederitz et al. (2001), Vymazal (2011) |
| Oxygen release—increased aerobic decomposition | Barko et al. (1991), Sorrell and Boon (1992) |
| Oxygen release—support of the deposition of heavy metals | Vymazal (2011) |
| Oxygen release—increased nitrification | Brix (1997), Yang et al. (2001), Fraser et al. (2004) |
| Carbon secretion—increased denitrification | Munch et al. (2005); Ruiz-Rueda et al. (2009) |
| Oxygen fluctuations have a limited effect in horizontal constructed wetlands | Vymazal (2011) |
| Release of antibiotics, phytometallophores, and phytochelatins | Drobot'ko et al. (1958), Seidel (1964), Seidel (1976), Vymazal (2011) |
| The roots' secretions encourage chelation of metals that reduces their toxicity | Vymazal (2011) |
| *Absorption* | |
| Nutrient storage | Gersberg et al. (1986), Wathugala et al. (1987), Tanner et al. (1995), Vymazal (2011) |
| Nutrient absorption by plants is marginal | Brix (1994), Geller (1997), Lantzke et al. (1998), Langergraber (2005), Langergraber and Simunek (2005), Brisson and Chazarenc (2009), Vymazal (2011) |
| Phytoremediation of metals | Salt et al. (1995), Weis and Weis (2004) |
| Phytoremediation of salts | Shelef et al. (2012) |
| *Microclimate conditions* | |
| Reducing the light that limits the growth of algae | Brix (1997) |

(*Continued*)

**TABLE 2.3 (*Continued*)**
**Possible Roles of Vegetation in Constructed Wetlands**

| Roles of Vegetation in Constructed Wetlands | Sources |
|---|---|
| Insulation from cold weather in the winter | Čížková-Končalová et al. (1996), Smith et al. (1997), Brix (1998) |
| Insulation from radiation in the spring | Haslam (1971), Brix (1994) |
| Reducing the speed of wind | Vymazal (2011) |
| Stabilizing the sediment surface | Vymazal (2011) |
| *Other roles* | |
| Pathogen removal | Wand et al. (2007) |
| Insect and odor control | Wood (1995) |
| Creating decorative gardens | Nelson et al. (2008), Tencer et al. (9002) |
| Increasing the variety of wildlife | Brix (1997) |
| Aesthetic appearance | Brix (1994), Wood (1995) |
| Bioindicators | Shelef et al. (2011) |
| *Plant production* | |
| Producing decorative plants | Belmont and Metcalfe (2003), Zurita et al. (2009) |
| Producing fibers for construction materials | Vymazal (2011) |
| Crops for bioenergy | Vymazal (2011) |
| Animal feed | No reference |

*Source:* Shelef, O. et al., *Water*, 5, 405, 2013.

and is collected in drains (Vymazal, 1998). Over the years, several modifications were added on the three basic designs (Figure 2.11).

### 2.3.2.3 Surface-Flow Constructed Wetlands

Constructed wetlands with surface water flow can be divided according to the chosen vegetation type: floating, submerged, or immersed. Water plants infuse oxygen into the rhizosphere area, thus encouraging aerobic decomposition processes. Floating plants (such as the water hyacinth) have an extensive root system that help filter suspended particles from the water and are used as a growth medium for organisms that decompose the organic matter. Coverage of the surface water with plants prevents sunlight penetration and the development of algae on the one hand, but on the other hand may inhibit the penetration of oxygen into the water. When there is no regular collection of the plants and they decay in the water, this may lower the oxygen concentrations in the water. Submerged plants need sunlight to penetrate the water in sufficient quantity for photosynthesis and therefore are not suitable for treatment where the water has high turbidity. In systems in which submerged bank plants are planted, the plants themselves are used as an adhesion substrate and as a factor that slows down the water flow and therefore encourages the deposition of suspended particles. Most of the oxygen in these basins enters the water from the surface (Vymazal et al., 1998).

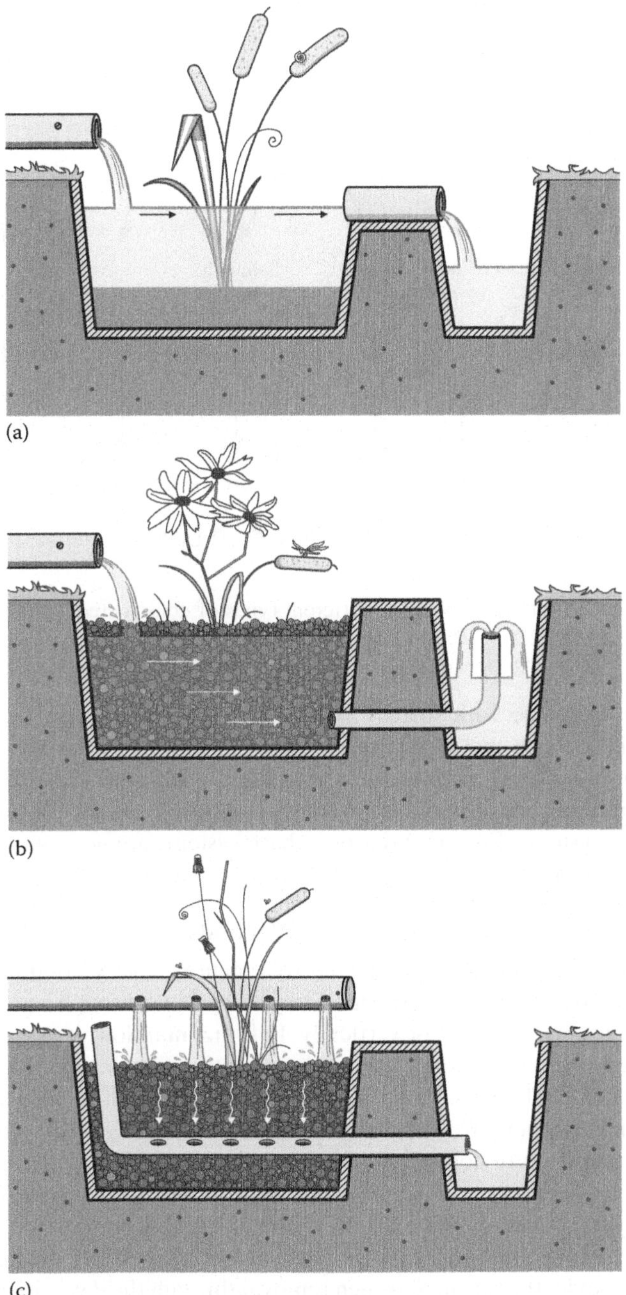

**FIGURE 2.10** (a) Constructed wetland with surface flow. (b) Constructed wetland with subsurface horizontal flow. (c) Constructed wetland with subsurface vertical flow. (Based on Vymazal, J. et al., *Constructed Wetlands for Wastewater Treatment in Europe*, Backhuys Publishers, Leiden, the Netherlands, 1998, 366pp.)

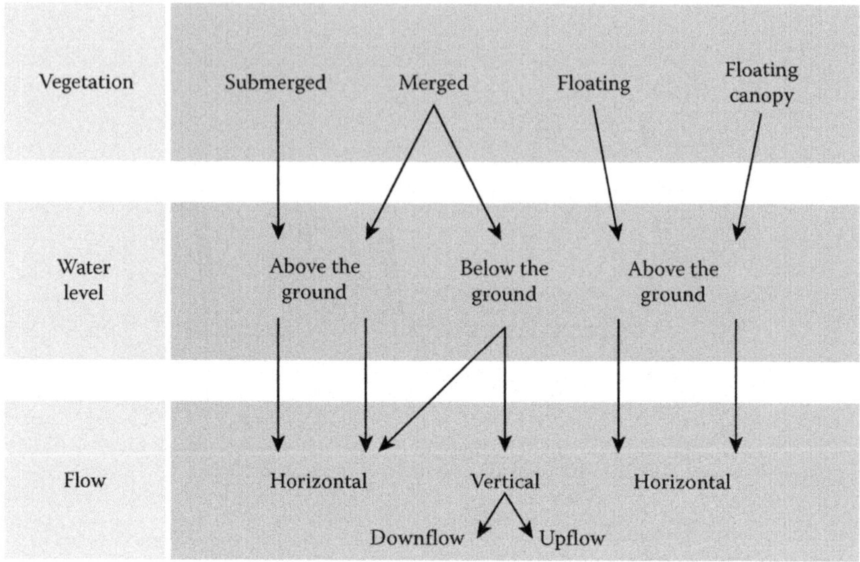


**FIGURE 2.11** Scheme describing the different types of constructed wetlands. (According to Vymazal, J., *Water*, 2(3), 530, 2010, doi:10.3390/w2030530.)

The main challenge with surface water flow systems (beyond the large size of the system) is the risk that they can provide a habitat for mosquitoes, especially when there is a high organic load and anaerobic conditions. The development of anaerobic conditions may also cause bad odors, which poses another problem, especially in home systems. For these reasons, these systems are not commonly used for greywater treatment.

### 2.3.2.4 Subsurface-Flow Constructed Wetlands

The subsurface-flow constructed wetland treatment is a very common wastewater treatment method used in home systems. Greywater flows through a porous substrate of different sizes, horizontally or vertically. In horizontal flow, water enters through an opening on one side of the basin and goes out on the other side at the same height; the water level does not rise above the substrate level. In vertical flow, water is streamed in the upper part of the basin and seeps through and exits at the bottom. Because the substrate is not saturated with water, vertical-flow systems are characterized by higher oxygen concentrations than those of horizontal-flow systems. The high oxygen concentrations enable faster decomposition of the organic matter and consequently can reduce the facility area needed. However, because conditions are not anoxic, there is no nitrogen removal through the denitrification process. Greywater is not rich in nitrogen as compared with full domestic sewage (where the main nitrogen source in wastewater is urine), and there is an advantage in keeping the nitrogen in the water. As a result, if treated greywater is used for irrigation, excess nitrogen is not a widespread problem.

A major drawback of constructed wetlands is that in order to achieve good results, a relatively large area should be used that limits its use for households. This is especially problematic in countries like Israel where space is an expensive commodity. To increase the effectiveness of treatment and save space, recirculating constructed wetlands were developed in which part or all of the entire volume of water is recirculated several times through the treatment system.

Because they are natural dynamic systems, constructed wetlands are highly complex, influenced by many factors, and characterized by high variability. This complexity makes it difficult to meet water quality standards under changing conditions over time. Several approaches to planning constructed wetlands have been suggested by various researchers, each based on a different model; some use regression equations, others on equations of zero order and first order (Reed et al., 1995; EPA, 2000; IWA, 2000), but no model accurately describes all of the existing data. Nevertheless, for most practical purposes if the quality and quantity of water to be treated are known, as well as the treated effluent quality required, it is possible to design the appropriate type of wetland and predict its performance. When designing wetlands, one must consider natural processes that create background concentrations of certain pollutants such as nitrogen and suspended material, which should be removed from the water. Environmental factors such as seasonal changes in temperature and precipitation also influence treatment efficacy outside the designer's control. To design a treatment process using constructed wetlands, several variables have to be determined:

1. *Pollutant concentration of the treated effluent required according to the relevant regulations*
2. *Background concentration of pollutants*
3. *Pollution load per unit area*
4. *Retention time* required for bacteria to break down the organic matter and nutrients (if required)
5. *Hydraulic conductivity and flow velocity at the facility*: qualities that affect the nature of the flow, sedimentation, and absorption of solids.

These variables dictate the size of the facility (i.e., volume, area, substrate size, and the amount of substrate) and its hydraulic properties, enabling the greywater to flow through without spilling and clogging. Ultimately this determines its effectiveness.

### Example 2.2: Designing a Horizontal-Subsurface Flow Constructed Wetland (based on EPA, 2000)

Let us suppose that we want to plan a constructed wetland for a neighborhood. The monthly maximal flux $(Q) = 50$ m$^3$/day and the maximal contaminant concentrations of the raw water are $BOD_5$ 165 mg/L and TSS 140 mg/L. According to regulation, the exiting concentrations must be 10 mg/L for $BOD_5$ and TSS.

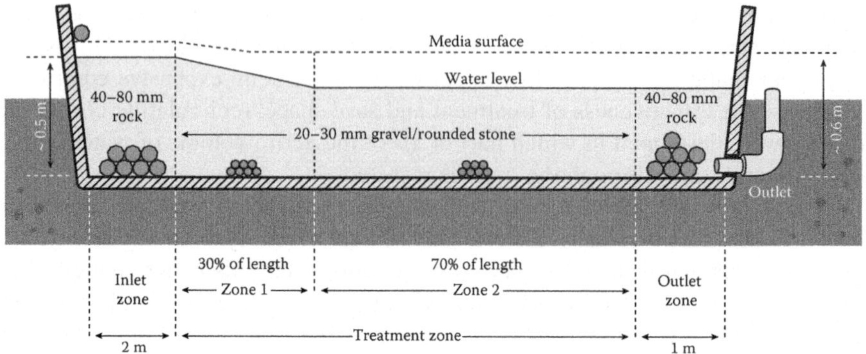


**FIGURE 2.12** Schematic cross section of constructed wetland with horizontal subsurface flow. (From EPA, *Constructed Wetlands Treatment of Municipal Wastewater*, 2000.)

There are two basic elements to consider in planning a constructed wetland:

1. The constructed wetland has four regions (Figure 2.12):
   Inlet zone, usually around 2 m in length, containing a substrate of pebbles 40–80 mm in diameter.

   *Zone 1 (first treatment area)*, consisting of 30% of the basin length (excluding the inlet and outlet zones). This is the area where the main treatment takes place so the hydraulic conductivity of the substrate in this region equals 1% of the *clean* hydraulic conductivity (as measured in the laboratory). In this example there would have to be crushed gravel with a diameter of 20–30 mm. This is due to a high solid and biomass load that clogs the pores of the substrate followed by reduced hydraulic conductivity.

   *Zone 2 (second treatment area)*, consisting of 70% of the basin length (with the same substrate as in Zone 1). The hydraulic conductivity in this area is about 10% of the substrate *clean* conductivity.

   Outlet zone of about 1 m in length, with a substrate identical to the one at the entrance.
2. The liquid flow in the horizontal constructed wetland can be calculated with Darcy's law.
   Here are the design stages:
   - Determine areal loading rate (ALR) in g/m$^2$ day (as for multiplication).
   - Determine constructed wetland (CW) width using Darcy's law.
   - Determine CW length and head loss of the initial treatment zone using Darcy's law.
   - Determine CW length and head loss of the final treatment zone using Darcy's law.
   - Determine bottom elevation using bottom slope.
   - Determine water elevations throughout the CW using head loss.
   - Determine water depth accounting for slope and head loss.
   - Determine media depth.
   - Determine number of cells.

*The following are the assumptions made for the design*:

a. ALR to $BOD_5$—6 gr $BOD_5$ per cubic meter/day; ALR to TSS—20 gr TSS per cubic meter a day
b. The use of gravel, 20–30 mm in diameter, with "clean" hydraulic conductivity of 100,000 m/day.
c. Bottom gradient 0.005%
d. Water depth at the entry to the basin: 0.4 m ($D_{WO}$)
e. Water depth at the beginning of the second area: 0.4 m ($D_{Wf}$)
f. Substrate depth: 0.6 m ($D_m$)
g. Maximal allowed head loss in the first treatment area: 10% of the substrate depth—0.06 m ($dh_i$).

For the course of design:

The cross-sectional area required to remove the two pollutants can be calculated by

$$A_S = \frac{Q \cdot C}{ALR} \tag{2.28}$$

where

$Q$ denotes the content discharge ($m^3$/day)
$C$ denotes the pollutant concentration ($g/m^3$)

For the $BOD_5$ removal:

$$A_S = \frac{50 \dfrac{m^3}{day} \cdot 165 \dfrac{g}{m^3}}{6 \dfrac{g}{m^2 \times day}} = 1375\ m^2 \tag{2.29}$$

For the TSS removal:

$$A_S = \frac{50 \dfrac{m^3}{day} \cdot 140 \dfrac{g}{m^3}}{20 \dfrac{g}{m^2 \cdot day}} = 350\ m^2 \tag{2.30}$$

3. We will use the larger area to ensure the removal of the two pollutants.
The required area for the first treatment area: 30% of 1,375 $m^2$ = 413 $m^2$
The required area for the second treatment area: 70% of 1,375 $m^2$ = 962 $m^2$

$$Q = K \cdot W \cdot D_{Wf} \cdot \frac{dh}{L} \tag{2.31}$$

where $L$ denotes the length of the first area

$$L = \frac{A}{W} \tag{2.32}$$

Subtracting these equations and isolating the width ($W$),

$$W^2 = \frac{Q \cdot A}{K_i \cdot dh_i \cdot D_{WO}} \tag{2.33}$$

By substituting the values,

$$W^2 = \frac{50\,\frac{\text{m}^3}{\text{day}} \cdot 413\,\text{m}^2}{100{,}000\,\frac{\text{m}}{\text{day}} \cdot 0.01 \cdot 0.06\,\text{m} \cdot 0.4\,\text{m}} = 860\,\text{m}^2 \tag{2.34}$$

$$W = 29\,\text{m}$$

This is the minimal width required to avoid overflow from the basin. The designer should plan a basin with a width greater than this value.

Calculating the length of the first treatment area,

$$L = \frac{A_{Si}}{W} = \frac{413\,\text{m}^2}{29\,\text{m}} = 14\,\text{m} \tag{2.35}$$

4. Calculating the length of the second treatment area,

$$L = \frac{A_{Sf}}{W} = \frac{962\,\text{m}^2}{29\,\text{m}} = 33\,\text{m} \tag{2.36}$$

$$\text{Total area} = \frac{50\,(\text{m}^3/\text{day}) \cdot 165\,(\text{g/m}^3) \cdot 2\,(\text{m}^2)}{60\,(\text{g/day})} = 275\,\text{m}^2$$

The total length of the facility can thus be calculated: 2 m at the entrance area + 14 m for the first treatment zone + 33 m for the second treatment area + 1 m for the exit area = 50 m long by 29 m wide. When planning constructed wetlands for wastewater treatment, it is customary to build two identical and parallel systems for backup and maintenance. In the treatment facilities for greywater of a single house or a small number of houses, building several parallel facilities would make the project costly. Therefore, several treatment facilities are not built in parallel, and in case of failure or closure of the facility for maintenance, greywater is discharged to the municipal wastewater collection system.

There are other methods for calculating the size of the required basin that introduce additional variables, such as temperature, into the calculation. Alternatively, when the goal is to remove a specific pollutant, such as TSS or nitrogen, the basin should be designed accordingly. (Additional equations for planning constructed wetlands can be found in Vymazal et al. (1998) and Reed et al. (1995).)

#### 2.3.2.5 Vertical Flow Constructed Wetland

As mentioned, in a vertical constructed wetland, the flow is not saturated, and the water enters both a liquid phase and a gas phase in the substrate. This allows oxygen to pass into the biofilm and thus increases the rate of biodegradation. The ability of air to penetrate through is influenced by the substrate particle size. In systems where a trickling filter is used, the system is open to the atmosphere at the top and bottom. This allows the transport of air into the substrate as a result of the temperature difference between the liquid and the air outside. Several other greywater treatment systems force air to flow through the substrate, for example, the passively aerated vertical bed (Green and Friedler, 1998; Alfiya et al., 2007).

A review of the performance of constructed wetlands with vertical flow found an optimal ratio of approximately 2 $m^2$ of basin area per person, or 60 g $BOD_5$ per day, or 120 g of COD per day (Molle et al., 2005).

**Example 2.3: Calculating the Area of a Vertical-Flow Constructed Wetland Basin (based on Molle et al., 2005)**

*The following are given*:

$Q = 50$ m³/day

$BOD_5 = 165$ mg/L

$COD = 342$ mg/L

$TSS = 140$ mg/L

This is based on the assumption that $A = 2$ $m^2$ per 60 g of $BOD_5$, or 120 g of $COD$.

Hence, the basin area equals

$Total\ area = 50(m^3/day) \times 165(g/m^3) \times 2(m^2)/60(g/day) = 275\ m^2$

Usually, the total area is divided into a number of basins with the parallel basins operating intermittently. In the preceding example, the area can be divided into five basins of 55 $m^2$ each (275/5 = 55), three of which will form the first stage of treatment and two of which will form phase B (Figure 2.13). In stage A, it is possible to operate one basin out of the three and replace it when necessary to prevent clogging. In stage B, the basins can be operated intermittently so that at any given time, one basin is active and the other is idle.

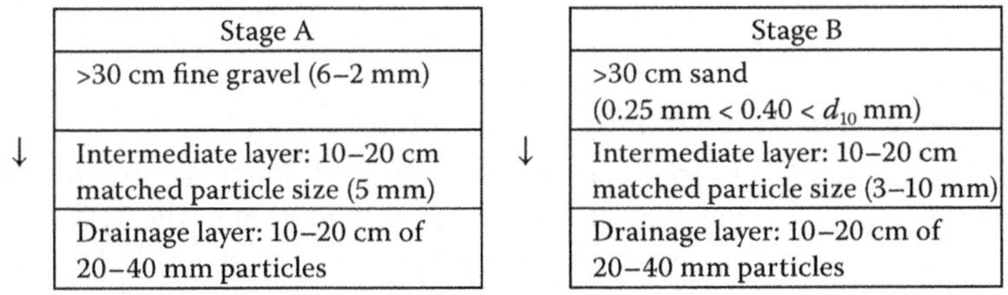

| | Stage A |
|---|---|
| | >30 cm fine gravel (6–2 mm) |
| ↓ | Intermediate layer: 10–20 cm matched particle size (5 mm) |
| | Drainage layer: 10–20 cm of 20–40 mm particles |

| | Stage B |
|---|---|
| | >30 cm sand (0.25 mm < 0.40 < $d_{10}$ mm) |
| ↓ | Intermediate layer: 10–20 cm matched particle size (3–10 mm) |
| | Drainage layer: 10–20 cm of 20–40 mm particles |

**FIGURE 2.13** Schematic structure of a constructed wetland with vertical flow. (Following Molle et al., *Water Sci. Technol.*, 51(9), 11, 2005.)

## CASE STUDY: AN EXTENSIVE TREATMENT SYSTEM

Extensive treatment systems are often designed to treat greywater in rural or otherwise remote areas with irregular power supply. In these cases, systems without pumps with low electrical consumption are required. These systems usually include a settling basin and a form of biological treatment.

For example, a treatment system in Kathmandu, Nepal, treats greywater from the kitchen, bathroom, and laundry of a house with seven occupants (Morel and Diener, 2006). Greywater is collected in a 500 L settling tank, and from there it flows into a regulation tank with a 200 L siphon that feeds a VFCW. The area of the constructed wetland is 6 $m^2$. The substrate consists of a 20 cm thick bottom layer of 20–40 mm gravel, and over that there is a 10 cm thick layer of 10 mm gravel. The top layer is 60 cm of coarse sand, and the plants *Phragmites karka* and *Cinna latifolia* were planted in the bedding. The treated greywater is collected into a 700 L container and is used for irrigation, car washing, and flushing toilets. The system treats about 500 L/day on average.

### Disadvantages

The system consists of several stages, and its space requirement is relatively high, so it is suitable for areas where land is not expensive (i.e., suburban or rural areas).

### Advantages

The quality of the treated greywater was stable over the monitoring period of the system (about 2 years). The system needed only a few maintenance operations, such as emptying the sludge from the settling tank once a year, checking the regulation tank, and annual pruning. On the other hand, there was no need to clean the substrate of the constructed wetland or replace the gravel and sand, and there was no need to clean the storage tank of the treated water.

The cost of the system was relatively low (at the prevailing prices in Nepal), at about 430 US$. Annual savings in expenditure on water was 40 US$ considering that the price of water in Nepal is 0.23 US$ $m^3$. Water savings pays for the cost of the system after 10 years. In countries with higher construction costs, it is reasonable to assume that the price of water is also higher, and therefore the savings on water expenditure will be higher.

The water quality from this system, and a comparison with other case studies, is displayed in Table 2.9.

#### 2.3.2.6 Systems of Recirculating Constructed Wetland

To overcome the need for large area required by traditional constructed wetlands, recirculation of all (or a portion) of the treated water through the system was suggested. These recirculating constructed wetlands are much more compact without compromising the treated water quality. Various designs have been developed including hybrid systems that combine different types of constructed wetlands (horizontal and vertical flow in series) to exploit the advantages of each treatment method.

One example of such a system is the recirculating vertical-flow constructed wetland (RVFCW). It consists of two tanks: the top tank that is in itself a VFCW and the bottom tank that is used as a reservoir. Greywater is passed back and forth between the top and bottom tanks. Once greywater enters, it is recirculated for 6–8 h after which it is pumped out for reuse (Figure 2.14).

This system has several advantages: the recirculation allows a longer retention time with a low footprint; the unsaturated flow through the substrate, and the fact that the substrate is open at the top and bottom, allows passive transport of air (oxygen) into the substrate, similar to the mechanism of oxygen transport in trickling filters; increased oxygen supply enhances the rate of organic matter decomposition, and it allows nitrification; and finally, the aerobic conditions of the facility prevent noxious odors (Gross et al., 2007b).

Mass balance can be used to characterize the system and predict its performance. The treatment element (source/sink) $r$ can be described by a first-order reaction model. In fact, the main treatment takes place in the constructed wetland bed and

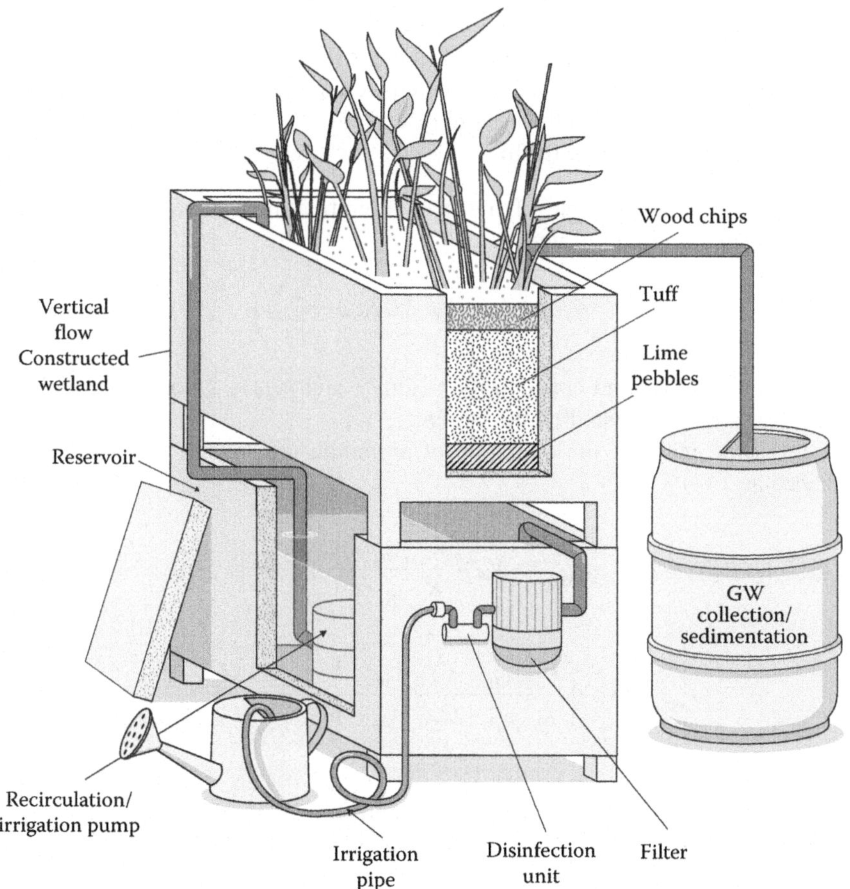


**FIGURE 2.14** RVFCW.

not in the reservoir; therefore, the mass balances have to be separated as follows (Sklarz et al., 2010):

For the mass balance of variable $i$ in the substrate,

$$V_R \cdot \frac{dC_{i.R}}{dt} = Q \cdot \left(C_{i,R} - C_{i,W}\right) + r_i \tag{2.37}$$

For the mass balance of variable $i$ in the storage tank,

$$V_R \cdot \frac{dC_{i.R}}{dt} = Q \cdot \left(C_{i,W} - C_{i,R}\right) \tag{2.38}$$

where

$C_{i,W}$ is the concentration of variable ($i$) in the substrate (g/m$^3$)
$C_{i,R}$ is the concentration of variable ($i$) in the reservoir (g/m$^3$)
$r_i$ is the first-order source/sink element of variable $i$
($g/h^1$) is related to removal or biodegradation
$i$ -TSS, $BOD_5$, $NH_4^+$, $NO_2^-$, and $NO_3^-$ in the different equations
$V_W$ is the water volume in the constructed wetland (m$^3$)
$V_R$ is the water volume in the bottom reservoir (m$^3$)
$Q$ is the recirculation rate (m/h)

The following is an example of the first-order treatment equations for TSS and ammonia oxidation into nitrate:

$$r_{TSS} = -k_{TSS} \cdot V_W \cdot \left(C_{TSS.W} - C_{TSS}^*\right) \tag{2.39}$$

$C_{TSS}^*$ is the background concentration (g/m$^3$), and $k_{TSS}$ is 22 on average (1/h) for domestic wastewater treated by the system.

In the same manner, the oxidation of ammonia into nitrite and nitrate can be described as

$$r_{NH_4^+} = -V_W \cdot \left(\frac{k_{max,NH_4^+} \cdot C_{NH_4^+,W}}{K_{S,NH_4^+} + C_{NH_4^+,W}}\right) \tag{2.40}$$

$$r_{NO_2^-} = V_W \cdot \left(\frac{k_{max,NH_4^+} \cdot C_{NH_4^+,W}}{k_{S,NH_4^+} + C_{NH_4^+,W}} - \frac{k_{max,NO_2^-} \cdot C_{NO_2^-,W}}{K_{S,NO_2^-} + C_{NO_2^-,W}}\right) \tag{2.41}$$

$$r_{NO_3^-} = V_W \cdot \left(\frac{k_{max,NO_2^-} \cdot C_{NO_2^-,W}}{K_{S,NO_2^-} + C_{NO_2^-,W}}\right) \tag{2.42}$$

The average reaction constants found for this system for domestic wastewater are summarized in Table 2.4.

**TABLE 2.4**
**Average Reaction Constants in an RVFCW for Greywater Treatment**

| Variable | Reaction Constant ($k$ 1/h) |
|---|---|
| TSS | 22 |
| $NH_4$[a] | 9 |
| $NO_2$[a] | 115 |

[a] Additional details about setting the reaction constants can be found in Sklarz et al. (2010).

## CASE STUDY: INTENSIVE COMMERCIAL TREATMENT SYSTEM

A company called Nubian markets a treatment system called Oasis GT600 in Australia and elsewhere in the world. The system includes a collection tank or ditch for greywater collected from the showers, baths/hot tub, handbasins, and laundry (NSW-Health, 2011).

The system consists of (1) a solid separation phase that includes a 1 mm mesh and settling tank, (2) a biological treatment phase that consists of an aerated and submerged biofilter, and (3) a UV disinfection.

The system's volume is 350 L and is capable of treating up to 50 L/h (1200 L/day). The system dimensions are 1900 mm × 450 mm × 2415 mm.

### Advantages

- The system's performance is strong and meets Australian regulatory requirements.
- The system has an aesthetically pleasing appearance and is wrapped in a plastic cover (resistant to sunlight radiation), which precludes unintentional, hazardous, or unpleasant access to greywater or to treated greywater.
- The area occupied by the system is relatively small, allowing its installation in the basement of a building or in a yard.

### Disadvantages

- The system is complex and requires maintenance by professional workers.
- The water quality from this system, and a comparison with other case studies, is displayed in Table 2.9.

### 2.3.3 Activated Sludge

The activated sludge process was developed in 1913 by Clark and Gage at the experimental station Lawrence in Massachusetts, United States. It was first implemented in full scale in 1914, in a municipal wastewater treatment facility in Manchester, England (Metcalf and Eddy, 2003). The name *activated sludge* was given to the process because it is based on the production and preservation of a mass of active microorganisms capable of stabilizing waste under aerobic conditions. The aeration tank provides contact time during which the incoming wastewater is aerated and mixed with the microbial suspension. Mechanical equipment or diffusers are used for mixing and introducing oxygen into the process. From there, the mixture of biomass and wastewater is transferred into a settling tank where the microbial suspension settles and returned to the aeration tank to further biodegrade the *new* incoming organic matter (Figure 2.15). A small part of the sludge by-product, which contains mainly nonbiodegradable solids and excess biomass, is removed from the system on a periodic basis. Inadequate removal will cause excessive accumulation and ultimately an increase in the solids' concentration in the effluent (Metcalf and Eddy, 2003).

Another feature of the activated sludge process is the formation of microbial flocs, typically sized to between 50 and 200 μm, which can be removed from the system by sedimentation. Usually, in the settling tank, approximately 99% of the TSS can be removed from the water.

The activated sludge process is commonly used to treat municipal wastewater, but almost never used for greywater treatment. As such, this chapter does not review the planning process of activated sludge for the treatment of greywater.

### 2.3.4 Rotating Biological Contactor

#### 2.3.4.1 Principles of the Process

In a treatment system using an RBC, a biomass is attached on a rotating round disk surface. The disks are located on horizontal axes in a tank, at a certain level of

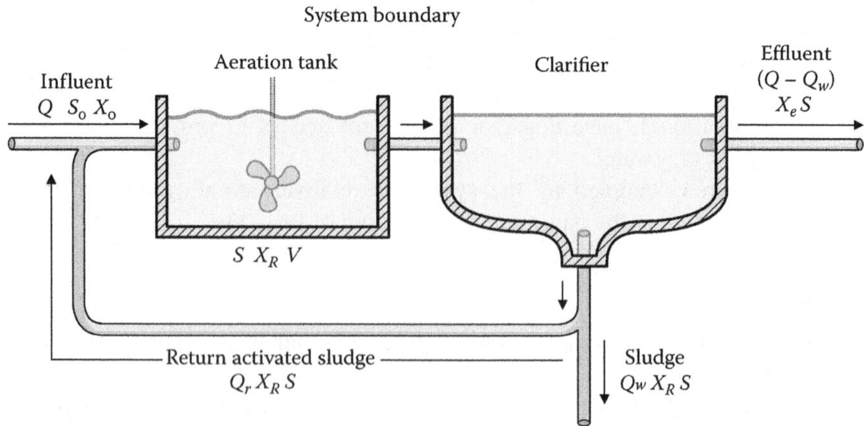


**FIGURE 2.15** Schematic description of the activated sludge process.

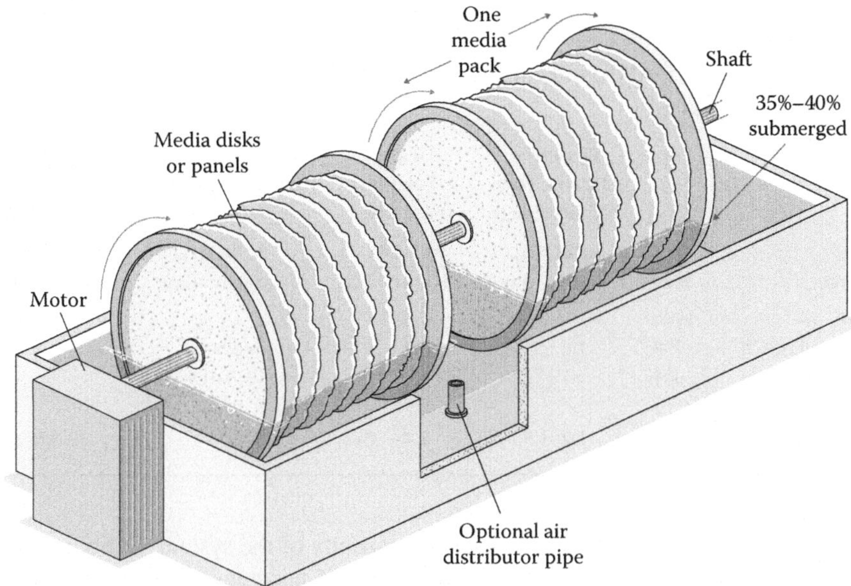


**FIGURE 2.16** Drawing of a motor-driven RBC.

immersion (Figure 2.16). The axes, which are driven mechanically or by compressed air, rotate at a speed of 1–2 rpm. In typical aerobic RBC systems, nearly 40% of the disks' area is immersed in the process container. The higher the degree of immersion (which can reach up to 85%), the lower the pressure on the axes and the disks that carry the biomass—but the aeration level of the biomass decreases as well.

The biomass that performs the treatment process consists of different bacteria and other microorganisms that stick to the rotating disks and form a biofilm. Typically (depending on the organic load), it takes 1–2 weeks for a biofilm to be well established on the disks. The biofilm tends to peel when the weight of the biomass on the disks is too heavy due to shear forces. The peeled-off biomass pieces create sludge, which must be removed by sedimentation. In general, this sludge has good settling properties. Biomass growth and sloughing are cyclical processes.

As the contactor rotates, oxygen dissolves on the part that has contact with air through diffusion on the fluid layer adjacent to the disk. The turbulence generated by the disk rotation introduces oxygen into the tank's liquid. The organisms that grow on the disk aerobically oxidize organic carbon (matter) from the wastewater through reduction of the dissolved oxygen (the electron acceptor) to $CO_2$. Some of the systems are also used to oxidize ammonia to nitrate (WEF and ASCE, 1998; Nolde, 2000).

### 2.3.4.2 Factors Influencing the Effectiveness of the RBC Process

#### *2.3.4.2.1 Organic Load*

The organic load in attached growth systems is expressed in relation to the substrate surface. The organic load in RBC should match the available surface area

**TABLE 2.5**
**Typical Organic Loads Used for Design of RBC for Treatment of Urban Wastewater**

| Source | Year | Dissolved Organic Load for First Stage $\frac{kg\ BOD_{dis}}{100\,m^2 \cdot day}$ | Organic Load for First Stage $\frac{kg\ BOD_{tot}}{100\,m^2 \cdot day}$ | Dissolved Organic Load Based on All Stages $\frac{kg\ BOD_{dis}}{100\,m^2 \cdot day}$ | Organic Load Based on All Stages $\frac{kg\ BOD_{tot}}{100\,m^2 \cdot day}$ |
|---|---|---|---|---|---|
| WEF | 1996 | 1–1.8 | 2–2.9 | 0.75–1 | 1.5–2 |
| WEF and ASCE | 1998 | 1.2 | 3.1 | 0.25–0.5 | |
| Rittmann and McCarty | 2001 | | 4.5 | 0.3–1.5 (0.5–0.8) | |
| Metcalf and Eddy | 2003 | 1.2–1.5 | 2.4–3.0 | 0.25–1 | 0.5–2 |

for biomass growth and the oxygen transfer capacity of the system. In high organic loads, the thickness of the biofilm increases, oxygen becomes a limiting factor, and a white-grey layer (which results from the growth of anaerobic filament organisms) is often generated. (See details in the section on dissolved oxygen concentration below). Problems can occur due to overload, and include bad odors, decrease in the effectiveness of the process and structural load causing the axes and the disks in the system to break. Typical values of organic loads used for the design of RBC for the treatment of wastewater can be found in the literature and are listed in Table 2.5. In spite of variation between sources, the magnitude remains the same.

Grady et al. (1999) updated an empirical model to anticipate the performance of RBC systems on the basis of $BOD_5$ concentration for design purposes:

$$\frac{S_e}{S_i} = e^{-0.4\left(A/Q\right)^{0.5}} \tag{2.43}$$

where

$S_e$ is the total concentration of $BOD_5$ in effluents after secondary sedimentation (mg/L)
$S_i$ is the influent $BOD_5$ concentration (mg/L)
$A$ is the disk (biofilm carrier) surface ($m^2$)
$Q$ is the feeding flow rate ($m^3$/day)

Using the equation, the required disk surface for designing a RBC can be found.

*Biofilm growth control*: The thickness of the biofilm is an important variable in the process; however, a distinction should be made between total biofilm thickness and the thickness of the active layer. The total biofilm thickness ranges between 70 and 4000 μm, depending on the hydrodynamic conditions. It was found that the thickness of the biomass layer in a heterotrophic biofilm in RBC systems ranges typically from 300 and 1400 μm, and the solid content in the biofilm composes on average 4.1%, of which about 3% is organic and 1.1% nonorganic. This suggests a VS/TS ratio of about 74% with the rest as water (Meng and Ganczarczyk, 2004). The $BOD_5$ removal rate within a thick biofilm is not very different from that in a thin biofilm because of the resistance to diffusion within the film. The thickness of the active portion that contributes to $BOD_5$ removal ranges from 20 to 600 μm. Oxygen concentration decreases with penetration of the biofilm depth. It has been found that there is a limit to the diffusion of oxygen and nutrients into the biofilm (Bungay et al., 1969). The authors reported that in an active layer with a thickness of 50–100 μm, the rate of $BOD_5$ removal was maximal, with no improvement using thicker biofilm. Venkataraman and Ramanujam (1998) reported a maximal $BOD_5$ removal with biofilm thickness of 70–100 μm, and Palma et al. (2003) reported a maximal $BOD_5$ removal with a biofilm thickness of 100–200 μm.

The biofilm is detached from the disks when it reaches a critical thickness and its weight becomes too heavy. To retain an effective biofilm thickness, it is possible to do one of the following: increase the speed of rotation (increasing the shear forces), reverse the direction of rotation of the disks, distribute the incoming discharge and load over a number of treatment stages, execute periodic starvation of an overloaded facility, or use a chemical treatment. The pieces of detached biofilm constitute the outgoing sludge from the facility and can be separated by settling. The settling properties of the RBC sludge are usually not an issue, allowing efficient separation from the supernatant.

*Concentration of dissolved oxygen*: As suggested earlier, diffusion of nutrients and oxygen into the biofilm dictates the removal of organic matter in RBC systems. The oxygen in the liquid is dissolved as a result of the disk rotation that lifts wastewater in the air and trickles it back toward the reactor. An acceptable concentration of dissolved oxygen to maintain aerobic biological process in RBC is 3–2 mg/L (WEF and ASCE, 1998).

It should be noted that low concentrations of dissolved oxygen in high-organic-load RBC result in the production of sulfide in the internal layers of the biofilm and consequently encourage the growth of filament organisms such as *Beggiatoa*, *Thiothrix*, and *Lepothrix*. The presence of these microorganisms creates bad odors and causes a significant decrease in the RBC's treatment efficiency. In addition, the biofilm changes in color from green-black to grey-white. Excess biomass does not sink well, and puts a strain on the hinges and the substrate disks, which may lead to their collapse.

*Rotation velocity*: Rotation velocity is an important variable that influences many factors in the system, such as the concentration of dissolved oxygen, uniform growth of biomass, and removal of excess biomass. A typical rotation speed ranges between 1 and 2 rpm (WEF and ASCE, 1998; Grady et al., 1999; Rittmann and McCarty, 2001), which is equivalent to a linear speed of 5.11–23 m/min in full-scale systems. Characteristic linear speed in a design is 18–20 m/min (Grady et al., 1999; Rittmann and McCarty, 2001).

*Direction of flow and distribution into stages*: The flow structure in an RBC requires primary sedimentation to remove any solids that may accumulate in the bottom of the cells. The flow in RBC can be perpendicular to the rotation direction of the disks, or in parallel, with no reported differences on the treatment efficiency (WEF and ASCE, 1998).

Usually, RBC is divided into stages, and each stage is in a separate cell. Each phase (cell) contains at least one rotating axis, and each axis holds several disks. It is acceptable to build RBC systems with 2–4 stages. Most of the treatment takes place in the first two stages, and the remaining steps are mainly used for polishing. With the progress of the liquid between stages, there is an exponential decrease in pollutant (e.g., $BOD_5$) concentration (Grady et al., 1999). Often, the residence time in RBC is designed for an hour, which corresponds to a 4.9 L tank volume for 1 m$^2$ of substrate surface in low density (WEF and ASCE, 1998). For example, the average hydraulic load of 0.4–012 m$^3$/m$^2$ a day is suitable for hydrocarbon $BOD_5$ removal; however, for maintaining nitrification, a typical hydraulic load is 0.1–0.04 m/day or even less. Reports can also be found citing a slightly different hydraulic load of 0.03–0.12 m/day (Metcalf and Eddy, 2003).

RBC's hydraulic load is defined as the relation between flux and the available surface area for biomass growth:

$$HL = \frac{Q}{A} \tag{2.44}$$

where

$HL$ is the hydraulic load (m$^3$/[m$^2$ × day])
$Q$ is the inlet flow rate (m$^3$/day)
$A$ is the total surface area of the disks (m$^2$)

It was found that for a hydraulic load of 40–200 L/(m$^2$ × day), an RBC unit of four stages produces effluent of better quality than that produced by a two-stage facility with the same surface area; the highest efficiency of $BOD_5$ removal is obtained in the first stage and decreases with the transition between stages. Distribution into stages is advantageous when the reaction is of first order. The more steps into which the process is divided, the closer it is to a plug flow–type reactor. Consequently, each stage has a biomass of different thickness and growth characteristics. In the first stage, the biomass will be heterotrophic since wastewater contains significant concentration of organic matter. In the last stage, once organic matter has decreased and ammonia concentration has increased, nitrification is usually the dominant microbial process. These conditions give the advantage to autotrophic nitrifying bacteria.

Other factors affecting the efficiency of the treatment in an RBC include temperature, which also affects the transition of oxygen and the biodegradation rate of the substrate. It also includes the degree of immersion, which is defined as the percentage of the disk diameter immersed in the liquid. Typical immersion ranges from 20% to 40%. A high degree of immersion reduces the oxidation ability because the retention time in the air is reduced and the biomass is mostly aerated while in contact with the air. In addition, a high degree of immersion consumes more energy due to greater friction (Rittmann and McCarty, 2001). In contrast, a low level of immersion puts a heavier load on the axes holding the disks.

RBCs are used to treat various types of wastewater including industrial wastewater, agricultural wastewater such as from aquaculture facilities, as well as domestic wastewater and greywater. In one case in Israel, RBC was used to treat greywater from 14 dormitory apartments of married students; some of them are with children (Friedler et al., 2005). The dorm facility consisted of two cells, 15 L each. The disks had a diameter of 0.22 m with a total area of 1 $m^2$. The average influx was 7.5 L/h, with a retention time of 2 h. The performance of the facility resulted in effluent of *excellent quality* (Halperin and Aloni, 2003).

RBC systems allow intensive treatment of greywater with low investment and maintenance compared to other intensive treatment. (See further expansion in Section 2.5 and Chapter 7 that discusses technological–economic aspects.)

### 2.3.5 Membrane Bioreactor

Membrane bioreactor (MBR) is a wastewater treatment system that combines biological treatment with the separation of solids through membrane filtration (Figure 2.17). The membrane creates a physical barrier and is based on its pore size allowing only certain very small colloidal particles to pass through where the rest remain in the brine. The permeate is therefore composed mainly of water, dissolved ions, and very small particles. The MBR system is an improvement on the regular activated sludge process since the biomass is separated from treated effluent by membranes rather than through sedimentation. The effective separation of biomass allows a TSS concentration of 15,000–10,000 mg/L to be reached in the biological reactor. This high concentration allows shorter hydraulic retention time in the reactor while maintaining a high sludge age, and therefore the volume of the reactor is small. The treated water allows unrestricted reuse for most purposes, with typical TSS

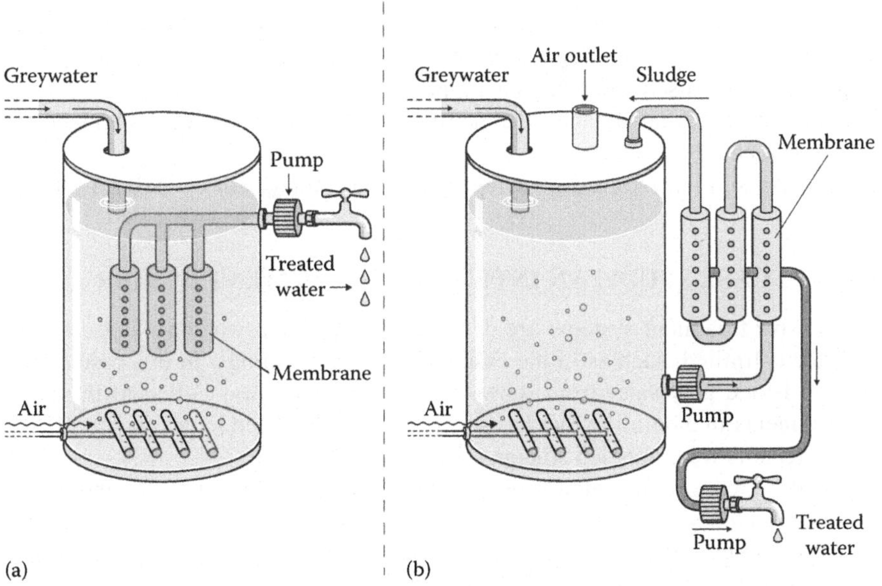


**FIGURE 2.17** Schematic diagram of MBR systems. (a) MBR with submerged membrane and (b) MBR with external side membrane.

**TABLE 2.6**
**Advantages and Disadvantages of MBR**

| Advantages | Disadvantages |
|---|---|
| Ability to treat a large volume of water with low hydraulic retention time | High construction cost |
| Relatively long solids retention time (SRT), leading to reduced production of sludge | Lack of knowledge about membrane longevity |
| Activation at low oxygen concentrations with potential for nitrification and denitrification simultaneously in systems designed for long SRT | High maintenance cost of the membranes (cleaning and replacing) |
| High-quality effluent in terms of turbidity, bacteria, TSS, and $BOD_5$ | High energy consumption |
| Compactness—low footprint | |

concentration of less than 5 mg/L and turbidity of less than 1 NTU. Moreover, membranes that are commonly used in MBR do not allow passage of bacteria since its pore size (~0.5–5 μm) is smaller than the diameter of the bacteria. Viruses can pass through this membrane because their size range is the same or smaller as that of the membrane pores, but it is rare for viruses to pass through the membrane, probably due to the biological biofouling layer that develops across it (e.g., Gilboa and Friedler, 2008).

The MBR allows the activated sludge process to be implemented compactly. Therefore, this implementation can be applied for on-site greywater reuse in densely populated areas or in public institutions such as sports facilities and office buildings. In addition, the membrane separation prevents flushing of the biomass as a result of momentary high discharge, which is typical in greywater production.

Despite its many advantages, the major drawback of the MBR is its high cost and high maintenance.

Table 2.6 summarizes the advantages and disadvantages of the MBR systems:

**CASE STUDY: AN INTENSIVE TREATMENT BY MBR**

Intensive treatment systems are designed to treat high water volumes where space is limited, such as in the case of high-rise buildings. In this case study from Israel, greywater from showers and handbasins was collected in seven apartments in a building that has seven floors and treated by an on-site MBR. The tenants were married students (Friedler et al., 2006).

In this case, the treated greywater was not recovered, but the amount of water treated was large, and the system was located in the basement of the residential building.

The greywater was collected in a 300 L regulation tank after coarse filtering with a 1 mm mesh. The MBR consisted of a process tank whose volume was 100 L. The retention time ranged between 5 and 8 h and sludge age ranged between 15 and 20 days. A centrifugal pump pushed liquid from the process tank into the membrane unit. The cross flow velocity through the membrane was 4 m/s. The pressure gradient across the membrane was set to 1 atmosphere. The filtrate was collected in a container, and the sludge was returned to the process tank. The total area of the membranes (8 units) was 0.34 $m^2$.

**Advantages**

- The quality of the treated greywater is very high.
- The area requirement is relatively low and therefore this treatment is suitable for high-rise buildings.

**Disadvantages**

- Relatively high construction and operation costs (energy and chemicals) of MBR systems relative to other systems and as a result the high annual cost

The water quality from this system, and a comparison with quality in other case studies, is presented in Table 2.9.

### 2.3.5.1 Membrane Configuration

#### *2.3.5.1.1 Sidestream Membrane*

The membrane is located outside the biological reactor in a separate cell. The wastewater stream is fed to the biological reactor, where the wastewater comes in contact with the biomass. The mixed liquid (wastewater and biomass) is pumped into the membrane unit, the concentrate is recirculated back to the biological reactor, and the effluent is discharged. In this system, the transmembrane pressure (TMP) and flow velocity are determined by the pump. Such membranes are more protected because they are in a separate container and their maintenance is simpler and more convenient.

#### *2.3.5.1.2 Submerged Membrane*

The membrane is submerged in the biological reactor, so there is no recirculation of wastewater. The effluent is separated from the biomass within the biological reactor. The TMP is determined by a pump drawing the effluent from the membrane. Some of the pressure is created by the water head (water pressure) above the membrane. As a result, the pump must overcome less head than in systems based on a sidestream membrane.

### 2.3.5.2 Membrane Characterization

This membrane should possess mechanical strength, allowing a high rate of filtration (high filtrate discharge) and a high degree of rejection (high degree of selectivity). Usually, filtrate discharge and selectivity each come at the expense of the other

**TABLE 2.7**
**Characteristics of the Different Membrane Configurations**

| Membrane Type | Application | Turbulence Degree | Cost | Area to Volume Ratio ($m^2/m^3$) | Advantages | Disadvantages |
|---|---|---|---|---|---|---|
| Pleased cartridge | MF (dead end) | Very low | Low | 800–1,000 | Strong and compact structure | Hard cleaning, easily blocked |
| Plate-and-frame | UF, RO | Medium | High | 400–600 | Detachable for cleaning | No counter current cleaning, complicated design |
| Spiral-wound | RO, UF | Low | Low | 800–1,000 | Strong and compact structure, low energy consumption | No counter current cleaning, complicated cleaning |
| Tubular | Cross flow TSS removal | High | High | 20–30 | Simple mechanical clean, can treat high TSS | Very high membrane costs |
| Capillary tube (inside-out) | UF | Medium | Low | 600–1,200 | Between tubular and hollow fiber | |
| Hollow fiber (outside-in) | MF, UF, RO | Very low | Very low | 5,000–40,000 | Counter current clean, compact, can treat high colloidal concentration | Sensible for high pressures |

*Note:* MF, micro filtration, UF, ultra filtration, RO, reverse osmosis.

because greater selectivity is achieved by small pores, which decreases membrane permeability resulting in a small filtrate discharge.

There are currently five configurations of membrane processes on the market. Each configuration has its own advantages and disadvantages. The configurations are based on a planar or cylindrical shape: capillary tube, plate and frame, spiral wound, and tubular and hollow fiber. Their characteristics are listed in Table 2.7.

### 2.3.5.3 Principles of the Process

#### *2.3.5.3.1 Types of Flow through the Membrane*

*Dead end flow.* This refers to a system that has vertical feed flow to the membrane. The particles are stopped on the membrane and form a *filter cake* on the surface.

At some point, the filter cake causes clogging of the membrane, and it is necessary to stop the filtration and clean the membrane.

*Cross flow*: The rejected phase flows in tangent to the membrane surface, and the passing phase flows through the membrane (perpendicular to the general flow direction). The large shear forces of the liquid flowing at a tangent to the membrane sweep the aforementioned biological biofouling layer, and therefore the cake layer remains thin. For this reason, it is possible to operate a system with a tangential flow in large flux for a long time.

*Completely mixed flow*: This is similar to the tangential flow, but the feed area and the filtration area are each mixed. Membrane clogging is reduced due to the large shear forces operating all the time.

#### 2.3.5.3.2 Pore Size

The types of membrane separation processes can be ranked according to the average pore size of the membrane. The classification of processes according to pore size is presented in Table 2.8.

#### 2.3.5.3.3 Filtrate Flow

The filtrate flow is expressed by the following equations:

$$J = \frac{\Delta p}{\mu R_{tot}} \tag{2.45}$$

$$\Delta p = \frac{P_f + P_r}{2} - P_p \tag{2.46}$$

$$R_{tot} = R_m + R_c \tag{2.47}$$

where

- $J$ is the flux passing through the membrane ($m^3/m^2/h$)
- $R_{tot}$ is the total resistance of the membrane ($m^2/kg/h$)
- $\mu$ is the liquid viscosity ($kg \times s/m^2$)
- $\Delta p$ is the TMP ($N/m^2$)
- $P_f$ is the feed pressure (bar or atm)
- $P_r$ is the pressure of returning flow (bar or atm)
- $P_p$ is the effluent pressure (bar or atm)
- $R_m$ is the clean membrane resistance ($m^2/kg/h$)
- $R_c$ is the cake resistance ($m^2/kg/h$)

#### 2.3.5.3.4 Membrane Resistance

The membrane resistance $R_{tot}$ is influenced by several factors: pore size, adsorption resistance of the membrane surface, the formation of biofouling, and concentration polarization.

#### 2.3.5.3.5 Shear Forces

During filtration, a filter cake is formed that clogs the membrane. A turbulent flow regime across the membrane can control the development of the filter cake by the

**TABLE 2.8**
**Types of Treatment Processes by Membrane Pore Size**

| Separation Type | Pore Size | MWCO | Material Rejected from the Membrane | Material That Passes the Membrane | Driving Force (TMP) | Main Usage |
|---|---|---|---|---|---|---|
| MF | 0.1–10 μm | 100,000–1E + 6 Da | SSs including bacteria | $H_2O$ and dissolved matter | 1–2 atm | Pretreatment for RO; liquor industry |
| UF | 5–100 nm | 5,000–300,000 Da | Colloidal matter above 1000 days | $H_2O$ and salts | 1–10 atm | Separation of oil emulsion, milk products, pretreatment for RO |
| Nanofiltration | 1–10 nm | 200–10,000 Da | Most of the bivalent ions above 300 days | $H_2O$ and most of the one-valent ions | 5–25 atm | Water softening; extracting and concentrating low-molecular-weight materials |
| RO | 1–10 A° | <200 Da | Dissolved and suspended matter with low molecular weight | $H_2O$ and little mineral salts | 10–80 atm | Water desalination, extraction of salts or sugars |

constant shear that results from the high flow speed. The greater the turbulence of the fluid, the thinner the layer that forms. The thickness of the filter cake is stabilized when a resistant state is reached.

##### *2.3.5.3.6 Clogging the Membrane Surface*

Membrane fouling or biological fouling is caused by solids and biomass (the latter of which grows in the form of a biofilm) accumulating on the membrane surface and increasing the resistance through the membrane until it becomes completely clogged.

##### *2.3.5.3.7 Concentration Polarization*

Concentration polarization is the accumulation of the solution at the membrane-fluid contact point and the creation of a boundary layer where the concentration is higher than in the feed solution. Concentration polarization can increase the transfer of material rejected by the membrane into the filtrate due to the increase in the concentration gradient close to the membrane surface. Therefore, to reduce the degree of concentration polarization, the turbulence around the membrane surface is increased.

#### 2.3.5.4 Energy Requirement

The primary energy requirements of MBRs are for creating a pressure gradient in the membrane and aeration of the biomass in the reactor. There are significant differences in the energy requirements of the different membrane configurations. In a submerged membrane, the TMP is low, ranging from 0.1 to 0.3 bars. In sidestream membranes, the TMP is high, ranging between 2 and 5 bars. The energy consumption of a sidestream membrane is usually higher by two orders of magnitude than that of a submerged membrane (Gander et al., 2000). When the membrane is submerged, more than 90% of the energy cost is for aeration, while in a sidestream membrane, the energy requirement for aeration is only about 20% of the total energy requirement. The cost of pumping the liquid in the sidestream membrane is about 60%–80% of the total cost, and together with the cleaning requirements, it leads to high operating costs as compared to the submerged membrane. So even though the acquisition and space costs are smaller, the operating costs are high for a sidestream membrane relatively speaking because the required membrane area is smaller.

**COMPARISON OF THE FOUR CASE STUDIES**

This chapter reviewed four different case studies representing four greywater treatment strategies: a physicochemical system that includes filtration and disinfection, an extensive system with biological treatment such as a constructed wetland, an intensive system with a MBR, and a commercial intensive system (Table 2.9).

Except for the physicochemical system, the quality of treated greywater produced by the systems was of high quality and allowed for effluent recovery.

In the physicochemical and extensive systems, the microbial quality of treated greywater was not reported. In these systems, a relatively high

**TABLE 2.9**
**Comparison of the Performance of the Facilities Reviewed**

| System | Filtration, Sedimentation, and Disinfection | | Primary Sedimentation Followed by VFCW | | MBR | | Sedimentation Followed by Biofilter and UV Disinfection (Nubian) |
|---|---|---|---|---|---|---|---|
| | Raw Greywater | Treated Greywater | Raw Greywater | Treated Greywater | Raw Greywater | Treated Greywater | Treated Greywater |
| Source | March et al. (2004) | | Morel and Diener (2006) | | Friedler et al. (2006) | | NSW-Health (2011) |
| TSS (mg/L) | 20–126<br>44 (±31) | 15–23<br>18.6 (±2.9) | 52–188<br>98 (±53) | 6–1<br>3 (±2) | 92 (±115) | 12 (8) | <20 |
| Turbidity (NTU) | 5–62<br>20 (±13) | 5.6–42<br>16.5 (±7.5) | | | 65 (±68) | 0.2 (±0.1) | |
| COD (mg/L oxygen) | 39–441<br>171 (±130) | 24–118<br>78 (±30) | 177–687<br>411 (±174) | 7–72<br>29 (±20) | 211 (±141) | 40 (±16) | |
| $BOD_5$ (mg/L oxygen) | | | 100–400<br>200 (±93) | 0–12<br>5 (±4.6) | 69 (±33) | 1.1 (±1.7) | <20 |
| Total nitrogen (mg/L nitrogen) | 4.7–40<br>11.4 (±9.4) | 3.2–11.8<br>7.1 (±2.9) | | | | | |
| Ammonia (mg/L nitrogen) | | | 4–26<br>13 (±8) | 0–2<br>0.5 (±0.6) | | | |
| Phosphorus (mg/L) | | | 1–5<br>3 (±1) | 1–4<br>2 (±1) | | | |
| Fecal coli (cfu/100 mL) | | | | | $10^5 \cdot 3.4$ ($10^5 \cdot \pm 4.2$) | 27 (±56) | <10 |

concentration of chlorine was maintained (above 1 mg/L chlorine). In the extensive system, the treated greywater can be disinfected.

The Nubian system includes disinfection, and the fecal coliform count is less than 10 cfu/100 mL. Treatment using MBR removes fecal coli, but the treated greywater requires additional disinfection. Other studies reported the disinfection of treated greywater using MBR with UV and chlorine (Friedler and Gilboa, 2010).

The treatment technologies presented earlier represent different types of greywater treatment suitable for different applications.

The basic treatment was intended to recover greywater in hotels or public buildings, but the reported quality of the greywater is not high enough and may create a health risk or odor nuisances if recovered for toilets flushing.

The extensive treatment is suitable for treatment in rural areas, suburbs, or developing countries. The quality of the treated greywater in this case is relatively high but should also be disinfected.

The intensive treatments are more expensive. They could be implemented in high-rise buildings where the available space for a treatment facility is limited and where the initial cost of facility construction is divided among a large number of tenants.

## 2.4 DISINFECTION

To meet sanitary standards for using greywater, disinfection of the treated water will usually be required. Microbial contamination can be reduced in two main ways: by physical removal of pathogens (e.g., membrane filtration) and by inactivation. Inactivation does not remove the pathogens from the water but rather prevents them from being viable (in other words capable of causing disease) (Crittenden et al., 2005). This can be achieved by chemicals such as chlorine, bromine, ozone, and hydrogen peroxide, which have the ability to oxidize and damage specific enzymes and/or the cell wall of the microorganisms; photochemistry, such as solar radiation and UV that damage the nucleic acids of the microorganisms (Acher et al., 1997); and the use of heat.

The disinfection capability of a compound is not always sufficient to allow its use as a disinfectant, and of course a large number of other factors have to be taken into account. The properties of an ideal disinfectant include the following: available in large quantities and at a reasonable price; not toxic to humans and animals; noncorrosive; nonstaining; not adsorbed by organic matter other than bacterial cells; able to help ameliorate odors from the water; capable of penetrating through surfaces; safe for transport, storage, treatment, and use; soluble in water or in cell tissue with a low loss of disinfection efficiency when unused; effective in higher dilution; and effective in a wide spectrum of air temperatures (Metcalf and Eddy, 2003).

### 2.4.1 Disinfectant Concentration and Contact Time

The most significant factors in the process of disinfection are the relative amount of the disinfectant (disinfection dose) and the contact time. In 1908, an empirical model known as Chick's law was published (Equation 2.48). The model was based on Harriet Chick's observation that for a given concentration of a disinfectant, the longer the contact time, the greater the kill (Metcalf and Eddy, 2003):

$$\frac{dN_t}{dt} = -k \cdot N_t \tag{2.48}$$

where

- $dN_t/dt$ is the rate of change in the organisms' concentration over time (number of bacteria/time)
- $k$ is the inactivation rate constant ($T^{-1}$)
- $N_t$ is the number of organisms in time $t$ (number of bacteria)
- $t$ is the time

Solving the following equation by assuming that $N_0$ is the number of organisms in time zero,

$$\ln \frac{N_t}{N_0} = -k \cdot t \tag{2.49}$$

or

$$\frac{N_t}{N_0} = e^{-k \cdot t} \tag{2.50}$$

The inactivation constant $k$ can be found empirically by substitution in the graph of $-\ln (N_t/N_0)$ values against contact time $t$ (minutes).

Around the same time, a model by Herbert Watson was published, associating the inactivation rate constant and the disinfectant concentration:

$$k = k' \cdot C^n \tag{2.51}$$

where

- $k$ is the inactivation rate constant (1/min)
- $k'$ is the mortality rate constant (1/min)
- $C$ is the disinfectant concentration (mg/L)
- $n$ is the dilution coefficient (depending on the water quality)

The combination of the two models is called the Chick–Watson model and it is used in the present day:

$$\ln \frac{N_t}{N_0} = -k' \cdot C^n \cdot t \tag{2.52}$$

The $n$ and $k'$ values can be found empirically using the survival data of organisms versus the different disinfectant concentrations and contact times.

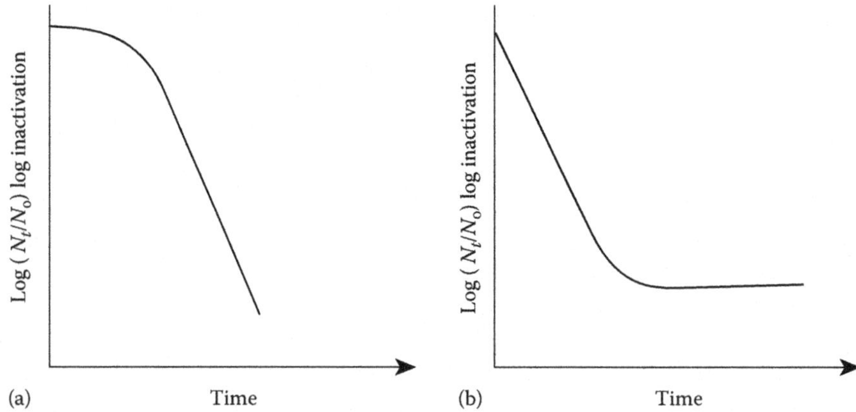


**FIGURE 2.18** Deviations from Chick's model: (a) *Shoulder* deviation caused by substances that react with the disinfectant. (b) *Tailing* deviation caused by organisms that find shelter from the disinfectant. (From Metcalf and Eddy, *Wastewater Engineering: Treatment and Reuse*, Tchobanoglous, G., Burton, F.L., and Stensel, H.D., eds., McGraw-Hill, New York, 2003.)

The following explanation is given for different values of $n$:

For $n=1$, the disinfectant concentration and contact time are of equal importance.
For $n > 1$, the disinfectant concentration is more important than the contact time.
For $n > 1$, the contact time is more important than the disinfectant concentration.

When the water contains substances that react with the disinfectant, there is a deviation from the model and a *shoulder* is formed at the beginning of the reaction (Figure 2.18a). Another deviation from the model is called *tailing*, which occurs when there are organisms that are bound to particles and protected by them and therefore continue their operation over time (Figure 2.18b).

## 2.4.2 Disinfection with Chlorine

The use of chlorine is the most common disinfection method due to its efficiency, simplicity of use, inexpensiveness, and long residual effect (because it prevents regrowth of microorganisms in water systems). The most common chlorine compounds in disinfection are chlorine gas ($Cl_2$), chlorine dioxide ($ClO_2$), and hypochlorite salts: sodium hypochlorite (NaOCl) (liquid) and calcium hypochlorite ($Ca(OCl)_2$) (powder/tablets).

Chlorine reacts with water and forms a strong acid (hydrochloric acid, $HCl^-$) and a weak acid (hypochlorite acid, HOCl):

$$Cl_2 + H_2O \rightarrow HOCl + HCl^- \tag{2.53}$$

Here is a reaction of hypochlorite salts (sodium and calcium):

$$Cl(OCl)_2 + H_2O \rightarrow 2HOCl + C_a(OH)_2 \tag{2.54}$$

$$NaOCl + H_2O \rightarrow HOCl + NaOH \tag{2.55}$$

The hypochlorite acid decomposes into hydrogen ions ($H^+$) and hypochlorite ions ($OCl^-$):

$$HOCl \rightarrow H^+ + OCl^- \tag{2.56}$$

Therefore, chlorination of water creates four configurations of chlorine: the chloride ion ($Cl^-$), chlorine ($Cl_2$), hypochlorite acid (HOCl), and the hypochlorite ion ($OCl^-$). Chlorine will hardly be present in the water after it transforms, and the two latter configurations are called free active chlorine. The form of active chlorine that is dominant in the water depends on the water's pH. A pH value higher than 7.5 causes a reduction in the hypochlorite acid concentration and thus reduces the disinfection efficiency. The optimal pH value for water disinfection is $5.5 < pH < 7.5$.

When there are nitrogenous compounds in the water, they react with the active chlorine to form chloramines, which also have a disinfection capacity. The chlorine found in these compounds is called chloramines, but it is less effective than hypochlorite acid:

$$NH_3 + HOCl \rightarrow NH_2Cl + H_2O \quad \text{Monochloroamine} \tag{2.57}$$

$$NH_3 + 2HOCl \rightarrow NHCl_2 + 2H_2O \quad \text{Dichloroamine} \tag{2.58}$$

$$NH_3 + 3HOCl \rightarrow NCl_3 + 3H_2O \quad \text{Trichloroamine} \tag{2.59}$$

Trichloroamine is not stable and decomposes into nitrogen gas and a chloride ion.

The reactions of chlorine in water determine the concentration of free chlorine, and they are usually described by the chlorination curve (Figure 2.19).

Keeping the concentration of active chlorine to disinfect the water is a complex operation because chlorine reacts not only with ammonia but also with other material such as metals and organic matter, since it is a strong oxidizer. The term *chlorination breakpoint* refers to the process in which chlorine is added to the water and reacts with all the compounds that can be oxidized, to the point where all the chlorine added will remain as active free chlorine. This is the only way to ensure effective disinfection (Figure 2.19).

When chlorine is added to water containing oxidative inorganic materials such as $Fe^{2+}$, $Mn^{2+}$, $H_2S$, or oxidative organic materials, these compounds react with the chlorine and reduce most of it to chloride ion (point A). After this immediate requirement is satisfied, the additional chlorine continues to react with ammonia to create chloramines (between points A and B). When the molar ratio of chlorine to ammonia

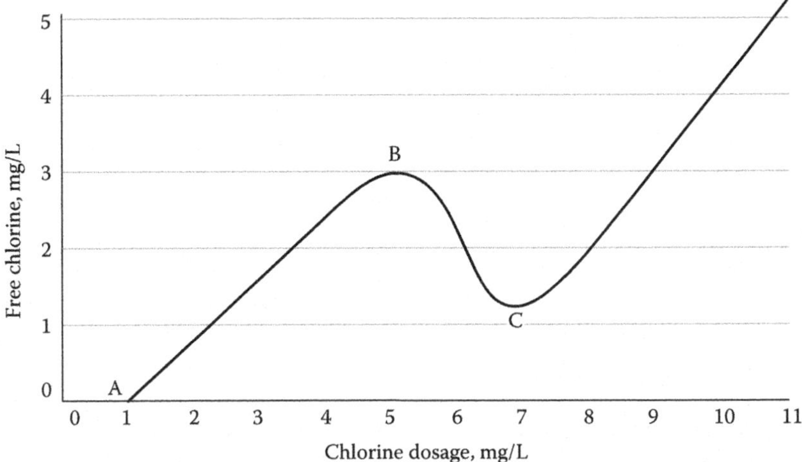


**FIGURE 2.19** Schematic description of the chlorination curve, including the *breakpoint*. Point A—all free active chlorine reacted with oxidative compounds. From point A to point B—the creation of chloroamine compounds. From B to C—destruction of chloroamine compounds. Point C—breakpoint, any chlorine now added will remain free active chlorine in the water. (From Metcalf and Eddy, *Wastewater Engineering: Treatment and Reuse*, Tchobanoglous, G., Burton, F.L., and Stensel, H.D., eds., McGraw-Hill, New York, 2003.)

is reduced, monochloramines and dichloramines are formed. With the increase in the chlorine concentration, from points B to point C, some of the chloramines become trichloramines. The rest of the chloramines are oxidized into nitrogen dioxide and nitrogen gas and chlorine is reduced to chloride ion; therefore, a decline in the concentration of active chlorine is observed. The addition of chlorine beyond point C (breakpoint) creates a proportional increase in the concentration of free active chlorine.

The amount of chlorine that reacts with substances present in the water up to the breakpoint is called the chlorine demand of the water. The chlorine demand is defined as follows:

$$\text{Chlorine demand} = \text{chlorine dose} - \text{concentration of active chlorine.}$$

#### 2.4.2.1 Factors Influencing the Disinfection Efficiency

The disinfection efficiency of chlorine is influenced by many factors such as the chemical characteristics of the water, extent of the mixing, concentration of particles in the water, the characteristics of the microorganisms, the concentration of active chlorine, and contact time. In general, high levels of disinfection are achieved when the concentration of active chlorine is high, the contact time is long, the pH is low, turbidity is low, and disturbing factors are absent. The disinfection efficiency varies from one organism to another according to their sensitivity to chlorine. In general, bacteria are the most sensitive to chlorine, followed by viruses, and finally protozoa.

Contact time is defined by the active chlorine concentration $C$ (mg/L) and contact time $T$ (min) with the water after the chlorination breakpoint. Contact time is usually calculated by mathematical linear integration of the disinfectant's concentration versus time, using the classical empirical model of Chick–Watson.

A significant disadvantage of chlorine use is the by-products generated by the reaction of chlorine with organic matter, creating toxic or carcinogenic compounds such as trihalomethanes (Rebhun et al., 1997). In addition, it was found that its effectiveness in the inactivation of protozoa (*Cryptosporidium parvum*, *Giardia lamblia*) is low (Crittenden et al., 2005), but these pathogens are not likely to be found in greywater.

In greywater that is treated well, the concentrations of organic matter and ammonia are low, and therefore the chlorine demand is usually low, standing at a few mg/L.

### 2.4.3 Disinfection Using Ozone

Ozone ($O_3$) is an unstable gas, formed when oxygen molecules are broken down into atomic oxygen (O). It is possible to produce ozone by electrolysis, which is a photochemical or radiochemical reaction using electrical discharge. At room temperature, ozone is a blue gas with a characteristic strong odor. Because of its smell, it is usually detected before it reaches concentrations that have deleterious health effects. When the gas reaches a concentration of about 20% in the air, it becomes volatile (about 240 g/m$^3$ air). In air, ozone is slightly more stable than in water, but in both cases it breaks down in minutes.

Ozone decomposition is described in the following series of equations:

$$O_3 + H_2O \rightarrow HO_3^{\,+} + OH^- \tag{2.60}$$

$$HO_3^{\,+} + OH^- \rightarrow 2HO_2 \tag{2.61}$$

$$O_3 + HO_2 \rightarrow HO^- + 2O_2 \tag{2.62}$$

$$HO^- + HO_2 \rightarrow H_2O + O_2 \tag{2.63}$$

The free radicals $HO^{\cdot}$ and $HO_2^{\cdot}$ are very strong oxidizers and are the active ingredients in disinfection. The prevailing conjecture maintains that bacteria mortality occurs as a result of damage to the cell wall (lysis).

In general, ozone is a reliable disinfectant and is considered more effective for virus disinfection than chlorine.

The presence of oxidative compounds in the water causes a "shoulder effect" in the ozone disinfection curve (similar to that of chlorine). Ozonation does not create dissolved solids and is not affected by ammonia ions or by the pH level of the water. It should be noted that in water containing significant halogen concentrations such as chloride or bromide, using excess oxidant may cause the formation of toxic by-products such as bromates. Therefore, it should be used with care and applied only when deemed suitable.

*The ozone disinfection model.* The model that describes disinfection with ozone is given here:

$$\text{For } U < q \quad \frac{N}{N_0} = 1 \tag{2.64}$$

$$\text{For } U > q \quad \frac{N}{N_0} = \left(\frac{U}{q}\right)^{-n} \tag{2.65}$$

where
- $N$ is the number of organisms after disinfection
- $N_0$ is the number of organisms before disinfection
- $U$ is the ozone dose consumed
- $N$ is the gradient of portion–reaction curve
- $q$ is the value of the breakpoint with the $x$-axis when $N/N_0 = 1$ or $\log N/N_0 = 0$

Since the efficiency of the ozone transfer from the production container to the reaction reactor (where disinfection takes place) is not 100%, the efficiency of ozone transfer should be considered when determining the required ozone dose according to the following equation:

$$D = U\left(\frac{100}{TE}\right) \tag{2.66}$$

where
- $D$ is the required dose of ozone (mg/L)
- $U$ is the consumed dose of ozone (mg/L)
- $TE$ is the efficiency of ozone transfer (%)

In recent years, the efficiency of ozone transfer to the solution has been increased, and it ranges in many cases between 80% and 90% following the use of designated injectors, such as those based on the Venturi principle and injection in a vacuum.

### 2.4.4 Disinfection Using Stabilized Hydrogen Peroxide

The compound hydrogen peroxide plus (HPP) is a combination of hydrogen peroxide with organic molecules. The organic molecules stabilize the compound, so that it remains active for a long time, thus adding a residual disinfectant effect that is not found in the regular peroxide compound ($H_2O_2$).

It was found that using HPP as a disinfectant of treated greywater is practically and economically feasible for small treatment systems. It has high disinfection efficiency, a residual effect, and no by-products that are known to be harmful to health. For these reasons, it is an acceptable alternative to the use of chlorine (Ronen et al., 2010).

### 2.4.5 Disinfection Using UV Radiation

A disinfection method that has been gaining popularity in recent years is by UV, which is radiation at wavelengths between 100 and 400 nm. It is customary to divide it into four categories:

UV-A—long wave lengths (315–400 nm)
UV-B—medium wavelengths (280–315 nm)
UV-C—short wavelengths (280–200 nm)
UV vacuum—(200–100 nm)

In general, the shorter the wavelength, the higher the energy causing more damage to the organism. On the other hand, the shorter the wavelength, the more efficiently it is absorbed in the water, and therefore wavelengths shorter than 200 nm cannot be used to disinfect water. Inactivation of bacteria occurs when the DNA absorbs radiation in short wavelengths up to about 300 nm. Therefore, the effective radiation range for water disinfection is between 200 and 300 nm (the peak of DNA absorption occurs at a wavelength close to 254 nm).

*The operation principle of a UV lamp*: Most UV lamps that are used to disinfect water operate on the vapor pressure of mercury. It is based on the operation principle that when voltage is applied to mercury vapors, the free electrons and ions are accelerated due to the electric field generated between the two electrodes. When the voltage is of sufficient magnitude, the accelerated electrons move to a higher energy level and collide with electrons and cations. As a result, they return to a lower energy level while releasing energy that is manifested in the creation of light of a wavelength corresponding to the 200–300 nm range.

There are two main types of UV lamps for disinfection of water: low-pressure (0.01–0.001 millibars) and medium-pressure lamps (3 to 1 bars) with high intensity. Low-pressure lamps are considered more effective because most of the light they emit is of a uniform wavelength (of 254 nm, the wavelength that hits the DNA molecule). Medium-pressure lamps emit light in a broad spectrum, most of which is ineffective for disinfection. Still, due to the fact that low-pressure lamps have significantly lower radiation power than medium-pressure lamps, in many cases, more low-pressure lamps are required than medium-pressure lamps to achieve the same disinfection result. It should be noted that in small disinfection facilities (for a single house or several houses), a single low-pressure lamp can be used. In recent years, low-pressure lamps with high intensity have been developed whose radiation power is higher than that of regular low-pressure lamps.

*The disinfection mechanism*: The photons emitted from UV lamps react directly with the organism's nucleic acids, and the radiation creates a double bond between two thymine molecules, thus preventing the multiplication of the DNA. This, in turn, prevents reproduction of microorganisms. Most viruses found in drinking water have only RNA (i.e., they contain the base uracil instead of thymine), and therefore they have low sensitivity to UV. The rotavirus and adenovirus are among the more resistant organisms to UV radiation.

*Regrowth*: Since UV radiation does not kill the organisms and after being exposed to radiation they are still metabolically active, some of the organisms have self-repair mechanisms by means of enzymes. There are primarily two such enzymatic repair mechanisms, one of which requires light in the visible spectrum to activate and is thus referred to as photoreactivation. The other does not require light for its activation and is therefore called dark repair mechanism.

*The UV dose required for disinfection*: As with chemical disinfectants, the efficiency of disinfection by UV depends on the organism's sensitivity to the disinfectant and on the disinfection dose actually absorbed by the organism. For a chemical disinfectant, the required disinfectant dose is calculated by multiplying the disinfectant concentration and contact time. Analogically, when using UV radiation, the disinfection dose is calculated as the product of the radiation intensity and the contact time to radiation:

$$D = I \cdot t \tag{2.67}$$

where

$D$ is the disinfection dose (mW × s/cm$^2$)
$I$ is the radiation intensity (mW/cm$^2$)
$t$ is the contact time (s)

Hence, it is possible to change disinfection dose by changing contact time, radiation intensity, or both.

Radiation intensity ($I$) is calculated by dividing the radiation power emitted from the lamp by the lamp's surface area:

$$I = \frac{P}{A} \tag{2.68}$$

where

$P$ is the actual power (mW)
$A$ is the surface area (circumference · lamp length, cm$^2$)

According to the Beer–Lambert law, radiation diminishes exponentially from the lamp's surface and is influenced by the distance from the lamp's surface and by the absorption rate of the transmission medium:

$$I_d = I_0 \cdot 10^{-\alpha \cdot d} \tag{2.69}$$

where

$I_d$ is the radiation intensity at distance $d$ from the lamp's surface (mW/cm$^2$)
$I_0$ is the initial radiation intensity on the lamp's surface (mW/cm$^2$)
$d$ is the distance from the lamp's surface (cm)
$\alpha$ is the absorption in 254 nm wavelength (a.u./cm)

**TABLE 2.10**
**Typical Absorption Values of Effluents of Different Qualities**

| Effluent Type | Absorption (Absorbance Unit/cm) |
|---|---|
| Primary effluent | 0.5–0.8 |
| Secondary effluent | 0.3–0.5 |
| Filtered secondary effluent | 0.2–0.4 |
| Secondary effluent after MF | 0.15–0.3 |
| Effluent after RO | 0.05–0.2 |
| Greywater effluent | |
| After treatment with RBC | 0.015–0.045 |
| After treatment with MBR | 0.05–0.045 |

*Sources:* Gilboa, Y. and Friedler, E., *Water Res.*, 42(4–5), 1043, 2008; Metcalf and Eddy, *Wastewater Engineering: Treatment and Reuse*, Tchobanoglous, G., Burton, F.L., and Stensel, H.D., eds., McGraw-Hill, New York, 2003.

Typical absorption values of effluents of different qualities are presented in Table 2.10.

Since the intensity of radiation diminishes with the distance from the lamp, the average radiation intensity in the system is often calculated for planning purposes (taking into account that the water flow in the disinfection reactor is turbulent):

$$I_{avg} = I_O \cdot \frac{\left(1 - e^{-\alpha \cdot d}\right)}{\alpha \cdot d} \tag{2.70}$$

where $I_{avg}$ is the average radiation intensity (mW/cm$^2$).

### 2.4.6 Factors Influencing the Disinfection Efficiency

The quality of water has a great impact on the efficiency of disinfection using UV radiation. Microorganisms can be shielded by SSs found in the water, allowing them full protection from radiation. In addition, the SSs may reduce the intensity of the radiation that hits the microorganisms by refraction, reflection, scattering, and absorption of the radiation.

Over time, the quartz sleeves surrounding the UV bulb are covered by chemical and/or biological fouling. Among the factors affecting the rate of fouling formation, and hence the frequency of cleaning required, are alkalinity, hardness, iron, and pH. The removal of fouling from low-pressure lamps can be achieved by washing with *chemical cleaning solutions* that require the operation of the system to be interrupted. In the case of medium-pressure lamps that experience much more fouling, frequent mechanical cleaning is required during operation, and thus does not call for interruption of the system.

Another factor affecting the efficiency of disinfection is the age of the lamp. Lamp intensity decays with time, and the decay is a function of the number of hours of work, the frequency and interval between operations, and the power (output) per unit length of the lamp. The end of life of the UV lamp is in effect when its power reaches about 80% of the initial intensity, which occurs after 6000–8000 h of operation in most cases.

The benefits of using UV radiation include saving the storage and dosing facilities required for chemical disinfection substance, avoiding by-products of known health risk, and preventing high inactivation efficiency. A dose of UV radiation of 70 mW/s/cm$^2$ has been found highly efficient for removing FC, *P. aeruginosa* sp., and *S. aureus* sp. However, even a high radiation dose (of 400 mW/s/cm$^2$) is not very efficient for complete removal of all bacteria as was demonstrated by heterotroph plate count and regrowth after UV disinfection (Gilboa and Friedler, 2008). Since UV disinfection (as well as disinfection by ozone and membrane filtration) has no residual effect, the microbial quality of the water after disinfection cannot always be guaranteed. In addition, the use of a UV lamp is often a more expensive option than chlorination.

## 2.5 COMPARING TECHNOLOGIES

### 2.5.1 Introduction

Greywater reuse poses a complex challenge since it is a local (on-site) treatment, usually with a limited budget. Also the water source often varies in quality and quantity, and it is used to meet a variety of needs such as irrigation, flushing, and rinsing. Moreover, regulation varies between states and countries as expressed by numerous laws, regulations, and recommendations across the globe. Consequently, there are uneven quality requirements for effluent, and the public often perceives greywater as a nonrisky source. As a result, there is a wide range of measures for recovery and/or greywater treatment, and, as expected, the quality of facilities is not uniform. Ideal treatment systems should be efficient, reliable, simple to operate and maintain, and relatively cheap. When treatment is intended to recover treated greywater in a dense urban area for toilet flushing, for example, a compact system with a low footprint is required. This requirement is much less important when it comes to recovery in suburban and rural areas; however, in these areas frequency and low maintenance are much more important. This chapter reviewed 33 articles and study results. They covered the quality of effluent from different greywater treatment systems and the relative removal rates achieved in experiments conducted in the laboratory or in the field. Reports from the literature were classified into physical–chemical treatment and biological treatment and were statistically compared. Despite the difficulty in conducting such a comparison, resulting from the variation between studies, some insights can be learned.

The physical–chemical treatment was divided into

1. *Deep media filtration* (Gerba et al., 1995; Šostar-Turk et al., 2005; Itayama et al., 2006; Pidou et al., 2008; Friedler and Alfiya, 2010; Godfrey et al., 2010; Mandal et al., 2010)
2. *Coagulation and flocculation* (Coag./Floc.) *with sedimentation* (Friedler and Galil, 2003; Šostar-Turk et al., 2005; Friedler et al., 2008; Pidou et al., 2008)

3. *Coagulation and flocculation* (Coag./Floc.) *with sedimentation and filtration* (Šostar-Turk et al., 2005; Friedler and Alfiya, 2010)
4. *Membrane filtration that includes UF, nanofiltration, and RO* (Ramon et al., 2004; Šostar-Turk et al., 2005; Friedler et al., 2008; Li et al., 2009; Hourlier et al., 2010)

Biological treatment was divided into

1. *Anaerobic treatment* (Elmitwalli and Otterpohl, 2007; Hernández-Leal et al., 2007; Hernández-Leal, 2010)
2. A combination of anaerobic and aerobic treatment
3. *Surface flow constructed wetland (SFCW)* (Gerba et al., 1995; Kadewa et al., 2010)
4. *Horizontal-flow constructed wetland (HFCW)* (Jenssen and Vrale, 2003; Morel and Diener, 2006; Paulo et al., 2009; Raude et al., 2009; Masi et al., 2010)
5. *Vertical flow constructed wetland (VFCW)* (Jenssen and Vrale, 2003; Gross et al., 2007a,b; Paulo et al., 2009; Weissenbacher and Müllegger, 2009; Zuma et al., 2009; Kadewa et al., 2010; Ronen et al., 2010)
6. A combination of constructed wetland with vertical flow and constructed wetland with horizontal flow (Jenssen and Vrale, 2003; Paulo et al., 2009)
7. *Rotating biological contactor (RBC)* (Friedler et al., 2005; Friedler and Gilboa, 2010)
8. *Sequential batch reactor (SBR)* (Hernández-Leal et al., 2007; Hernández-Leal, 2010)
9. *Membrane biological reactor (MBR)* (Kovalio, 2004; Lesjean and Gnirss, 2006; Merz et al., 2007; Scheumann and Kraume, 2009; Friedler and Gilboa, 2010; Huelgas and Funamizu, 2010; Kraume et al., 2010)

### 2.5.2 Comparison of Effluent Quality from Different Physical–Chemical Treatment Systems

The average concentration of the constituents in effluent produced by the reviewed physical–chemical systems is summarized in Table 2.11.

In general, the physical properties of greywater can be determined by measuring the concentrations of TSS and turbidity. These were found to be relatively low, and their average concentrations for all systems were 16 mg/L and 7.3 NTU, respectively. However, when examining the variables that characterize dissolved organic matter, the performance of physical treatment systems was less impressive. For example, COD and $BOD_5$ concentrations were on average 119 and 27 mg/L (Table 2.11). Greywater generally contains lower nutrient concentrations (nitrogen and phosphorus) than the total domestic wastewater, so relatively low concentrations were measured in the effluent too. The average concentrations of total nitrogen, ammonia, soluble reactive phosphorus, and total phosphorus in all physical–chemical treatment systems are 9, 2.1, 0.91, and 1.3 mg/L, respectively. No significant difference was found between the concentrations of nutrients in the effluent in the various systems ($p > 0.05$).

**TABLE 2.11**
**Summary of Effluent Quality from Various Physical–Chemical Treatment Facilities**

| | TSS | Turbidity | COD | BOD | AS | TN or TKN | TAN | $PO_4$ | TP |
|---|---|---|---|---|---|---|---|---|---|
| Type of Treatment | mg/L Avg. (SD) | NTU Avg. (SD) | mg/L Avg. (SD) | mg/L Avg. (SD) | mg/L (SD) | mg/L Avg. (SD) | mg/L Avg. (SD) | mg-p/L Avg. (SD) | mg/L Avg. (SD) |
| Coag./floc. + SED | 6.2 (1.7) | 6.6 (3.0) | 209 (100) | 51 (35) | 10 (0) | 10 (7.6) | 1.6 (0.7) | 0.1 (0) | 1 (0) |
| Coag./floc. + SED + Filt. | 4.7 (0.4) | 3.9 (0) | 42 (30) | 27 (24) | 0.5 (0) | 2.6 (0) | 2.3 (0) | 1 (0) | |
| Filtration—media | 25.3 (43) | 18 (27) | 140 (113) | 32 (26) | 0.5 (0) | 14 (13.6) | 1.5 (0.6) | 0.7 (0.8) | 0.5 (0.7) |
| Membrane | 8.5 (7.4) | 1.1 (0.5) | 92 (78) | 14. (27) | 5.9 (4.9) | 4 (6.5) | 3 (5.9) | 2 (3.4) | 1.9 (3.2) |
| All (average) | 16 (29.7) | 7.3 (16) | 120 (98) | 27 (29) | 5.3 (4.9) | 9 (10.4) | 2.1 (3.2) | 0.9 (1.8) | 1.3 (2.2) |

*Notes:* The no. of observations ranges between 2 and 13. (The results are in mg/L unless stated otherwise.)

TSS, total suspended solids; COD, chemical oxygen demand; BOD, biological oxygen demand; AS, anionic surfactants; TN, total nitrogen; TAN, total ammonia nitrogen; $PO_4$, soluble reactive phosphorous; TP, total phosphorous; SED, sedimentation.

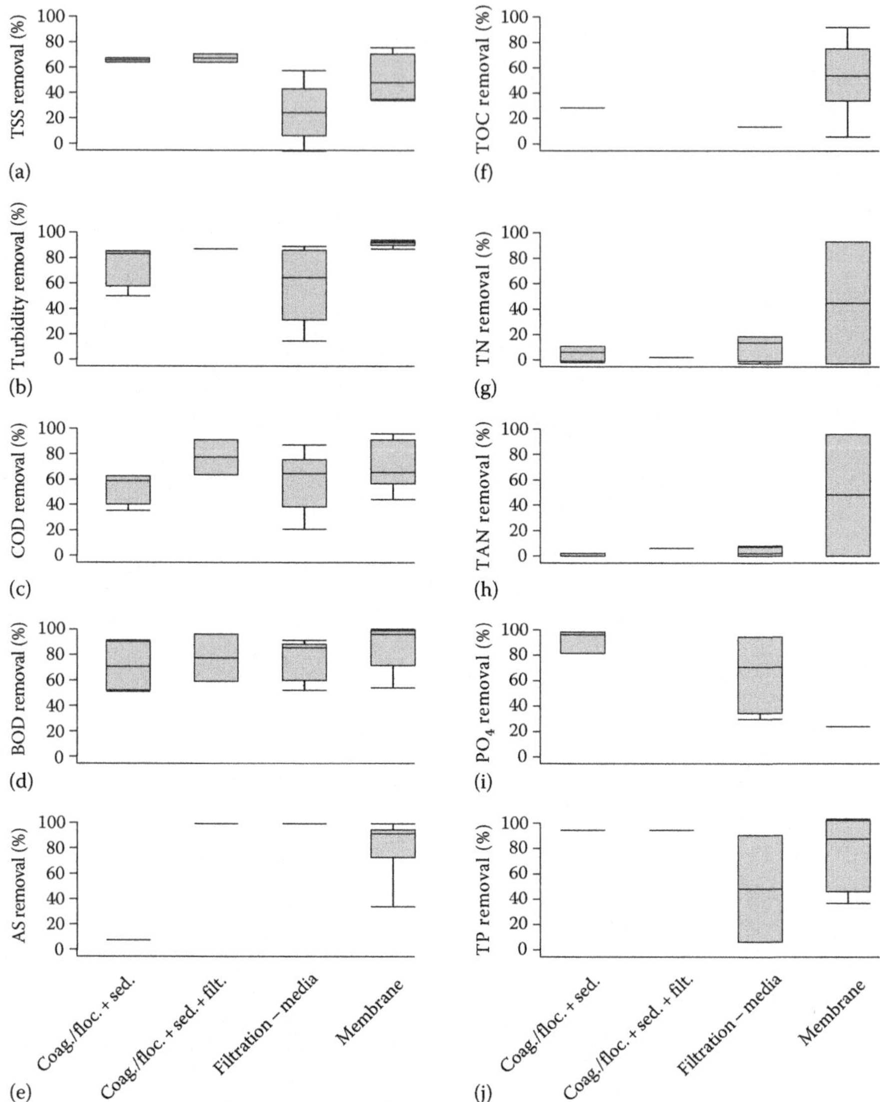


**FIGURE 2.20** Percent removal of (a) turbidity, (b) TSS, (c) COD, (d) $BOD_5$, (e) anionic surfactants, (f) total organic carbon, (g) total nitrogen, (h) total ammonia, (i) dissolved phosphorus, and (j) total phosphorus from various physical greywater treatment systems.

As mentioned, physical–chemical treatment is more efficient for treating suspended material such as TSS and turbidity, but less efficient for the removal of dissolved material such as total organic carbon or nitrogen (Figure 2.20). For example, the removal efficiency in the media filtration systems was the lowest, while in the coagulation/flocculation systems, it was the highest (and there were fewer data for membrane systems). A similar trend was also seen in the treatment of turbidity where the removal rate in the membrane systems was the highest. As expected, the removal efficiency of COD and $BOD_5$ of physical and chemical systems was relatively low, ranging from 55% to 87% of the COD and from 56% to 95% of the $BOD_5$. In filtration systems, coagulation and flocculation with sedimentation had a lower removal percentage than in coagulation and flocculation systems with filtration and membranes.

As noted, the removal of dissolved material is generally low in the simple physical systems (except for membrane treatment, depending on the membrane type). For example, one report found that the flocculation system did not remove anionic surfactants at all (Šostar-Turk et al., 2005), while in membrane systems, an average of 78% of the anionic surfactants was removed. Coagulation/flocculation with filtration systems removes over 90% of the surfactants, but there is only one report on a single type of treatment (Šostar-Turk et al., 2005). Removal of the different nitrogen forms is relatively low except when filtration is membranal, but the range reported is wide (0%–99%). The removal percentage of dissolved phosphorus and total phosphorus is higher and stands on average at about 70%, a rate resulting mainly from chemical deposition with the solids.

### 2.5.3 Comparison of Effluent Quality of Different Biological Treatment Systems

Biological treatment is more complex than physical treatment. It usually includes physical elements such as filtering and sedimentation, biological degradation of organic matter, and sometimes also biological transformations/removal of nutrients such as phosphorus and nitrogen. Sometimes, the biological treatment is also designed to remove micropollutants, such as hormones and pesticides. In addition, biological treatment is divided into treatment in an aerobic (with oxygen) or anaerobic (without oxygen) environment or a combination of the two. In this chapter, we compared a small number of the principles of common biological treatments for greywater. The average concentration of constituents in the effluent of the biological systems reviewed is summarized in Table 2.12. It was found that the average concentration of SSs and turbidity of the effluent in biological systems is 12 mg/L and 11.3 NTU, respectively, and the quality of the effluent from the MBR is the highest.

The average concentration of COD and $BOD_5$ (representative analyses of organic matter) in the treated greywater of all the systems reviewed is 99 and 15.1 mg/L, respectively. It should be noted that the variance between the different systems is very high. In terms of COD, the anaerobic facilities yield effluent of a lower quality than the rest of the facilities ($p < 0.05$), because anaerobic is less efficient than aerobic digestion (Table 2.12). In the reports on anaerobic facilities, the COD concentration in treated water ranged between 255 and 474 mg/L. Anaerobic methods are often used as initial treatment when the organic load is high. This is usually the case for dark greywater since it requires a relatively small energy investment (as compared to aerobic treatment). A complimentary aerobic treatment is then followed, treating the effluent to the required quality. Anaerobic treatment can also be used for energy production, such as biogas, but this process is not common in greywater treatment because the amounts of gas generated do not justify the involved costs.

**TABLE 2.12**
**Summary of the Average Concentrations of TSS, Turbidity, COD, $BOD_5$, TOC, and AS in Greywater Treated in Various Biological Treatment Systems**

| | TSS | Turbidity | COD | BOD | AS | TN or TKN | TAN | $PO_4$ | TP | TOC |
|---|---|---|---|---|---|---|---|---|---|---|
| Type of Treatment | mg/L Avg. (SD) | NTU Avg. (SD) | mg/L Avg. (SD) | mg/L Avg. (SD) | mg/L Avg. (SD) | mg/L Avg. (SD) | mg/L Avg. (SD) | mg-p/L Avg. (SD) | mg/L Avg. (SD) | mg/L Avg. (SD) |
| Anaerobic | | | 364 (104) | | 35 (1.8) | 33 (1.4) | 4.9 (2.8) | 7.1 (0.3) | 7.1 (1.4) | |
| Anaerobic + aerobic | | | 100 (0) | | 1.3 (0) | 26 (0) | 0.4 (0) | | 5.8 (0) | |
| SFCW | 9.5 (5.2) | 3.8 (2.6) | 14 (9.3) | 1.4 (2) | 2.7 (2.6) | | | | | 6.5 (2.8) |
| HFCW | 26.1 (22.5) | 58 (79) | 193 (238) | 53 (95) | 24 (0) | 5.6 (3.3) | 2.1 (0.3) | | 1.6 (2.4) | 0 (0) |
| VFCW | 8.7 (5.9) | 11 (11) | 58 (57) | 11 (15) | 0.9 (1.1) | 5.8 (4.6) | 1.3 (0.8) | 0.1 (0) | 2.4 (2.9) | 7.2 (0.5) |
| VFCW + HFCW | 9.8 (0) | 26 (0) | 94 (0) | 17 (14) | | 2.1 (0.6) | 1.5 (1.1) | | 1.2 (1.6) | 0 (0) |
| RBC | 12 (5.7) | 1.3 (0.7) | 44.3 (3.8) | 4.2 (2.2) | | | | | | 0 (0) |
| SBR | | | 91 (9) | | | | | | | |
| MBR | 6.5 (7.8) | 0.3 (0.2) | 21 (10) | 2.1 (1.7) | 0.2 (0.2) | 4.8 (4.9) | 0.6 (1) | 0.8 (0.1) | 2.3 (1.7) | 0 (0) |
| All (average) | 12 (11) | 11 (26) | 99 (134) | 15 (41) | 6.7 (12.4) | 12 (12) | 1.7 (2.1) | 2.9 (3.5) | 3.3 (2.8) | 6.8 (1.7) |

The average concentration of anionic surfactants in greywater treated by biological treatment systems stands at 6.7 mg/L, with a median of 1.1 mg/L, which means that most of the reported concentrations are below average. The highest concentrations were observed in greywater treated in anaerobic facilities and in constructed wetlands with horizontal flow.

The average concentrations of total nitrogen, ammonia-N, total phosphorus, and dissolved phosphorus in all biological facilities surveyed are 12, 1.7, 3.3, and 2.9 mg/L, respectively (Table 2.12).

### 2.5.4 Removal Efficiency of Pollutants by Different Biological Treatment Systems

The average percent removal of TSS and turbidity in all biological systems is 82% and 87%, respectively. Although not statistically significant, it was observed that membrane treatments that combine aerobic and anaerobic processes and combined HFCW/VFCW yielded a solid removal of over 90%, higher than HFCW and RBC, which resulted in an average solid removal of 72%.

As expected, removal of organic matter was relatively high, averaging 79% and 91% for COD and $BOD_5$, respectively.

The ratio of COD removal in the anaerobic treatment systems surveyed was the lowest among the biological systems and averaged 51% ($p<0.05$). In addition, removal by MBRs was higher than in both VFCWs and anaerobic reactors ($p<0.05$). In terms of $BOD_5$, there were no statistically significant differences ($p>0.05$) between the systems.

Significant removal of anionic surfactants averaging 80% was found in all types of treatment that contained an aerobic stage. The average percent removal in anaerobic systems was 20% and in the SFCWs about 55% and was measurably lower than the other systems ($p<0.05$).

The average removal of total nitrogen and total ammonia of all the systems was 49%. The average removal of total nitrogen through MBRs and systems that combine HF/VF constructed wetlands flow was approximately 78% and 76%, respectively. This average is higher as it combines aerobic and anaerobic processes, which likely results in coupled nitrification and denitrification. Percent removal of either anaerobic or combined anaerobic followed by aerobic treatment systems was much lower, averaging about 6% and 2%, respectively ($p<0.05$).

The average removal of dissolved phosphorus and total phosphorus in all the systems surveyed stands at 28% and 47%, respectively. The highest removal of total phosphorus (75%) was observed in systems that combine constructed wetlands with vertical and horizontal flow, and the lowest (13% and 33%) were observed in anaerobic and SBR systems ($p<0.05$), respectively (Figure 2.21).

### 2.5.5 Comparison of Biological Treatment and Physical and Chemical Treatment

The comparison of the two types of treatment (biological and physical–chemical) indicates trends and differences in treatment principles. In both types of

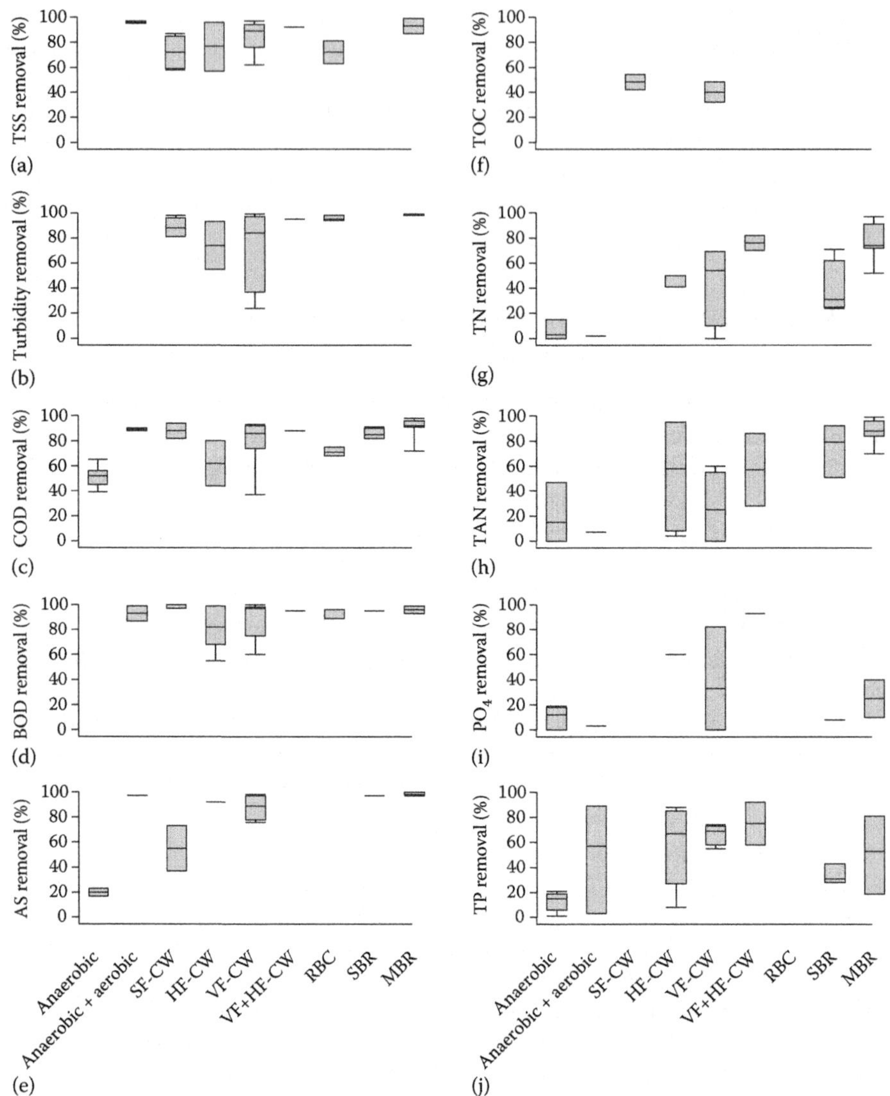


**FIGURE 2.21** Percent removal of (a) TSS, (b) turbidity, (c) COD, (d) $BOD_5$, (e) anionic surfactants, (f) total organic carbon, (g) total nitrogen, (h) total ammonia, (i) dissolved phosphorus, and (j) total phosphorus from various biological greywater treatment systems.

treatment, the concentration of solids in effluents was relatively low (Tables 2.11 and 2.12), but solids removal in biological treatment was higher (averaging 82%) than that in physical treatment which averaged 60% (Figure 2.22). Solids removal was effective in biological treatment because it usually included an element of physical filtration or sedimentation. Moreover, most of the solid material in greywater is organic and biodegradable. Further, it is clearer that the percent removal

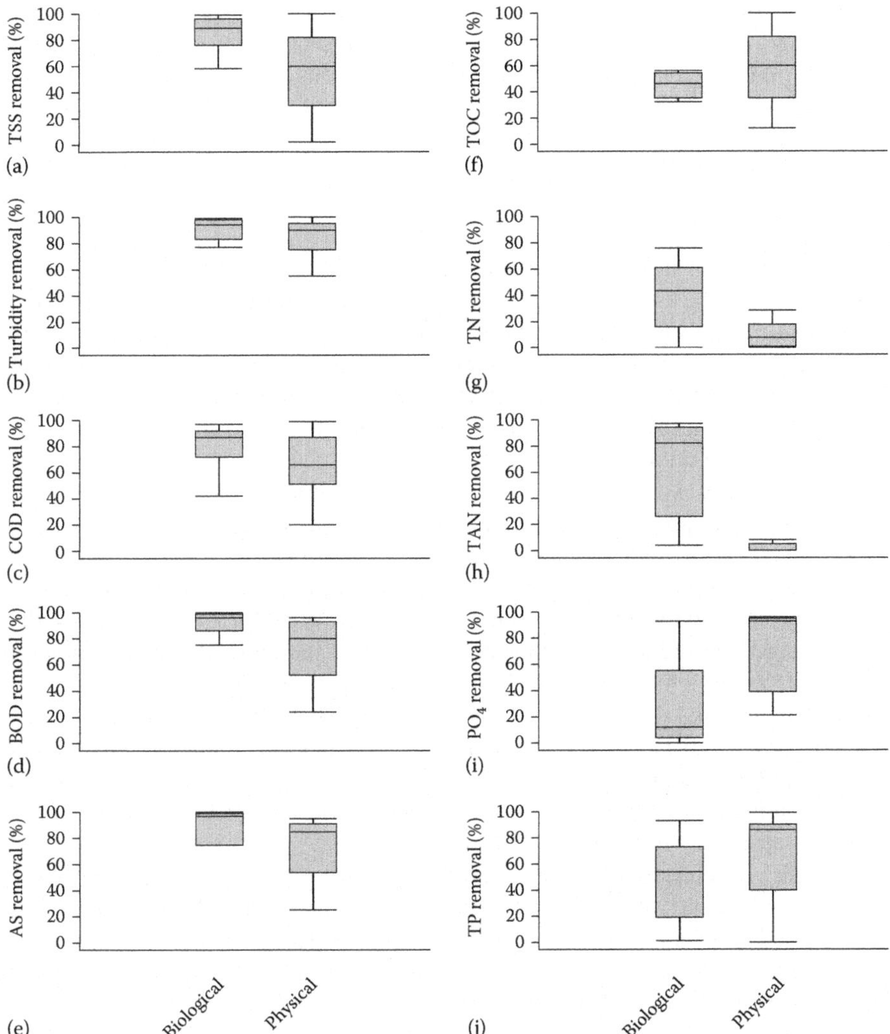


**FIGURE 2.22** Comparison biological and physical–chemical treatments in terms of the percent removal of (a) TSS, (b) turbidity, (c) COD, (d) $BOD_5$, (e) anionic surfactants, (f) total organic carbon, (g) total nitrogen, (h) total ammonia, (i) dissolved phosphorus, and (j) total phosphorus. (The numbers on the graph are Student's test $p$-value.)

of organic matter, as expressed by $BOD_5$ and COD, is higher on average in biological treatment than in physical–chemical treatment ($p < 0.05$). This is expected, as biological treatment is meant to biologically degrade dissolved and suspended organic material.

The average removal of total nitrogen in biological treatment averaged approximately 49% as compared to 23% in physical–chemical treatment. Since most of the nitrogen is dissolved, it is not treated effectively by physical treatment systems but is removed by the nitrification/denitrification coupling in biological facilities. Average phosphorus removal in physical treatments was approximately 69%, as compared with 28% in biological treatments. This is because phosphorus is chemically bound and sinks with the solids and is actually not treated biologically in the greywater treatment systems.

## 2.6 SUMMARY

A variety of greywater reuse technologies have been developed that are based on chemical, physical, and biological solutions. These innovations were motivated by a combination of factors including a need to treat different qualities and quantities of water, a need to treat water for various end uses, the lack of regulatory uniformity, and the need for simple and effective solutions. In the preceding chapter, we discussed the variation in treatment principles and discussed their pros, cons, and effects on effluent quality.

As expected, it was found that the TSS and turbidity (which are variables representing physical properties) are treated fairly well by physical treatment systems. Interestingly, comparable—and at times better—removal was also achieved with biological systems since most have a filtration or sedimentation component. Efficient removal of SSs and turbidity is also important for disinfecting greywater since their low concentration reduces the chlorine demand and increases the disinfection efficiency of UV. The ability of physical systems to treat dissolved organic matter is very limited. In contrast, the removal of dissolved organic matter in biological systems is relatively high. It is important to note that in anaerobic conditions, treatment of organic matter including surfactants is less efficacious than in aerobic conditions. However, it must be remembered that creating aerobic conditions is usually accompanied by an energy investment.

Nutrient concentration (nitrogen and phosphorus) is relatively low in greywater relative to full wastewater and hence also in treated greywater. As a result, in many cases treated greywater can be used as a substitute for fertilizer when used for irrigation. Phosphorus removal will usually be higher with physical treatment than with biological treatment because it often precipitates as salt (e.g., calcium or aluminum phosphates). However, the rate of nitrogen removal is higher in biological treatment because of the nitrification/denitrification processes.

# 3 Greywater Usages

## 3.1 INTRODUCTION

Treated greywater can be used for various nondrinking uses, such as flushing toilets, watering gardens (both private and public), washing streets and cars, washing laundry, and firefighting. This chapter focuses on flushing toilets and watering gardens, which are the most common uses with the highest potential for saving (potable) water.

## 3.2 GREYWATER REUSE FOR FLUSHING TOILETS

The development of urban centers is accompanied by an increase in population density and a reduction of the areas available for the construction of new buildings. Consequently, most of the construction in cities constitutes high-density high-rise buildings. Beyond the environmental benefits of such construction, in these buildings, the potential for potable water savings as a consequence of greywater reuse is between 10% and 30% of the urban water demand (Friedler and Galil, 2003; Campisano and Modica, 2008).

The amount of water required for flushing toilets is approximately 30% of the total domestic water consumption (see Figure 1.1). For example, in Israel, the consumption of water for toilet flushing is about 55 liters per person per day (LPD). This consumption is distributed throughout the day with two peaks in the morning and evening (Figure 3.1). In contrast, the flow of treated greywater is constant thanks to the equalization tank installed at the beginning of the treatment process, which regulates the incoming flow (see Chapter 2 where treatment is discussed). Therefore, it is necessary to install a second storage tank for treated greywater in order to regulate between the treated greywater production (constant) and its use for flushing toilets (varying with time).

Engineering challenges posed by greywater reuse for flushing toilets include preventing inadvertent cross-connections (i.e., unwanted connection of greywater line with potable water line; see Chapter 4, which deals with risks and risk assessment) and choosing the right volume of the storage tank for treated greywater. The volume of the storage tank depends on the distribution of water consumption for toilet flushing and on the yield of treated greywater. These flow rates depend on the nature of water use by the occupants (see Chapter 1) and on the size of the reuse project (number of individuals to be served). The larger the number of individuals, the more stable is the water consumption for toilet flushing. Although water use habits are quite repetitive, daily changes are possible in the distribution of water use of a small reuse project. Therefore, there is some uncertainty in determining the needed

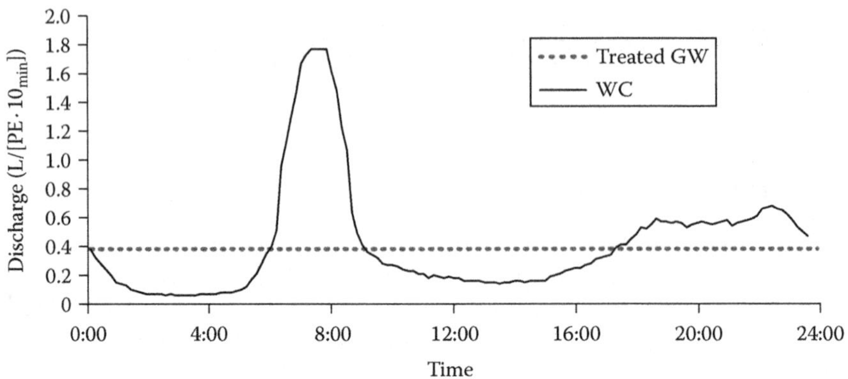


**FIGURE 3.1** Diurnal variation of water consumption for toilet flushing (WC) and flow of treated greywater. (Based on data from Butler, D. et al., *Water Sci. Technol.*, 31(7), 13, 1995; Penn, R. et al., *Urban Water J.*, 9(3), 137, 2012.)

storage volumes. This uncertainty is usually overcome by applying safety factors (i.e., adding some proportion to the calculated volume).

The choice of storage capacity affects the degree of water savings. With the increase in storage tank volume, the level of water savings for flushing toilets increases because the volume of water lost through overflow is minimized (Campisano and Modica, 2008). Nevertheless, when choosing a storage tank, one must also include economic considerations. The larger the tank, the more expensive it is and the larger the allocated space it requires, and space may be a valuable commodity in dense urban areas.

Uncertainty also has an economic aspect. Larger safety factors will increase the cost of water storage and, thus, affect the economic viability of the reuse project. If it is expected that in a small number of cases there will be lack of water for toilet flushing, the shortfall can be supplied by the supplementation of freshwater (ensuring that there is no cross-connection) instead of by increasing the storage. Alternatively, a large storage tank can be used, and then the excess water can be used for other purposes (garden irrigation, washing cars, etc.). The problem of determining the volume of the storage tank cannot be solved analytically because of the nature of the pattern of water consumption for toilet flushing. Several methods based on the storage of potable water for urban supply have been published; of them, two are discussed below: *mass curve analysis* (Ripple, 1883) and *sequent peak analysis* (Thomas and Burden, 1963).

### 3.2.1 Graphical Calculation of Storage Volume (Ripple, 1883)

In the graphical calculation method, the cumulative treated greywater consumption and the cumulative treated greywater production are plotted against time. See Figure 3.2 as an example. In this figure, three time periods are depicted: I, 0:00–7:30

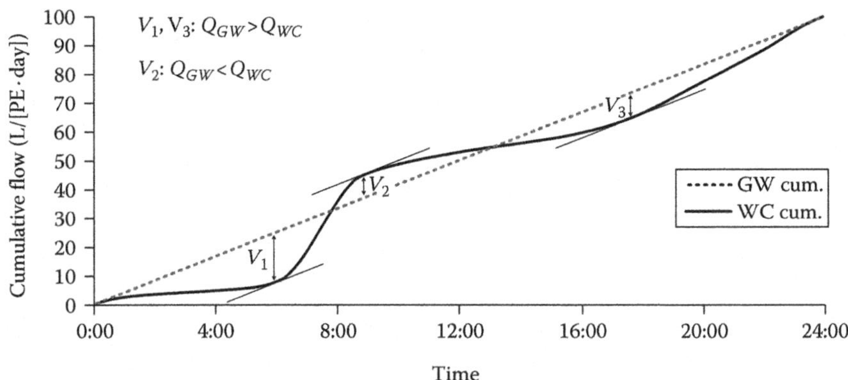


**FIGURE 3.2** Cumulative volume of consumption for flushing toilets and supply of treated greywater. (Based on the diurnal patterns presented in Figure 3.1.)

(the cumulative inflow is generally greater than the cumulative outflow); II, 7:30–13:00 (the cumulative inflow is generally smaller than the cumulative outflow); and III, 13:00–24:00 (the cumulative inflow is generally greater than the cumulative outflow). For each of these periods, the required storage volume is the maximum difference between the supply of greywater and its demand (i.e., $V_1$ for period I, $V_2$ for period II, and $V_3$ for period III). Since both I and III are periods of excess supply, only the greater value of $V_1$ and $V_3$ is considered ($V_1$ in our case). Thus, the overall required storage volume can be calculated using the following equation:

$$V_{Tot} = (V_1 + V_2) \cdot 1.33 \quad (3.1)$$

where

$V_{Tot}$ is the total storage volume (L/person)
$V_1$ is the maximum excess volume, period I (L/person)
$V_2$ is the maximum missing volume, period II (L/person)

The safety factors is selected depending on the degree of uncertainty and required reliability to supply treated greywater for the reuse system (0.1–0.33). The value of 0.33, suggested by Ripple (1883), is used in planning potable water supply systems that require higher reliability (greywater reuse systems can be supplemented with potable water).

### 3.2.2 Sequent Peak Method (Thomas and Burden, 1963)

The sequent peak method was suggested by Thomas and Burden (1963). Below is an application of the method for determining the required storage volume for reuse of greywater in toilet flushing (Penn et al., 2012). In this example, an iterative simulation of the liquid volume in the storage tank is performed using the equation

$$V_{(i)} = \begin{pmatrix} V_{(0)} & V_{(i-1)} + \left(Q_{(i)}^{GW} - Q_{(i)}^{WC}\right) \cdot t_{(i)} \geq V_{(0)} \\ 0 & V_{(i-1)} + \left(Q_{(i)}^{GW} - Q_{(i)}^{WC}\right) \cdot t_{(i)} < 0^{**} \\ V_{(i-1)} + \left(Q_{(i)}^{GW} - Q_{(i)}^{WC}\right) \cdot t_{(i)} & \text{else} \end{pmatrix} \quad (3.2)$$

where

$V_{(i)}$ is the greywater volume in the tank at the end of period $i$ (L/person)
$Q_{(i)}^{GW}$ is the greywater flow rate during period $i$ (L/[person × 10 min])
$Q_{(i)}^{WC}$ is the flow rate demand for flushing toilets during period $i$ (L/[person × 10 min])
$t_{(i)}$ is the length of the time step $i$ (in this example, 10 min)

In each run, the initial condition of the tank is full. **When a negative volume is obtained, $V_{(0)}$ is recalculated:

$$V_{(0)}^{*} = V_{(0)} + \left|V_{(i-1)} + \left(Q_{(i)}^{GW} - Q_{(i)}^{WC}\right) \cdot t_{(i)}\right| \quad (3.3)$$

The iterative simulation is run until no negative volume is obtained.

## 3.3 GREYWATER REUSE FOR GARDEN IRRIGATION

### 3.3.1 Introduction

Both the need for garden irrigation and the required irrigation volume in a given area depend on several factors: precipitation rate and distribution, type of soil, and type of irrigated vegetation. Greywater is produced throughout the year so it can supplement the amount of water required for plants beyond the quantity provided by precipitation, especially in areas with precipitation of less than 250 mm per year (arid regions). Elsewhere, greywater will be discharged to the sewerage system (if it exists) in the wet season or used as excess irrigation and during the dry season will irrigate the garden. Intelligent design of irrigation can create water savings, increasing plant health and reducing (potential) environmental pollution while maintaining public health.

This chapter presents practical guidelines and recommendations that can be used in designing a greywater reuse system for garden irrigation.

### 3.3.2 Irrigation Efficiency

Irrigation efficiency is the ratio between the total amount of water irrigated and the amount of water absorbed by the plant. This ratio varies between different plants and also depends on the soil type and the irrigation method.

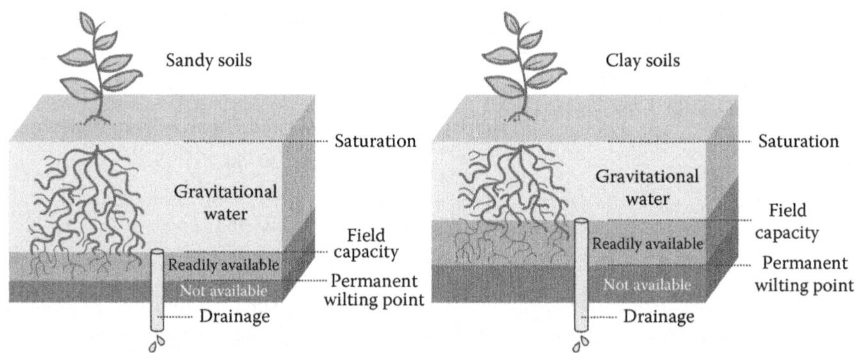


**FIGURE 3.3** Gravitational water and field capacity of sandy and clay soil. (From Doneen, L.D., *Irrigation Practice and Water Management*, Food and Agriculture Organization of the United Nations, Water Resources Development and Management Service, Rome, Italy, FAO Irrigation and Drainage Paper, no. 1, 1971, 84pp.)

Plants have root systems used for actively absorbing water from the soil. Both the spatial distribution of the root system in the soil and the root properties affect the efficiency of water absorption. Water's availability to plants also depends on the soil's water-holding capability (i.e., water-holding capacity of the soil). When the holding capacity is low, the water drains fairly quickly from the root zone or evaporates into the atmosphere. Following irrigation, it is customary to divide the irrigated soil into different areas where the water availability to a plant varies (from top to bottom): the saturated zone, the gravitational water zone, the field capacity zone, and the zone below the field capacity where the water is no longer available to the plant (permanent wilting point) (Figure 3.3). Gravitational water is defined as water found in the soil in an amount that exceeds its ability to adhere to the soil particles. This water flows with the help of gravitation to deeper layers. The moisture content in the field capacity zone is defined as the maximum water content of the wet soil after the gravitational water has drained, before run-off has been formed. The gravitational water and field capacity amounts depend on the type of soil. For example, in heavy soils, the field capacity is higher than in light soils (Table 3.1).

Irrigation efficiency also depends on the irrigation method. For example, irrigation by sprinklers is less efficient than drip irrigation, since some of the water in sprinkling reaches areas where no roots exist (Table 3.2). In addition, the evaporation rate of water irrigated by sprinkling is higher than that in a drip irrigation system.

### 3.3.3 Irrigation Volume and Frequency

Several important considerations must be made in planning an irrigation system: the penetration rate of the water into the soil, rate of evaporation, and type of vegetation.

**TABLE 3.1**
**Water-Holding Capacity of Various Soils**

| Type of Soil | Water-Holding Capacity/Field Capacity (mm Water/m Soil) |
|---|---|
| Coarse sand | 30–60 |
| Loam | 60–120 |
| Silt | 120–165 |
| Clay | 135–180 |

*Source:* Doneen, L.D., *Irrigation Practice and Water Management*, Food and Agriculture Organization of the United Nations, Water Resources Development and Management Service, Rome, Italy, FAO Irrigation and Drainage Paper, no. 1, 1971, 84 p.

**TABLE 3.2**
**Efficiency Coefficient of Irrigation**

| Irrigation Method | Efficiency Coefficient of Irrigation | Notes |
|---|---|---|
| Dripping | 0.93–0.95 | |
| Sprinkling | 0.8 | Lower in the presence of wind |

*Source:* Based on Be'er, R. et al., *Planning Low Water Consumption Gardens and Landscape*, 2nd edn., Water Authority, the Department to Promote Water Savings (Hebrew), 2009.

A plant's water requirement actually equals the water's evapotranspiration from the plant. The evapotranspiration depends on abiotic factors and most of all on temperature, sun radiation, wind, and relative humidity (Figure 3.4), as well as the plant's physiological properties and the leaves' surface area. Hence, the irrigation volume and frequency should be equal to the amount of water that has evaporated from the plant and can be determined according to irrigation efficiency.

#### 3.3.3.1 Calculating Irrigation Volume and Frequency

The daily evaporation values of water in different geographical regions and in different seasons can be found in tables for evaporation from Class A Pan (e.g., Table 3.3).

To calculate the amount of water that evaporates from a plant, abiotic factors that are reflected in Class A Pan evaporation and the correction coefficient that is dependent on the plant type must be considered. This is summed up in the following equation:

$$\text{Evapotranspiration} = (\text{Evaporation}) \cdot (\text{Crop factor}) \tag{3.4}$$

Table 3.4 demonstrates the evapotranspiration coefficients of different plants.

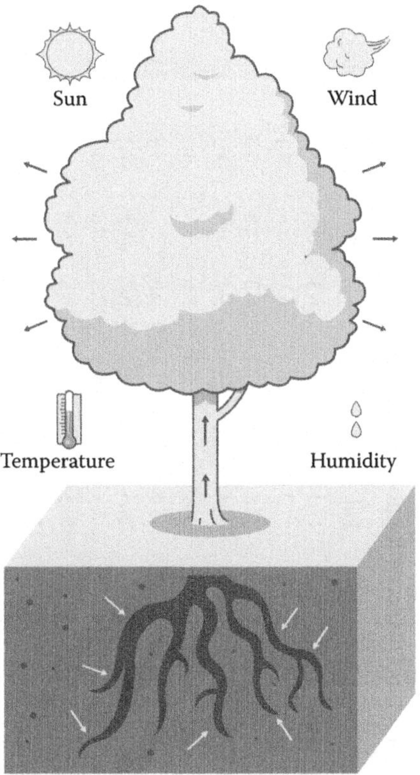


**FIGURE 3.4** Factors influencing evapotranspiration from plants. (From Doneen, L.D., *Irrigation Practice and Water Management*, Food and Agriculture Organization of the United Nations, Water Resources Development and Management Service, Rome, Italy, FAO Irrigation and Drainage Paper, no. 1, 1971, 84pp.)

The plant's water requirement can be calculated as follows:

$$NWR = TAWC \cdot DRZ \cdot MAD \tag{3.5}$$

where

*NWR* is the net water requirement, the total amount of water required by the plant (mm)

*TAWC* is the total available water capacity to the plant ($mm/m_{soil}$)

*DRZ* is the design root zone (m)

*MAD* is the management allowable deficit (depends on soil and the method of irrigation, usually assumed to be in the range of 40%–60%)

The *NWR* can also be calculated by the amount of water that the plant transpires and water loss from the soil (*MAD*):

**TABLE 3.3**
**Daily Evaporation Values from Class A Pan (Monthly Average) in Different Regions in Israel (mm/Day)**

| Region and Station | Geoclimate | Jan. | Feb. | Mar. | Apr. | May. | Jun. | Jul. | Aug. | Sep. | Oct. | Nov. | Dec. |
|---|---|---|---|---|---|---|---|---|---|---|---|---|---|
| Central coastal plain (Beit Dagan) | Med. | 2.3 | 2.7 | 3.8 | 5.3 | 6.5 | 7.4 | 7.5 | 6.7 | 6.3 | 4.8 | 3.3 | 2.1 |
| Central coastal plain (Negba) | Semiarid | 1.9 | 2.5 | 3.4 | 5.2 | 6.3 | 7.2 | 7.3 | 6.8 | 6.1 | 4.6 | 3.1 | 2.1 |
| Southern coastal plain (Habsor farm) | Arid savanna | 2.4 | 2.9 | 4.1 | 6.1 | 7.1 | 7.8 | 7.7 | 7.1 | 6.2 | 4.8 | 3.5 | 2.5 |
| Western Galilee (Eilon) | Med. | 2.2 | 2.4 | 3.4 | 5.1 | 6 | 6.4 | 6.1 | 5.6 | 5.2 | 4.5 | 3.3 | 2.3 |
| Judean Mountains (Rosh Tzurim) | Semiarid | 2.5 | 2.8 | 4.2 | 5.7 | 6.5 | 6.9 | 7.1 | 6.5 | 5.6 | 5 | 4 | 2.8 |
| Northern Negev (Beer Sheba) | Arid savanna | 2 | 2.6 | 3.8 | 5.9 | 7.6 | 8.6 | 8.4 | 7.8 | 6.6 | 5 | 3.4 | 2.1 |
| Hula Valley (Dafna) | Semiarid | 1.8 | 2.5 | 3.4 | 5.3 | 7.4 | 8.6 | 8.6 | 7.8 | 6.7 | 4.9 | 3.1 | 1.9 |
| Southern Golan Heights (Avney E Tan) | Med./semiarid | 2.4 | 2.7 | 4 | 5.7 | 7.8 | 8.6 | 9 | 8.2 | 7.3 | 5.8 | 4.2 | 2.9 |
| Beit Shean Valley (Havat Eden) | Arid savanna | 2 | 2.5 | 4 | 6.3 | 9.1 | 10.5 | 11.3 | 10.3 | 8.6 | 6.3 | 3.5 | 2 |
| Arava (Yotvata) | Extremely arid | 3.4 | 4.5 | 6.8 | 9.9 | 12.4 | 14.1 | 14 | 13.1 | 10.6 | 7.9 | 5.5 | 3.6 |

*Source:* Data taken from Israel Meteorological Service (Hebrew). http://ims.gov.il/IMS/Meteorologika/evaporation+Tub/monthly+data/.

**TABLE 3.4**
**Evapotranspiration Coefficients (CFs) for Different Types of Plants**

| | Evapotranspiration Coefficient | |
|---|---|---|
| **Type of Plant** | **Doneen et al. (1971)** | **Israel Water Authority (2009)** |
| Lawn | 0.4–0.6 | 0.45 regular maintenance<br>0.55–0.6 high maintenance |
| Flowers | 0.75–1.0 | 0.55–0.7 seasonal flowers<br>0.5–0.6 roses |
| Fruit trees | 0.8–1.0 | |
| Bushes | 0.3–0.4 | 0.3–0.4 young bushes and trees up to their establishment (3–4 years)<br>0.35–0.4 water-consuming established bushes and trees<br>0–0.1 water-saving established bushes and trees<br>0.1 water-saving established bushes and trees<br>0.2 water-saving established bushes and trees |

$$NWR = \frac{\sum_m \left(E_{pan} \cdot N_{days}\right) \cdot CF}{\sum_m N_{days} \cdot MAD} \tag{3.6}$$

where

$E_{pan}$ is the daily evaporation from Class A Pan (mm/day, monthly average) (e.g., Table 3.3)
$m$ is the number of irrigation months
$N_{days}$ is the number of days in a month
$CF$ is the crop factor, evapotranspiration coefficient (Table 3.4)

Hence, the amount of water for irrigation can be calculated:

$$GWR = \frac{NWR \cdot A}{1000 \cdot EF} \tag{3.7}$$

where

$GWR$ is the gross water requirement for irrigation (L/day)
$EF$ is the irrigation efficiency (—) (Table 3.2)
$A$ is the area of the plot ($m^2$)

The required amount of water should be irrigated at a suitable rate and frequency to allow optimal penetration of the water into the soil. Accordingly, the optimal method of irrigation should be selected.

Irrigation duration can be estimated according to the following equation:

$$DI = \frac{GWR}{AR} \tag{3.8}$$

where

*DI* is the duration of irrigation (h)
*AR* is the application rate (mm/h)

Irrigation frequency can be calculated according to the following equation:

$$I = \frac{NWR}{DE \cdot CF} \tag{3.9}$$

where

*I* is the time between consecutive irrigations (day)
*DE* is the evaporation rate from Class A Pan (mm/day) (Table 3.3)

In Israel and other countries, professional publications are available in which the estimated recommended amounts of water for irrigation of different crops in different seasons and geographical regions are provided (Table 3.5). These can be used instead of the evapotranspiration coefficient. There are also publications of recommended intervals between irrigations (Table 3.6).

Where irrigation with greywater is concerned, the amount of water available for irrigation is limited by the amount of greywater produced, which in turn depends on the number of household residents, nature of use patterns, and collection efficiency (see Chapter 1). The amount of water available for irrigation should be calculated up front if garden irrigation is to be exclusive with greywater or if the area planted will be determined by the amount of treated greywater available for watering. The following is an example of calculations to determine the amount of water required to irrigate a garden of 100 $m^2$ located on the coastal plain of Israel.

**CALCULATION EXAMPLE**

Suppose one wants to calculate the amount of greywater available for the irrigation of a 100 $m^2$ private garden of a home on the coastal plain of Israel in which five people live. Suppose that 30% of the garden area is designed to contain flowers, 30% shrubs, and 40% trees. In this example, the contribution of rainfall is zero since the calculation is performed for the dry season in Israel (May–October).

The potential of greywater production per capita is 100 LPD (see Chapter 1). Therefore, the amount of water available for irrigation is 500 L/day.

The total amount of water required by vegetation (NWR) during the irrigation months (May–October) for the different plants in the garden can be calculated according to Equation 3.5.

The average amount of water required for the flowers can be determined as follows:

$$\frac{(65\cdot 31+\cdots+4.8\cdot 31)\cdot 0.85}{(31+\cdots+31)}=5.6\frac{\text{mm}}{\text{day}} \tag{3.10}$$

The same calculation is performed for bushes and trees:

$$\frac{(65\cdot 31+\cdots+4.8\cdot 31)\cdot 0.35}{(31+\cdots+31)}=1.2 \tag{3.11}$$

$$\frac{(65\cdot 31+\cdots+4.8\cdot 31)\cdot 0.9}{(31+\cdots+31)}=3.0 \tag{3.12}$$

From this, we calculate the average daily water volume required for the plants:

$$\frac{5.6}{1000}\cdot 30+\frac{2.3}{1000}\cdot 30+\frac{5.9}{1000}\cdot 40=0.470\frac{\text{m}^3}{\text{day}}=470\frac{1}{\text{day}} \tag{3.13}$$

If the water loss from the soil is 40%, irrigation has also to replace the water evaporating from the ground:

$$\frac{470}{0.6}=783\frac{1}{\text{day}} \tag{3.14}$$

If we choose a drip system with efficiency of 95% as a method of irrigation, the actual amount of water is calculated as follows:

$$\frac{783}{0.95}=825\frac{1}{\text{day}} \tag{3.15}$$

In summary, the amount of greywater produced by a family of five individuals sums up to approximately 500 L/day. This amount is about 60% of that required for irrigation of a 100 $m^2$ garden. That is, 60% of the water consumption for garden irrigation can be supplied by the family's treated greywater, hence saving 60% of the potable water required for irrigating the garden during the dry period (May–October).

### 3.3.4 Irrigation Methods

The most common irrigation methods in home gardening are sprinklers, drip (subsurface or on surface), flood irrigation, and hosing. The choice between methods

**TABLE 3.5**
**Water Consumption in Gardens in Israel (L/($m^2 \times$ day))**

| | Spring | | | | Summer | | | | Autumn | | | |
|---|---|---|---|---|---|---|---|---|---|---|---|---|
| **Region and Climate** | **Lawn High Maint.** | **Lawn Medium Maint.** | **Bushes and Trees** | **Flowers, Vegetables, and Roses** | **Lawn High Maint.** | **Lawn Medium Maint.** | **Bushes and Trees** | **Flowers, Vegetables, and Roses** | **Lawn High Maint.** | **Lawn Medium Maint.** | **Bushes and Trees** | **Flowers, Vegetables, and Roses** |
| Coastal plain (med.) | 3.0 | 2.5 | 2.0 | 3.5 | 4.0 | 3.0 | 2.5 | 4.0 | 2.5 | 2.0 | 1.5 | 2.5 |
| Interior lowlands (med.) | 3.5 | 3.0 | 2.0 | 3.5 | 4.0 | 3.5 | 3.0 | 4.5 | 3.0 | 2.5 | 2.0 | 3.0 |
| Mountain area (med./semiarid) | 3.5 | 3.0 | 2.0 | 4.0 | 4.5 | 3.5 | 3.0 | 5.0 | 3.0 | 2.5 | 2.0 | 3.5 |
| Interior valley and central Negev (semiarid) | 3.5 | 3.0 | 2.0 | 3.5 | 4.5 | 3.5 | 3.0 | 6.0 | 3.0 | 2.5 | 2.0 | 3.0 |
| Arava (extremely arid) | 5.5 | 4.5 | 3.5 | 8.0 | 7.5 | 7.0 | 4.5 | 11.0 | 4.5 | 4.0 | 3.0 | 8.0 |

*Source:* Based on *The Gardener's Journal*, The Gardening and Landscaping Organization in Israel, 2000 (Hebrew).

**TABLE 3.6**
**Range of Recommended Time Intervals between Consecutive Irrigations**

| Type of Plant | | Possible Interval between Irrigations (days) Depending on the Type of Soil (Light/Medium/Heavy) |
|---|---|---|
| Lawn | Regular maintenance | 5–14 |
| | High maintenance | 5–21 |
| Flowers | Seasonal flowers | 3–7 |
| | Roses | 7–10 |
| Bushes | Young bushes and trees up to their establishment (3–4 years) | 7–14 |
| | Water-consuming established bushes and trees | 10–20 |
| | Water-saving established bushes and trees | Once in the summer |
| | Water-saving established bushes and trees | 2–3 times in the summer |
| | Water-saving established bushes and trees | 3–5 times in the summer |

*Source:* Israeli Water Authority, *Gardens and Landscape Planing*, Beer, R., ed., Israeli Water Authority, pp.61, 2009.

is rarely the result of formal calculation; it is usually based on plant type, climate, soil type, water quality, convenience, and the owner's resources (financial and otherwise). Due to concern about the exposure of humans and animals to greywater, it is usually desirable to water the garden with a surface or subsurface drip system in which the irrigation equipment is clearly marked. This is common practice. For example, the use of purple piping is an internationally agreed sign for effluent irrigation. Moreover, with drip irrigation, the uniformity and irrigation efficiency of water distribution are high. It should be remembered that drip irrigation requires the added expenditure on pipeline infrastructure. Table 3.7 presents the advantages and disadvantages of the different irrigation methods.

### 3.3.5 Time Interval between Consecutive Irrigations

As mentioned, in an optimally planned garden, it is important to determine the required irrigation duration and frequency. Greywater is produced every day, and storing it is not recommended as its quality will decline within a relatively short time (depending on the level of treatment). For this reason, in most cases, plants are irrigated with the greywater every day as opposed to two or three times a week as is common in gardens irrigated with freshwater. It is possible to increase irrigation efficiency in one of two ways: (1) store treated greywater of high quality in a way that ensures its quality is not compromised for approximately 2–3 days or (2) in larger gardens, it is possible to divide greywater irrigation between different regions and thus water a different area each day. This technique may increase the ability of the soil to store water and to further treat the water while reducing the risk of environmental pollution.

**TABLE 3.7**
**Advantages and Disadvantages of Various Irrigation Methods**

| Method | Advantages | Disadvantages |
|---|---|---|
| On-surface drip irrigation | Lower irrigation rate | Not suitable for all applications. |
| | Uniform distribution | Sensitive to water quality. |
| | High efficiency | Requires pressure above a certain minimum. |
| | Easy to install | Requires investment in pipeline infrastructure. |
| | Low weed growth | |
| | Aids in prevention of plant diseases | |
| | Easy to fertilize | |
| | Flexible irrigation regime | |
| | Possible to irrigate with high-salinity water by flushing (excess irrigation) | |
| | Not sensitive to pressure changes (only self-regulating drippers) | |
| | Reduces the exposure to water | |
| Subsurface drip irrigation | Low irrigation rate | Expensive. |
| | Uniform distribution | Not suitable for all applications. |
| | High efficiency | Requires professional knowhow. |
| | Low growth of unwanted weeds | Requires pressure above a certain minimum. |
| | Improves prevention of plant diseases | Impossible to monitor clogged drippers. |
| | Easy to fertilize | Leakages and bursts cannot be easily detected. |
| | Prevents damage to piping (by animals, people, or elements) | Problematic maintenance. |
| | Flexible irrigation regime | Requires investment in pipeline infrastructure. |
| | Possible to irrigate with high-salinity water by flushing (excess irrigation) | |
| | Not sensitive to pressure changes (only self-regulating drippers) | |
| | Reduces direct exposure to water significantly | |
| Sprinkling | Suitable for all applications | Low efficiency. |
| | Easy to operate | Easy to make mistake in planning. |
| | More expensive than flood irrigation but cheaper than dripping | Increases diseases and the growth of fungi. |
| | | Exposure to large amounts of water and spray may endanger public health. |

## 3.4 SUMMARY

The most common ways to use treated greywater are reuse in toilet flushing and reuse for watering gardens. Where reuse for toilet flushing is concerned, there are engineering challenges involved: to prevent inadvertent cross-connections between the potable water system and the treated greywater system and to design a storage tank of optimal volume. For garden irrigation, the reuse of greywater should be planned with an eye to the quantities and methods of irrigation to be used. Irrigation may create water spray (aerosols) or water impoundment on the soil surface and surface runoff. Moreover, irrigation water may percolate through the soil and reach the groundwater and contaminate it. For these reasons, the rate of irrigation, its frequency, and the quantities of water supplied should conform to the requirements of the vegetation, soil, and the environment, and not be in excess. It should be ensured that irrigation will not compromise public health or damage the environment, for example, by treating the water to safe levels.

Proper planning of greywater reuse for garden irrigation, as described in this chapter, will significantly decrease potential hazards to public health and the environment. The risk associated with greywater irrigation is discussed in Chapter 4.

# 4 Risk Assessment and Management

## 4.1 INTRODUCTION

Risk assessment is a process by which specific risk factors are identified, and an attempt is made to assess the potential damage they may cause under known conditions, with consideration of uncertainties (Wu and Farland, 2007). The framework of risk assessment has been developed mainly since the second half of the twentieth century. Originally, in the health field, the purpose of risk assessment was to understand and evaluate public health risks resulting from the chemical, food, and pharmaceutical industries (Parkin, 2007). With time, frameworks were evolved to assess environmental and ecological risks (e.g., environmental risk assessment). To enrich the risk analysis paradigm, social input, policy feasibility, and economic considerations must be considered, as well as the best means for risk communication. In addition, provisions to avoid or reduce risks should be considered if risk is to be incurred over time. This process in its entirety comprises the risk management framework.

The risk management process is complex and is often used as a significant aid in decision making. Optimal risk management of greywater use considers and balances protection of public health, environmental aspects, and other goals such as saving natural resources and maintaining economical feasibility. In this chapter, we discuss the risks associated with the use and management of greywater, how to assess this risk, and how to manage it (Figure 4.1).

Each of the steps shown in Figure 4.1 consists of substages that together help estimate risk and provide decision makers with a tool for setting an optimal policy:

1. *Hazard identification*: The object of assessment (i.e., greywater system or scheme) is divided into steps in the form of a flow chart to allow identification of all risk factors and their methods of transfer.
2. *Setting tolerable risk levels and associated targets*: Identifying acceptable safety levels as zero risk is usually unfeasible.
3. *Risk assessment*: Conducting a full risk assessment, quantitative where possible, specifically for each hazard.
4. Measures to achieve the targets that were identified in the preceding steps: In other words, specifying the required means and barriers to achieve the desired safety targets.
5. *Critical control points (CCPs)*: Finding the critical points that should be monitored to track the performance of the system or reuse scheme and its compliance with the safety targets and the means to mitigate the problems where necessary.

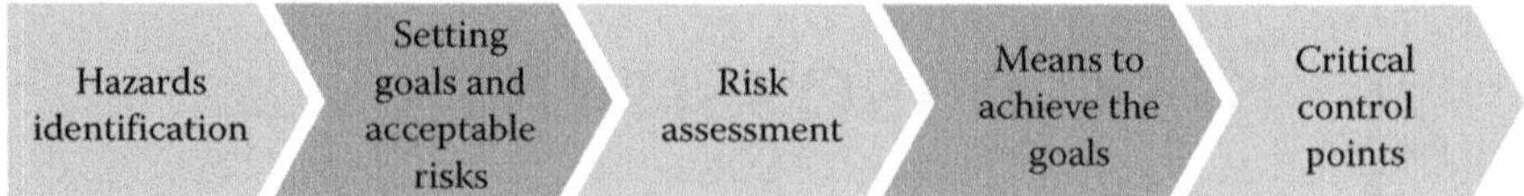


**FIGURE 4.1** The five steps of risk management.

## 4.2 HAZARD IDENTIFICATION

A general flow chart of greywater reuse scheme is presented in Figure 4.2. This chart facilitates the identification of the risk factors in each stage. The details are as follows:

### 4.2.1 Scope of the Assessment

#### 4.2.1.1 Single Household

Sanitation and environmental hazards may result from greywater use. These can be caused by odors, leaks, clogging, etc., and may be incurred due to poor maintenance or lack of professional knowledge. These can also occur from biological or mechanical failure of the system due to extreme changes in the water quantity or quality, for example, when the residents are absent for a long certain period or when a temporary overload of detergent occurs. However, there is a low probability of the spread of sanitary or environmental damage beyond the borders of the household. It is also worth noting that in a single household, there are many ways one can be exposed to pathogens (surfaces, contact, aerosols, etc.), and greywater is usually not the main route.

#### 4.2.1.2 Multiple Households

In this situation, managing the risk from a greywater system is somewhat more complicated. First, in the event of a failure in the system, the damage would affect a larger population than in a single household, and the risk of diseases being spread would be higher. Second, a greywater system can be a path through which pathogens are spread between people who do not come into direct daily contact with each other. Finally, risk is more likely due to potential for a "tragedy of the common" situation

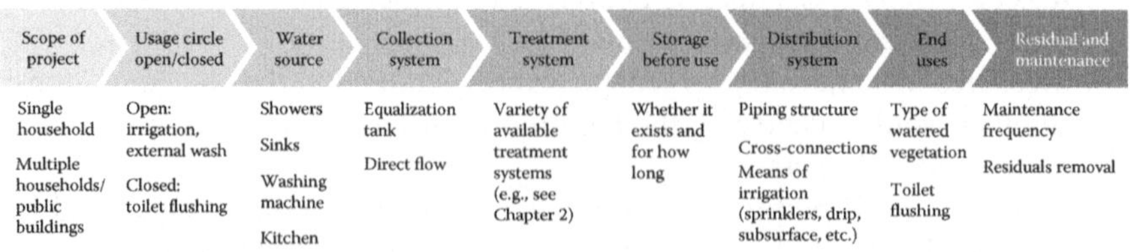


**FIGURE 4.2** General flow chart of a program to reuse greywater. (Based on NSW, *Management of Private Recycled Water Schemes*, New South Wales, Australia, Water for life, 2008.)

(Hardin, 1968). In such a case, there is no way of tracing the source of a given contribution to the communal water system, so there is less personal responsibility and a higher probability that users will allow unwanted materials into the system.

#### 4.2.1.3 Public Buildings

A public building is similar to multiple households in terms of risk from a system failure. However, there is a more active and knowledgeable control over a greywater system in most public buildings as compared to private homes because a technician or a professional company is responsible for the management and maintenance of the system.

### 4.2.2 Open/Closed Cycle of Use

#### 4.2.2.1 Open-Circuit Use

In greywater systems that transport water to the environment (outside the user's area), the sanitary and environmental risks may be wider-spread, for example, through groundwater contamination or disease transmission by animals (Figure 4.3a).

#### 4.2.2.2 Closed-Circuit Use

This refers to situations in which water is recycled for limited use in the private realm where contact with humans and the environment is minimal, and the water is afterward transferred to the central sewage treatment system (such as with toilet flushing). In these uses, the main risk is from cross-connection and the entry of greywater into drinking water. The environmental impact is lower than with open circle usage; however, it is still necessary to ensure that no other materials that will interfere with subsequent treatment enter the water (Figure 4.3b).

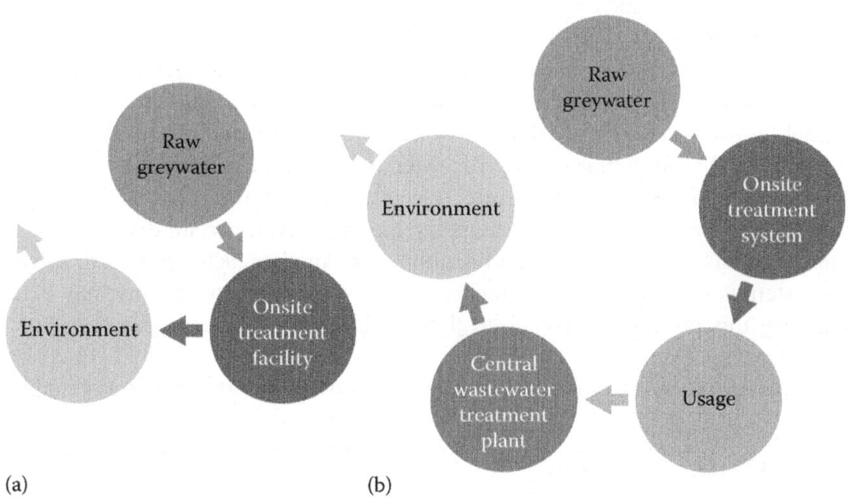


**FIGURE 4.3** Schematic drawing of (a) open and (b) closed use cycles.

### 4.2.3 Water Source

Water sources differ in their concentrations of pollutants and hence in expected risks. The identification and characterization of pollution sources in the various water sources is a critical step in the process of evaluating risks, and it should be conducted carefully and thoughtfully. For example, water from washbasins may contain opportunistic pathogenic microorganisms originating in the skin and mucous membranes (but usually without pathogens from fecal origin). It may also contain pollutants such as cosmetics, detergents, paints, thinners, and other materials related to household work or hobbies. Bath and shower water is expected to contain the same pathogens found in washbasins, but also those of fecal origin. Laundry water is expected to contain high concentrations of surfactants and salts, as well as pathogenic microorganisms mainly of fecal origin. Kitchen water is likely to contain high concentrations of oils, organic matter, detergents, and microorganisms originating from food handling such as *Salmonella*. (For further details on the characteristics of various water streams, see Chapter 1.)

### 4.2.4 Collection System

Using a storage tank to collect greywater can be considered a primary treatment in addition to being used to regulate the volume of water entering the treatment system. However, it is also associated with certain disadvantages and risks like noxious odors, regrowth of pathogens, and the need for periodic maintenance such as cleaning accumulated sludge. In some cases, greywater can be streamed directly into a treatment system (without the use of a storage tank).

### 4.2.5 Treatment and Disinfection System

The correct treatment and disinfection systems are determined by the characterization of the risk associated with the water source, as presented in the previous section. One of the most significant factors that affect risk is the quality of the water exiting the treatment system. Therefore, each system must be examined individually to identify potential failure points (Figure 4.4), as well as the risks resulting from these failures. For example, the microbial quality of water is greatly affected by the presence or absence of disinfection. Water disinfection will increase the microbial quality significantly and therefore reduce the sanitary risk. However, certain disinfectants may themselves cause environmental and health risks. Another factor related to the treatment system is water storage after the treatment, since the water quality may deteriorate during this period mainly due to the development of microorganisms. For this reason, if treated water is stored, the use of a disinfectant with a residual effect should be considered, or the disinfection unit should be positioned after the storage tank.

### 4.2.6 Distribution System

Cross-connections have the potential to be critical failure points in the cases of unprofessional system construction. In the case of failure, high exposure to

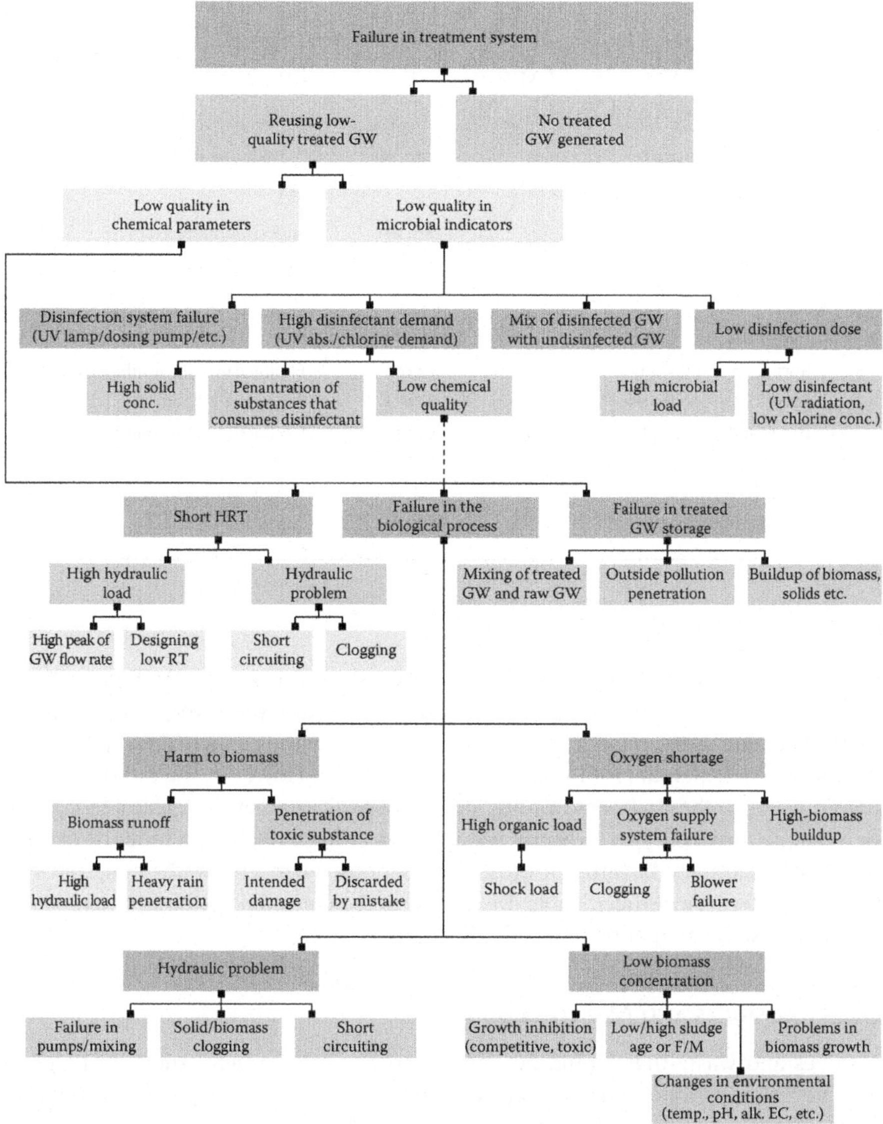


**FIGURE 4.4** Flow chart designed to identify failures in a hypothetical biological treatment system. (Based on Alfiya, Y. et al., *Wat. Sci. Technol.*, 67(6), 1389, 2013.)

greywater may occur leading to a high sanitary risk. Another failure point could be in the structure of the pipeline itself, which can lead to clogging or standing water (stagnation). Furthermore, the method of irrigation has a significant effect on the degree of exposure to water. Irrigation with sprinklers creates aerosols and high human exposure potential, while drip irrigation (and to an even greater extent subsurface irrigation) reduces the exposure significantly.

**TABLE 4.1**
**Common Methods of Exposure Following the Use of Greywater for Irrigation and Flushing Toilets**

| Source of Exposure | Amount | Means of Exposure |
|---|---|---|
| *Toilet flushing* | | |
| Toilet flushing—splash | Small | Ingestion/inhalation |
| Toilet flushing—aerosols | Small | Ingestion/inhalation |
| Cross-connections | Large | Ingestion |
| *Irrigation* | | |
| Sprinkling–splashing | Large | Contact/ingestion/inhalation |
| Sprinkling—aerosols | Small | Ingestion/inhalation |
| Break in irrigation piping—splash | Large | Contact/ingestion/inhalation |
| Surface irrigation | Large | Contact/ingestion |
| Unauthorized uses | Large | Contact/ingestion/inhalation |
| Cross-connections | Large | Ingestion |

### 4.2.7 End Uses

The various end uses also affect the level of risk. For example, greywater use for irrigating food crops involves higher potential exposure and sanitary risk than that of watering decorative plants. The use of water for flushing toilets does not create as great an environmental risk impact as does irrigation. However, the proximity of the treated water to the daily environment of humans is another dimension to consider, for example, proximity to airborne water droplets from flushing a toilet or exposure to potential sanitary hazards of cross-connections when greywater penetrates into drinking water (see earlier). Table 4.1 describes the common methods of exposure in greywater use for irrigation and toilet flushing.

### 4.2.8 Residues and Maintenance

Residues and waste that should be eliminated from the system, such as sludge and used filters, constitute another point of exposure. If incorrectly handled, they may lead to serious health and environmental risks. Another significant factor in exposure to raw and treated greywater is the nature and frequency of required system maintenance. Maintenance personnel may be exposed to greywater systems on a daily basis, so the risk level for this group in particular should be examined.

## 4.3 SETTING HEALTH AND ENVIRONMENTAL TARGETS AND TOLERABLE RISK LEVELS

It can be assumed that any action involves some risk, so a target of zero risk is never realistic. Therefore, before beginning the assessments of specific risks, the criteria for high and low risks must be defined. For example, a 1:1000 risk of mild intestinal disease is not qualitatively as severe as a 1:1000 risk of mortality; however, quantitatively, they are the same. Obviously, comparison between the two would require a normalization procedure. Similarly, in environmental risk assessment, data should be normalized, and goals defined in a way that enables various risk factors to be evaluated on the same scale, allowing comparison between them.

### 4.3.1 Health Targets

For health risk assessments, we will use the disability-adjusted life year (DALY) index.

The DALY index is a quantitative measure used to estimate the amount of healthy life lost as a result of death, illness, or any health malady (Homedes, 1996). It is used to measure a population's overall health by combining years lost as a result of death (years of life lost [YLL]) and the years of having to live with a health limitation (years lived with disability [YLD]). The index is normalized, so it offers the advantage of being usable to compare disparate health situations, such as terminal illness versus a three-day stomach virus. DALY is also used as an important tool in determining the maximal risk tolerated and can be used to articulate health goals when making public health management decisions. For example, the World Health Organization (WHO) stated that a DALY value of $10^{-6}$ per person-year (p-y) is the highest tolerable risk of water-transmitted diseases (WHO, 2008). That is, the maximum risk from water-related diseases should be no more than the loss of approximately one year of a healthy life in a population of one million people. Applying this logic, we can calculate the maximum tolerable risk of infection according to the characteristics specific to that disease. More information on the calculation methods of the DALY index and its application in decision making can be found in various literature sources (Homedes, 1996; WHO, 2006a, 2008).

Based on the goal of $10^{-6}$ DALY/p-y, the tolerable infection rates of several waterborne pathogens were calculated (Table 4.2).

**TABLE 4.2**
**Calculating the Maximum Tolerable Risk of Infection by Waterborne Pathogens**

| Pathogen | Taxa | DALY Loss per Case of Disease | Tolerable Disease Risk P.Y Equivalent to $10^{-6}$ DALY Loss P.Y | Disease/Infection Ratio | Tolerable Infection Risk |
|---|---|---|---|---|---|
| Rotavirus | Virus | 1.40E–02 | 7.14E–05 | 0.05 | 1.43E–03 |
| Campylobacter | Bacteria | 4.60E–03 | 2.17E–04 | 0.7 | 3.11E–04 |
| Cryptosporidium | Protozoa | 1.50E–03 | 6.67E–04 | 0.3 | 2.22E–03 |

*Source:* Drechsel, P. et al., Eds., *Wastewater Irrigation and Health: Assessing and Mitigating Risk in Low-Income Countries*, IDRC, Sri Lanka, 2010.

### 4.3.2 Environmental Targets

While health targets are direct and fairly obvious, environmental goals are more difficult to set and compare. This is because research on the subject of the environmental impact of greywater is extremely limited, so there are no agreed upon quantitative indices (such as DALY) of environmental effects. Questions arise, for example, "Is it better to determine maximal concentrations of specific substances that may cause damage to the environment?" "And should target values be established for general indices associated with health of the soil, vegetation, and water bodies?" "Or do substances that create immediate but reversible damage to vegetation have to be assigned the same weight as materials that are harmless in the near future, but which accumulate in the long run to cause permanent damage?"

In the absence of quantitative measures, only few qualitative environmental risk assessments were developed for greywater use such as in Australia (NRMMC and EPHC, 2006). In this example, estimates are based on the matrix in Figure 4.5, and their goal is to avoid risks that are rated above a *low* risk level. However, the disadvantage of such a framework is that certain risks may not be captured. For example, pollution due to greywater is usually local and therefore will be rated as having a minor environmental impact at the most even though it is possible that many small facilities will produce significant cumulative damage. To counter such deficiencies, it is important to develop a quantitative model for assessing environmental risks originating from the reuse of greywater.

## 4.4 RISK ASSESSMENT

Risk analysis of greywater should include health and environmental risk assessments, each of which evaluates distinct pathogens and pollutants. One of the first articles to propose such a combined framework for risk assessment of greywater use describes a qualitative health risk assessment framework weighing exposure and water quality data (Table 4.3), as well as a matrix for rating various environmental hazards according to their risk level (Dixon et al., 1999b). The proposed matrix is reminiscent of the one shown in Figure 4.5. Over the years, there has been an increasing tendency to quantitatively focus on the health aspect issues while neglecting environmental aspects.

The following sections present an assessment of health risks based on a method called quantitative microbial risk assessment (QMRA) (Haas et al., 1999). In addition, we offer a quantitative outline to examine environmental risks.

### 4.4.1 Health Risks

There are two main approaches to conduct health risk assessment: one based on epidemiological research and the other uses a more theoretical evaluation through QMRA. In epidemiological studies, a comparison is made of the effects of disease among populations with different exposures to the risk factor.

| Probability / Consequences | | Insignificant | Minor | Moderate | Major | Catastrophic |
|---|---|---|---|---|---|---|
| | | No detectable environmental impact | Potentially harmful to local ecosystem with contained onsite impact | Potential damage to regional ecosystem with primarily local contained impacts | Potentially lethal for local ecosystems, potential for off-site impacts | Potentially lethal to regional ecosystems or threatened Species. Vast on-site and off-site impacts |
| Almost certain | Event is expected to occur often (several times per year) | Low | Medium | High | Very high | Very high |
| Likely | Event will probably occur once within 1 to 5 years period | Low | Medium | High | Very high | Very high |
| Possible | Event might occur every 5 to 10 years | Low | Medium | Medium | High | Very high |
| Unlikely | Event could occur once in 20 years or in unusual circumstances | Low | Low | Medium | High | High |
| Rare | Event may occur only in rare circumstances once every 100 years | Low | Low | Low | Medium | High |

**FIGURE 4.5** A matrix for ranking environmental risk. (Based on NRMMC and EPHC, *National Guidelines for Water Recycling: Managing Health and Environmental Risks (Phase 1)*, Natural Resource Management Ministerial Council, Environment Protection and Heritage Council, Australian Health Ministers' Conference, 2006.)

**TABLE 4.3**
**Conceptual Analysis of Risk Range of Greywater Uses**

| | Low Risk | Medium Risk | High Risk |
|---|---|---|---|
| Population | Small/single household | | Large/several households |
| Exposure | Without contact (subsurface irrigation) | With contact (toilet flushing, washing) | Swallowing (drinking) |
| Dose–response | <1 virus for sampling, <1 bacteria for sampling | | 1<virus for sampling, $10^6$<bacteria for sampling |
| Delay before use | Immediate use | Use within hours | Use within days |

*Source:* Dixon, A.M. et al., *Water Environ. J.*, 13(5), 322, 1999b.

It is worth mentioning that no significant epidemiological studies have yet been conducted on the use of greywater. One study to be presented later (Fernandes et al., 2007) seems to deal with greywater, but it is actually about partially treated surface water. Nevertheless, we can learn from studies on the application of poor quality water for domestic use with low exposure. For example, a retrospective epidemiological study was conducted in Sydney, Australia, where there are dual reticulation systems in homes; part of the domestic water consumed (for toilet flushing, watering gardens, washing cars, and other external uses) is recovered wastewater (Sinclair et al., 2010). The recovered wastewater is of high microbial quality (less than 1 fecal coliform/100 mL, 10 coliforms/100 mL, 1 protozoa/50 L, and 2 viruses/50 L). According to risk assessment, the levels should not pose a risk detectable in epidemiological research. However, the risk assessments do not take into account unauthorized uses of water such as filling swimming pools, so it is possible that the real risk is much higher; the article claims that about 16% of swimming pool owners in the areas with a dual reticulation system fill their pools with the treated wastewater. The study was based on calls to a general physician about problems that could be linked to treated wastewater, such as gastrointestinal and respiratory problems as well as skin irritation. The study involved 14,458 people from a region with a normal water supply and 21,370 people from a region with double piping. The study found no significant difference between the regions in complaints related to the target illnesses. Based on these results, the researchers concluded that it is reasonable to assume that use of reclaimed water in well-managed dual reticulation systems is generally a safe option.

In contrast to the Australian study in which using a dual reticulation system demonstrated a positive outcome, a Dutch study demonstrates the potential for a negative outcome with a system of similar purpose. In December 2001, an accidental cross-connection was made between the drinking water of a neighborhood in Holland and partially treated surface water system, which was supposed to be used only for low-exposure purposes. As a result, the residents, whose water was polluted with water of poor quality, suffered from a high rate of diarrhea in comparison with those residents whose water was of normal quality (Fernandes et al., 2007). In another study in which this same case was examined, it was found that at the time, there were no binding regulations for water quality for nondrinking purposes (Oesterholt et al., 2007).

Where epidemiology is based around clinical outcomes rather than the specific contaminants causing the problem, a QMRA allows an assessment of the degree of risk caused by exposure to a particular pathogen. For example, a quantitative assessment of risk on the same aforementioned cross-connection event found that in the region where the failure occurred, the concentration of *Norovirus* was roughly 30 times higher than the allowable concentration (the maximum risk level required for drinking water according to the Dutch law in this case is below $10^{-4}$) (Oesterholt et al., 2007).

The QMRA method is an indirect measure of risk, based on mathematical models and data from experiments. The QMRA process is divided into the four general risk assessment steps (Figure 4.6):

1. Hazard identification: Defining risk factors and/or identifying indicators of the principal risks and evaluating their frequency in the relevant environment
2. Exposure assessment: Assessment of human pathways of exposure to the target hazards, appraisal of exposure duration and frequency, and identification of exposed populations
3. Dose–response model: Defining the mathematical relationship between dose of exposure and the probability of infection
4. Risk characterization: Processing the data collected from all earlier QMRA stages to assess the risk level and comparison with target levels

It was found that when realistic values are used in QMRA (using a Monte Carlo simulation with 10,000 repetitions), the risk values obtained are similar to those obtained in parallel epidemiological studies (Mara et al., 2007).

In recent years, it seems that there is a trend among policy makers to add the QMRA method to their toolbox. Thus, in 2006, a policy document for safe use of wastewater, excreta, and greywater was published by the WHO (2006a), as well as guidelines for using greywater in Australia (NRMMC and EPHC, 2006). Both use the QMRA platform as a tool for setting policy. Here too, we will use QMRA to present the risks involved in using greywater.

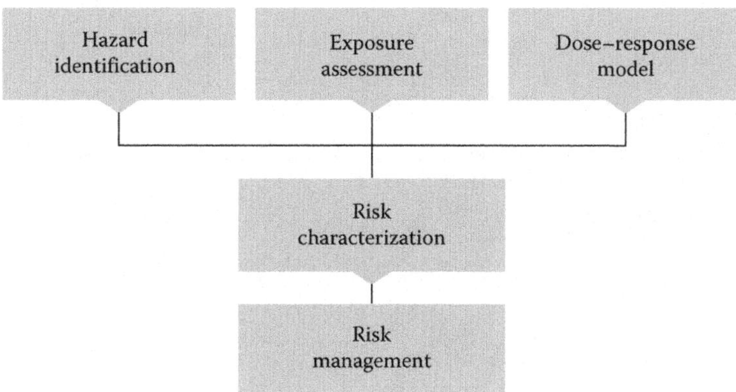


**FIGURE 4.6** Flow chart of risk assessment according to the QMRA method.

#### 4.4.1.1 Hazard Identification

Greywater is often considered harmless because it does not contain human secretions. However, it is known today that greywater may pose a health risk (Table 4.4) if used improperly (Dixon et al., 1999b; Eriksson et al., 2002; Ottoson and Stenstrom, 2003; Friedler, 2004; Friedler et al., 2006; Morel and Diener, 2006). Pathogenic organisms may reach greywater from three main sources: fecal contamination, food handling, and opportunistic pathogens on the skin or in mucous tissues (such as the mouth and nose).

Since the identification of all possible pathogens and their concentration in greywater are not practical, it is necessary to use indicator organisms. Although greywater usually contains low levels of fecal contamination, the vast majority of literature on this topic refers to the presence of a variety of gastrointestinal pathogens (Ottoson and Stenstrom, 2003; Birks and Hills, 2007) as traditionally fecal contamination is a main factor in wastewater quality assessment. The most common indicator of fecal pathogens is fecal coliforms. Although there are many reports that show high levels of fecal coliforms in greywater (Christova-Boal and McFarlane, 1996; Eriksson et al., 2002; Ottoson and Stenstrom, 2003; Birks and Hills, 2007), the relevance of this indicator in assessing the microbial quality of greywater is debatable (Dixon et al., 1999b; Ottoson and Stenstrom, 2003). For example, it was found that in some cases, given the right conditions, fecal coliforms are able to reproduce in an aqueous

**TABLE 4.4**
**Pathogen and Indicator Concentrations, as Found in Several Studies**

| Microorganism | Kitchen | Laundry | Washing | Light | Mixed |
|---|---|---|---|---|---|
| *Indicators* | | | | | |
| Total coliforms log10/100 mL | | 1.7–5.8[a] | 1.7–7.4[a] | 2.7–7.4[e] | 7.2–8.8[a] |
| *E. coli* log10/100 ml | | | 0.5–4.4[a] | | 2.0–6.0[a] |
| Fecal coliforms log10/100 mL | 4.8–6.1[a] | 1.4–6.6[a] | 1–6.6[a] | 1.0–6.9[e] | 3.0–8.0[a,e] |
| Fecal enterococci log10/100 mL | | 1.4–3.4[a] | 1–3.4[a] | 1.9–3.4[e] | 2.4–4.6[a,b,e] |
| *Pathogens* | | | | | |
| *Salmonella* spp. | | | Detected[d] | | |
| *Campylobacter jejuni* | | | 0[d] | | |
| *P. aeruginosa* log10/100 ml | | | 0–2[c] | 2.6–3.5[e] | 2.3–4.3[c,e] |
| *S. aureus* log10/100 mL | | | 5[c] | 4–5.7[e] | 4–5.7[e] |
| *Legionella pneumophila* units per 100 mL | | | 0[d] 2–[c] | 2.2[e] | 2.2–2.9[e] |
| *Cryptosporidium parvum* units per 100 L | | | 0[d] | | 0–8.3[e] |
| *Giardia* units per 100 mL | | | 0.1–1.5[d] | | 0–0.12[b] |
| *Coliphage* log10/100 mL | | | | | 0–3[c] |

[a] Winward et al. (2008).
[b] Birks et al. (2004).
[c] Gilboa and Friedler (2008).
[d] Birks and Hills (2007).
[e] Friedler et al. (2011).

environment outside the body. In addition, fecal coliforms are not unique to human feces. As a result, these assessments may be biased by an overestimation of human pathogens. However, it is also possible that the high reproductive ability of *E. coli* bacteria in raw water is balanced by their sensitivity to treatment (Ottoson and Stenstrom, 2003), and therefore their presence is indeed capable of giving a reliable estimate of the concentration of fecal pathogens after treatment. In summary, fecal contamination does exist to a certain degree in greywater and may pose an unnecessary risk (Ottoson and Stenstrom, 2003). Since the database for the relation between *E. coli* and fecal contamination is large, and as long as no other indicator is proven more effective, the use of *E. coli* as an indicator organism is appropriate. It can provide useful information about the microbial quality of greywater.

In contrast to risk from fecal contamination, only a few studies have investigated the risk resulting from bacteria associated with food handling, such as *Salmonella*, in a greywater use scenario. One study estimated that there will be a 4% increase in the number of infections with *Salmonella* spp. among people who use greywater without disinfection, as opposed to an increase of 0.001% among people who use disinfected greywater (Diaper et al., 2001a). A number of studies examined the risk resulting from the presence of these bacteria in regular wastewater (Gerba et al., 2008; Carlander et al., 2009), and their results can be extrapolated (until a direct study will be performed) for an assessment of the risks inherent in greywater.

Opportunistic pathogens often associated with the skin, nose, and ears, such as *Staphylococcus* and *P. aeruginosa*, may be found in greywater (Gross et al., 2007; Gilboa and Friedler, 2008; Winward et al., 2008). In a study published in 2010, concentrations of *S. aureus, P. aeruginosa, and E. coli*, as well as the presence of heterotrophic bacteria, were analyzed in toilet bowls and flush water under three scenarios: *regular* scenario where toilets used for excretion and flushed by freshwater, toilets flushed by treated greywater, or toilets flushed by treated and UV-disinfected greywater. The study found that concentrations of fecal coliforms and *S. aureus* were higher by about two orders of magnitude in toilets that used freshwater as compared to toilets that used greywater after UV disinfection. Hence, the researchers concluded that the additional risk in using treated and disinfected greywater, represented by fecal coliforms and *S. aureus*, is negligible. Concentrations of heterotrophic bacteria were found in the same order of magnitude in toilets where freshwater and disinfected greywater were used, a fact that increases only slightly their expected concentration in toilets that use greywater. Of the four microbial indices measured, only the concentration of *P. aeruginosa* was higher in greywater than in freshwater (where they did not exist at all). To assess the resulting risk, the researchers used the minimum value of infection due to oral exposure to *P. aeruginosa* as found in the literature ($10^6$ bacteria) and estimated that one must swallow about 83 L of greywater to cause infection (Table 4.5). Therefore, they concluded that the concentration of *P. aeruginosa* does not pose a significant sanitary risk in the case of flushing toilets with treated and disinfected water (Friedler and Gilboa, 2010).

As for nonpathogenic health risks, in most cases, public health will not be affected by chemical pollutants in greywater. However, if the water is used to irrigate edible

**TABLE 4.5**
**Volume of Greywater Required for Infection with *S. aureus* and *P. aeruginosa***

| | Exposure | Infection Dose (Number) | T. Greywater (l) | T. Greywater after UV (l) | Potable Water (l) |
|---|---|---|---|---|---|
| *P. aeruginosa* | Oral | Traumatized, $10^6$ | 238 | 83 | ∞ |
| | | Healthy, $10^8$ | $238 \times 10^2$ | $83 \times 10^2$ | ∞ |
| *S. aureus* | Skin | Traumatized, $10^2$ | 1.8 | ∞ | 0.19 |
| | | Healthy, $10^6$ | $1.8 \times 10^4$ | ∞ | $1.9 \times 10^3$ |

*Source:* Reprinted from *Sci. Total Environ.*, 408(9), Friedler, E. and Gilboa, Y., Performance of UV disinfection and the microbial quality of greywater effluent along a reuse system for toilet flushing, 2109–2117, April 1, 2010, from Elsevier.

plants or enters the drinking water supply, these factors could have an impact on health (Aertgeerts and Angelakis, 2003).

#### 4.4.1.2 Exposure Assessment

The extent of exposure is a major factor in determining the risk level. Exposure assessment should take into consideration the possible means of exposure, its frequency, and its extent. For example, exposure to pathogens in greywater could be through the digestive tract if a person swallows greywater, eats food irrigated with greywater, ingest aerosols, puts hands that touched greywater in their mouth, and so on. Other means of exposure are through dermal (skin) contact with the water or via inhalation (respiration system) through aerosols.

Many factors influence the extent of exposure of users to greywater including the type of greywater use (flushing toilets, irrigating, car washing, etc.), duration, and nature of maintenance work required for the treatment system. Even in water designated for irrigation, there is a great variation in the extent of risk depending on the irrigation technique and time of day, time spent in the irrigated area, frequency and duration of work in the garden, and the type of vegetation irrigated with greywater. In addition, the individual precaution that users take when in contact with greywater also has an effect. Table 4.6 presents different exposure scenarios with their associated water volumes, as found in the literature.

#### 4.4.1.3 Dose–Response Model

There are several possible outcomes of an exposure event to pathogenic microorganisms (see Figure 4.7). A certain percentage of those exposed will be infected by the pathogen. Infection means that a microorganism is surviving and reproducing in the host's body. Out of those infected by the pathogen, a certain percentage will develop a disease. Disease is evidence of the host body's negative reaction to the presence of the pathogen. Of those who become ill, a certain percentage will recover, a certain percentage will develop chronic disease, and a certain percentage will die of

**TABLE 4.6**
**Possible Exposure Scenarios and an Evaluation of the Associated Water Volumes Based on Sources Found in the Literature**

| Exposure Scenario | Water Volume |
|---|---|
| Accidental swallowing[a] | 100 mL |
| Indirect swallowing caused by contact with irrigated lawn and vegetation[a] | 1 mL |
| Swallowing of water spray from the irrigation system[a] | 0.1 mL |
| Swallowing/breathing of aerosols following toilet flushing[b] | 0.1 mL |
| Swallowing of soil soaked with greywater[c] | 10–100 mg<br>Equivalent to 0.01–0.1 mL water |
| Eating vegetables exposed to greywater[a,d] | 5 mL for lettuce, 1 mL for other vegetables |
| Contamination of drinking water through cross-connections[e] | Up to 500 mL in one time |
| Exposure due to contact with skin (without swallowing)[e] | A few liters |

[a] NRMMC and EPHC (2006).
[b] Frwtrell and Kay (2007).
[c] Mara et al. (2007).
[d] Shuval et al. (1997).
[e] The authors' assessment.

the disease (Figure 4.7). It should be noted that in epidemiological data, there is an inherent bias toward the most serious outcome, since infection without disease or mild disease is not necessarily reported and/or treated. In addition, a single death will get more attention than many cases of recovery. In dose–response assessment, the focus is mostly on the probability of being infected (*Pi*). From the perspective of public health, the bias is justified because focusing on infection is a conservative element in the analysis (Haas et al., 1999).

The dose–response model describes the relationship between the extent of exposure to the microbial organisms and the probability of infection. In the most general sense, the dose–response model is a mathematical function that includes the dose and yields the probability of a specific adverse impact, which is represented by numbers between 0 (no impact) and 1 (unavoidable impact) (Haas et al., 1999).

The required variables for the dose–response model are the dose ($d$) and the constants associated with the specific characteristics of each pathogen such as the infection constant ($\alpha$) and the concentration at which half of the population will be infected by the pathogen ($N_{50}$). Table 4.7 presents two common models by which the dose–response relationship of the major pathogens associated with wastewater is described together with their respected constant values.

The models can be used to assess the risk to individuals or to the population, from a single exposure event or from multiple exposures. When the model is for an individual, the result will show the probability of that person being infected with the specific pathogen. For example, $P_i(d) = 0.04$ means there is a 4% chance

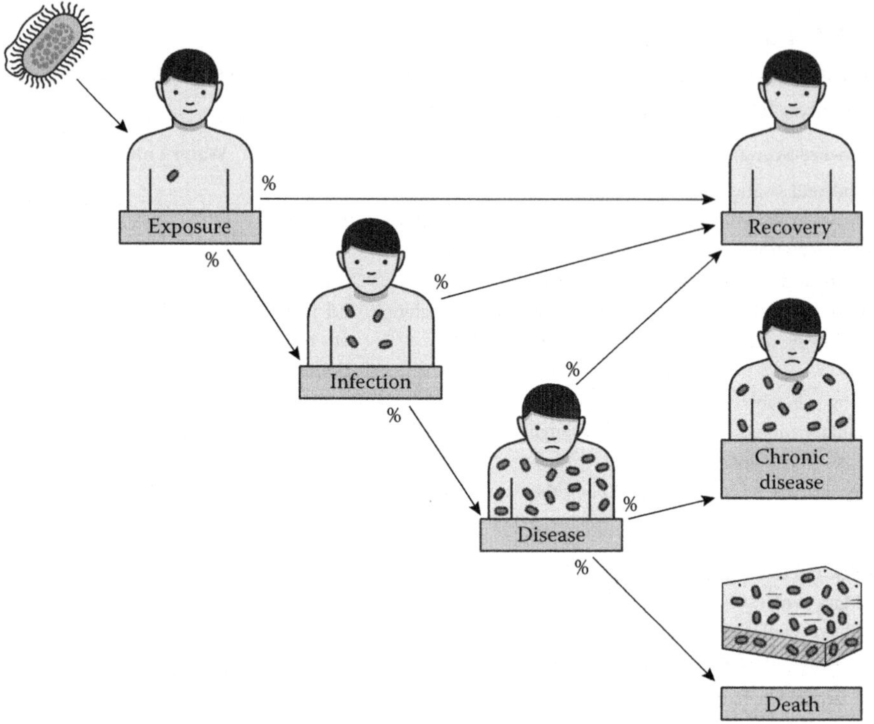


**FIGURE 4.7** Possible outcomes of an event of exposure to pathogenic microorganism. (Based on Haas, C.N. et al., *Quantitative Microbial Risk Assessment*, John Wiley & Sons, New York, 1999.)

that the person will be infected by exposure to concentration $d$ of the pathogen. For a population, the result will show the number of people in the population that is expected to be infected as a result of identical exposure to the pathogen. For example, $P_i(d)=0.04$ means that 4 people out of a 100 exposed will be infected by the pathogen.

#### 4.4.1.4 Risk Characterization

At this stage, the data collected in the previous steps are entered into the dose–response equations, and different risk assessments for various exposure scenarios are obtained. The risk assessments are compared with the targets.

### 4.4.2 An Example of Using QMRA to Assess Microbial Risks in Using Greywater (Taken from Maimon, 2010; Maimon et al., 2010)

#### 4.4.2.1 Hazard Identification

Of the fecal pathogens that are expected to be found in wastewater, viruses are a significant factor since they are secreted in very high concentrations by infected individuals and are potentially infectious even at a very low concentration. In addition,

**TABLE 4.7**
**Common Models and Constant Values of Major Pathogens Associated with Wastewater**

| | | |
|---|---|---|
| *Exponential model* | | |
| $P_i(d) = 1 - \exp(-\alpha d)$ | $\alpha$ | |
| *Cryptosporidium* | 0.057 | |
| *Giardia* | 0.020 | |
| *Beta-Poisson model* | | |
| $P_i(d) = 1 - [1 + (d/N_{50})(2^{1/\alpha} - 1)]^{-\alpha}$ | $\alpha$ | $N_{50}$ |
| *Rotavirus* | 0.253 | 6.17 |
| *Campylobacter* | 0.14 | 890.38 |
| *Vibrio cholerae* | 0.25 | 243 |
| *Shigella* | 0.27 | 1480 |
| *Salmonella* nontyphoid | 0.21 | 49.78 |
| *Salmonella* typhoid | 0.175 | $1.11 \times 10^6$ |

*Source:* CAMRA, Dose response assessment, http://wiki.camra.msu.edu/index.php?title=Dose_response_assessment, 2012.

viruses generally have high survivability even in the environment outside the body (Gerba et al., 1996; Ottoson and Stenstrom, 2003; WHO, 2006a). The rotavirus (for which there is a dose–response model) is a common virus that causes high incidence of intestinal disease (Gerba et al., 1996). In a risk assessment analysis, it presented the highest risk out of all the pathogens tested (Ottoson and Stenstrom, 2003). Because the methods for measuring rotavirus concentrations in greywater are not as simple and straightforward as those used to quantify *E. coli*, various studies have tried to find a correlation between rotavirus concentrations and the concentrations of *E. coli* in wastewater. The WHO (2006a) proposed that in domestic wastewater, there are between 0.1 and 1 units of rotavirus to every $10^5$ *E. coli*. In the national guidelines for water recycling in Australia (NRMMC and EPHC, 2006), an average value of 8000 virus units per liter of domestic sewage is proposed, containing an average of $10^7$ *E. coli* per 100 mL (or in other words, eight rotavirus units to $10^5$ *E. coli*). Due to concerns that using *E. coli* as an indicator would lead to overestimations, the use of coprostanol as a chemical measure of fecal contamination was examined (Ottoson and Stenstrom, 2003). Coprostanol is a metabolite of cholesterol secreted in feces and is used as an indicator for fecal contamination because of its very slow rate of decomposition. Using coprostanol and epidemiological studies, researchers reached an estimate of about 0.17 rotavirus units/mL in a neighborhood system for greywater recycling. Interestingly, though their study found that fecal contamination was overestimated by using *E. coli* as compared with chemical indices, the evaluation results of rotavirus concentrations according to the chemical index were similar to the concentration that would have been found by using the rotavirus/*E. coli* ratio estimated by the WHO (2006a). If the concentrations of *rotavirus* in the system are estimated according to three different methods, we obtain comparable concentrations.

#### 4.4.2.2 Exposure Assessment

A general exposure assessment for home garden irrigation is presented in Table 4.8 without details regarding different exposure barriers (Maimon, 2010).

#### 4.4.2.3 Dose–Response Model

There are two relevant equations to find the risk from exposure to rotavirus. One is the beta-Poisson model for a single exposure (Table 4.7 and Equation 4.1). The other yields the risk of infection as a result of several exposure events per year (Equation 4.2):

$$P_i(d) = 1 - \left[1 + \left(\frac{d}{N_{50}}\right) \cdot \left(2^{1/\alpha} - 1\right)\right]^{-\alpha} \tag{4.1}$$

$$P_{i(A)}(d) = 1 - \left[1 - Pi(d)\right]^n \tag{4.2}$$

where

$d$ is the number of rotavirus units to which there is exposure in each event
$n$ is the number of exposures per year, $\alpha = 0.253$, and $N_{50} = 6.17$ (Table 4.7)

#### 4.4.2.4 Risk Characterization

The probability of being infected by rotavirus due to cumulative annual exposure $P_{i(A)}(d)$ to different *E. coli* concentrations (adding up the three exposure scenarios presented in Table 4.8) according to the rotavirus/*E. coli* ratio of $1.7 \times 10^{-5}$ (Table 4.9) is shown in Table 4.10.

Compared with the target set using the DALY index, when the *E. coli* concentration is lower than 100 CFU/100 mL, the risk arising from the use of greywater for irrigation does not exceed the limit that was set and can therefore be defined as reasonable.

**TABLE 4.8**
**Exposure Assessment of Greywater Used for Watering Garden**

| Exposure Scenario | Water Volume (mL) | Frequency (Exposures per Year) |
|---|---|---|
| Accidental swallowing | 100 | 1 |
| Routine swallowing as a result of use and maintenance | 1 | 37 days |
| Swallowing of soil soaked with greywater | 10–100 mg | 5 days |

*Source:* Maimon, A., Safe reuse of greywater for irrigation in private households, MSc Thesis, Albert Katz International School for Desert Studies (AKIS), Ben Gurion University of the Negev, Sede boqer campus, Israel, 2010.

**TABLE 4.9**
**Rotavirus Concentrations according to Different Estimates of the Ratio Rotavirus/*E. coli***

| Estimated Concentration of Rotavirus per mL[a] | Rotavirus/*E. coli* Ratio | Source |
|---|---|---|
| 0.1–0.01 | $1\times10^{-5}$ to $1\times10^{-6}$ | WHO (2006a) |
| 0.17 | $1.7\times10^{-5}$ | Ottoson and Stenstorm (2003) |
| 0.8 | $8\times10^{-5}$ | NRMMC and EPHC (2006) |

*Source:* Maimon, A. et al. *Environ. Sci. Technol.*, 44(9), 3213, 2010.
[a] According to $10^6$ *E. coli*/100 mL.

**TABLE 4.10**
**Probability of Being Infected by *Rotavirus* due to Cumulative Annual Exposure $P_{i(A)}(d)$**

| *E. coli*/100 mL Concentration | Rotavirus Concentration Estimated for 100 mL | $P_{i(A)}(d)$* |
|---|---|---|
| 10,000 | 0.27 | 0.12 |
| 1,000 | 0.027 | 0.014 |
| 100 | 0.0027 | 0.0014 |

* Probabilities are rounded.

### 4.4.3 Environmental Risks

Certain greywater components can adversely affect soil characteristics and vegetation health (Eriksson et al., 2002; Cordy et al., 2004; Gross et al., 2005; Wiel-Shafran et al., 2005; Travis et al., 2008; Misra and Sivongxay, 2009). In most cases, the most significant impact, and therefore also the highest risk, will be in the immediate irrigated environment, that is, the home garden. Sometimes, the damage is cumulative in nature and can be detected only after several years of use (Travis et al., 2008).

In some cases, the impact of greywater use can be local and limited to the irrigated plot only, but it can also spread beyond the boundaries of the private plot to the environment outside causing groundwater or surface water pollution. Empirical data are lacking as to the scope and degree of these effects and the environmental risks of using greywater.

#### 4.4.3.1 Damage to Vegetation

There are conflicting reports about the impact of greywater irrigation, raw or treated, on the growth of plants: a negative effect of irrigation with greywater on growth was observed in some (Garland et al., 2000; Misra et al., 2010), others showed no decrease in growth (Pinto et al., 2010, Travis et al., 2010; Rodda et al., 2011;

Alfiya et al., 2012), and others reported increased biomass due to the nutrients contained in the treated greywater (Alfiya et al., 2012).

It has been found that phytotoxicity can occur following greywater irrigation due to the accumulation of salts such as sodium, chloride, and boron (Gross et al., 2005; Parks and Edwards, 2005; Misra and Sivongxay, 2009; Pinto et al., 2010). It has been reported that phytotoxicity can also follow irrigation with different concentrations (usually hundreds of mg/L) of surfactants (Wiel-Bubenheim et al., 1997; Wiel-Shafran et al., 2005; Eriksson et al., 2009; Rodda et al., 2011). The salt concentrations that will affect a plant depend on the type of plant. Table 4.11 lists the threshold values of soil salinity and in irrigation water of different crops. Because there are almost no values in the literature for decorative plants, values for edible plants are presented here as an example. There are halophyte plants capable of absorbing salts and, if the plants are harvested, some of the salts in the treated greywater can thus be removed (Shelef et al., 2012a).

Another environmental risk results from the accumulation of pathogenic bacteria on leaves and roots, which may affect the plants. These may also cause an additional health risk to humans and animals (as mentioned in Section 4.4.1) (Walker et al., 2004; Jackson et al., 2006; Finley et al., 2009). In summary, irrigation with raw greywater may damage plants, but such damage is mainly observed in particularly susceptible plants. Irrigation using treated greywater of high quality is not expected to harm plants.

#### 4.4.3.2 Damage to Soil

Because biological treatment does not remove salts, it is feared that the use of treated greywater will change soil properties and reduce its fertility. Additionally, oils and grease may build up in the soil and make the soil water repellent (Travis et al., 2008; Travis et al., 2010). Oil and grease can also exist in the ground as colloidal particles, imbibing at faster rate than would be expected by advective flow (Travis et al., 2011). This phenomenon may increase the transport of pollutants, which are adsorbed on the oil particles in the soil profile all the way down to groundwater. Another class of organic materials that might accumulate in greywater irrigated soils and can reduce hydraulic conductivity (Figures 4.8 and 4.9) is surfactants (Wiel-Shafran et al., 2006; Misra et al., 2010). Cosmetics and pharmaceuticals found in greywater (Ericksson et al., 2009) are organic micropollutants that might affect soil biota. Water salinity and the sodium adsorption ratio (SAR) may further affect soil sorptivity (Figure 4.10).

Table 4.12 summarizes reports on the possible effects that greywater irrigation has on plants, soil, and the environment. Most of the reports concern short-term experiments of limited scope, so it is difficult to evaluate the long-term effects of greywater irrigation. However, decades of experience in agricultural irrigation with treated wastewater show that monitoring the quality of the irrigation water and soil quality could prevent future damage.

Risk assessment framework may be useful for evaluating the knowledge that exists today and helping identify the information gaps that need to be bridged. Considering what is already known, it is possible to establish safeguards to reduce risks.

#### 4.4.3.3 Hazard Identification

A variety of environmental risk factors that are related to the use of greywater for irrigation are summarized in Table 4.13.

**TABLE 4.11**
**Threshold Values of Salts in Irrigation Water and in Soil Emulsion for Various Crops**

| | Crop | Threshold Concentration in Irrigation Water | | Electric Conductivity |
|---|---|---|---|---|
| | | Boron | Chloride | dS/M |
| Irrigation water | Grapefruit | 0.3–1.0 | 120–250 | 1.2 |
| | Orange | | 120–250 | 1.1 |
| | Avocado | 0.3–1.0 | 20–120 | |
| | Peach | 0.3–1.0 | | 1.0–1.1 |
| | Persimmon | 0.3–1.0 | | |
| | Grapes | 0.3–1.0 | | 1.0 |
| | Apple | 0.3–1.0 | | |
| | Sweet potato | 1.0–2.0 | | |
| | Tomato | 1.0–2.0 | 250–400 | 1.7 |
| | Potato | 1.0–2.0 | | |
| | Carrot | 2.0–4.0 | | |
| | Onion | 2.0–4.0 | | |
| Soil solution | Grass (Bermuda) | | 2450 | 6.9 |
| | Grapefruit | | 1775 | 1.8 |
| | Lemon | 0.3–0.5 | 1065 | |
| | Orange | | 710 | 1.7 |
| | Mandarin | | 1775 | |
| | Vine | | 2130–2840 | 1.5 |
| | Persimmon | 0.50–0.75 | | |
| | Avocado | | 355–532 | |
| | Deciduous (peach, apricot, plum, and almond) | | 532–1775 | 1.5–1.7 |
| | Tomato | 4–6 | | 2.5 |
| | Carrot | | | 1.0 |
| | Onion | | | 1.2 |

*Sources:* Data retrieved from Ayers, R.S. and Westcot, D.W., Water quality for agriculture, Food and Agriculture Organization of the United Nations, 1985; Maas, E.V., Salt tolerance of plants, In *Handbook of Plant Science in Agriculture*, Christie, B.R. (ed.), CRC Press, Boca Raton, FL, 1990a, pp. 57–75; Maas, E.V., Crop salt tolerance, In *Agricultural Salinity Assessment and Management*, Tanji, K.K. (ed.), American Society for Civil Engineering, NY, 1990b, Chapter 13; Maas, E.V. and Hoffman, G.J., *J. Irr. Drain Div-ASCE*, 103, 115, 1997.

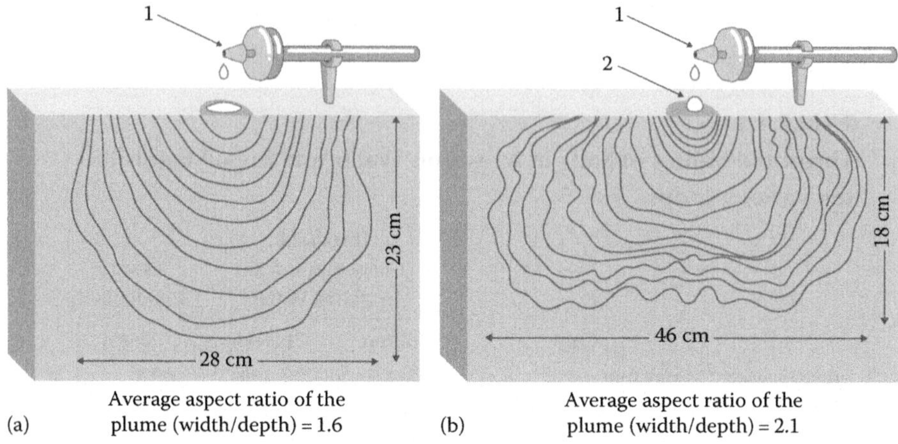


**FIGURE 4.8** Images from a 2D chamber used to compare pattern of water flow in the soil. (a) Irrigation with potable water. (b) Irrigation with water containing detergent at a concentration of 53.6 mg/L of anionic surfactant (expressed as MBAS). The lines on the cell are the wetting front. Point 1 is the location of the dripper and point 2 is the characteristic shape of a water droplet on a hydrophobic surface. (From *Ecol. Eng.*, 26(4), Wiel-Shafran, A., Ronen, Z., Weisbrod, N., Adar, E., and Gross, A., Potential changes in soil properties following irrigation with surfactant-rich greywater, 348–354, 2006, from Elsevier.)

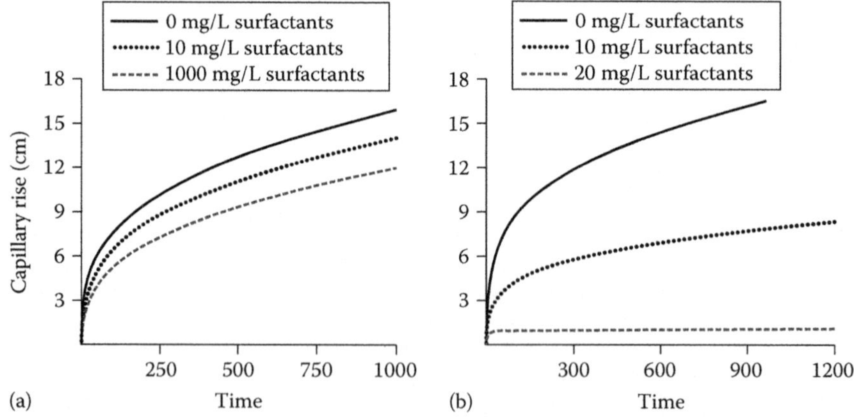


**FIGURE 4.9** (a) Capillary rise of detergent solution at different concentrations in wet sand containing 10% water. (b) The effect of detergent solution mixed with sand (final water content 10%) on the capillary rise of potable water. (From *Ecol. Eng.*, 26(4), Wiel-Shafran, A., Ronen, Z., Weisbrod, N., Adar, E., and Gross, A., Potential changes in soil properties following irrigation with surfactant-rich greywater, 348–354, 2006, from Elsevier.)

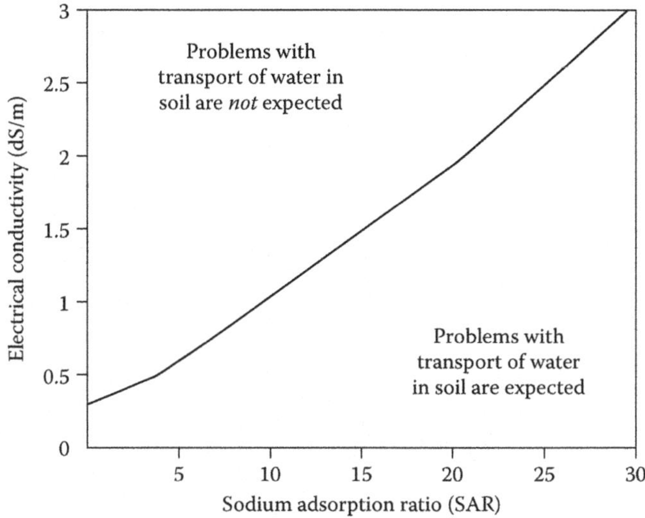


**FIGURE 4.10** *Threshold value* of soil permeability to water and its dependence on SAR and salt concentration. (From Rhoades, J.D. and Loveday, J., Salinity in irrigated agriculture, in Steward, B.A. and Nielsen, D.R., Eds., *Irrigation of Agricultural Crops*, ASCE (Monograph 30), American Society of Agronomists, pp. 1089–1142, 1990.)

#### 4.4.3.4 Exposure Assessment

Naturally, soil and biota in the immediate environment irrigated with greywater will be the most vulnerable to the risk factors in greywater. Sensitive areas (as well as groundwater or surface water bodies and other external environment) may be exposed to pollutants from greywater. However, it is difficult to assess the degree of exposure because the concentrations of pollutants in greywater are not uniform, and many factors may influence them (Eriksson, 2002; Eriksson et al., 2002, 2008; Gross et al., 2007; NSW-DEUS, 2007). One way to characterize the nature and extent of exposure is to distribute a combine consumption survey with water analyses to assess the expected concentration in the irrigated area. The result is referred to as predicted environmental concentration (PEC) (Eriksson, 2002).

#### 4.4.3.5 Dose–Response Model

Due to the large number of pollutants in greywater, it is difficult to define a safe concentration for each of them. Therefore, as in the microbial risk assessment, it would be more practical to adopt an approach that locates the most problematic pollutants (e.g., those whose PEC was high) and determine the highest concentration at which they are not expected to affect the environment adversely. This is sometimes called a predicted no effect concentration (PNEC) (Eriksson, 2002). For example, the PNEC was calculated for 32 XOCs compounds in a 2002 study by Eriksson. Other researchers worked on establishing the concentrations at which certain substances made an impact on the environment. For example, it was found that anionic surfactants (such as MBAS) already lowered the capillary rise in sand at concentrations of 10 mg/kg (soil)

**TABLE 4.12**
**Summary of Reports on Effect of Treated Greywater Irrigation**

| | Type of GW | Test | Findings |
|---|---|---|---|
| *Effect on plants* | | | |
| Eriksson et al. (2006) | Raw GW | Algal growth inhibition | Phytotoxic effect of laundry and kitchen GW. |
| Eriksson et al. (2006) | Raw GW | Willow cuttings, short-term acute bioassay | Phytotoxic effect of laundry and kitchen GW. |
| Garland et al. (2004) | Surfactants AE and CAPB | Hydroponic wheat growth | Slight (~20%) reduction of growth. |
| Garland et al. (2000) | Surfactant Igepon (1 g/$m^2$ growing area/d) | Hydroponic lettuce growth | Reduced growth. |
| Weil-Shafran et al. (2005, 2006) | Laundry GW | Lettuce plants | Chlorosis. |
| Pinto et al. (2009) | Synthetic GW | Plant growth | Small differences—not statistically nonsignificant. |
| Travis et al. (2010) | Raw artificial GW and treated artificial GW | Plant growth | No significant differences between treatments. |
| Misra et al. (2010) | Laundry GW and surfactant solution (high and low conc.) | Tomato plant biomass and leaf area | Biomass: GW ≥ TW ≥ LC ≥; HC; significant leaf area in GW over other. |
| Rodda et al. (2011) | GW | Vegetable growth | Growth of plants irrigated with GW was the same as growth of plants irrigated with tap water + nutrient solutions. |

(*Continued*)

**TABLE 4.12 (*Continued*)**
**Summary of Reports on Effect of Treated Greywater Irrigation**

| | Type of GW | Test | Findings |
|---|---|---|---|
| Rodda et al. (2011) | GW | Micronutrient content | Plants irrigated with GW showed higher levels of Na and metals. |
| Bubenhiem et al. (1997) | Igepon TC-42 (anionic surfactant) solution | Phytotoxicity | 250 mg/l and greater, which showed phytotoxicity in lettuce plants. |
| Alfiya et al. (2012) | Tap water, raw GW and treated GW | Phytotoxicity effect on ryegrass and biomass yield | No phytotoxicity found of all treatments. Treated GW yield higher biomass. No differences in biomass yield were found between tap water and raw GW. |
| *Effect on soil* | | | |
| Misra and Sivongxay (2009) | Laundry GW | Saturated hydraulic conductivity | Irrigating with high SAR reduces hydraulic conductivity. |
| Weil-Shafran et al. (2006) | Laundry GW and surfactant solutions | Capillary rise | Reduction of the capillary force. |
| Weil-Shafran et al. (2006) | Laundry GW | Sand infiltration rate | Caused soil to become hydrophobic. |
| Gross et al. (2005) | 3 years of raw GW irrigation | EC and boron (B) accumulation in loess Negev soil | Some accumulation of B. |
| Travis et al. (2008) | Dark raw GW | O and G accumulation | O and G accumulated up 20 cm depth, but no correlation was found with water repellency. |
| Pinto et al. (2009) | Synthetic GW | Soil pH and EC | Significant pH and EC increase. |
| Travis et al. (2010) | Raw artificial GW and treated artificial GW | Water repellency: $\zeta$—potential and WDPT | Correlation between WDPT with O and G and surfactants. Sand and loam soils irrigated with raw GW developed significant water repellency than soil irrigated with FW and TGW. |

(*Continued*)

**TABLE 4.12 (*Continued*)**
**Summary of Reports on Effect of Treated Greywater Irrigation**

| | Type of GW | Test | Findings |
|---|---|---|---|
| Misra et al. (2010) | Laundry GW and surfactant solutions | Soil water retention | GW and surfactant solutions caused a reduction in soil and water retention. |
| Rodda et al. (2011) | GW | Accumulation of metals and sodium in soil | EC, Na, and metals were accumulated with time. |
| Pathogens | | | |
| Finley et al. (2009) | SH and WM GW | FC and FS on edible plants | GW-irrigated food crops do not necessarily correlate to higher levels of FC and FS. |
| Gross et al. (2005) | 3 years of raw GW irrigation | Survival of FC in loess Negev soil | FC did not survive in soil. |
| Jackson et al. (2006) | GW | Pathogen bacteria on vegetables (carrots, spinach, onions, and peppers) | No significant difference between irrigating with GW and TW. |
| Travis et al. (2010) | Raw artificial GW and treated artificial GW | *E. coli* in loess and loam soils | Soils that were irrigated with raw GW had significantly more *E. coli* counts. |
| Walker et al. (2004) | *P. aeruginosa* strains PA01 and PA14 | Pathogenicity in vitro and in soil | Infect roots of *Arabidopsis* and sweet basil and causing mortality of the plants. |
| Alfiya et al. (2012) | Tap water GW and treated GW | FC accumulation in soil | No accumulation of FC found in all the three treatments. |

**TABLE 4.13**
**Environmental Risk Factors of Using Greywater for Irrigation**

| Parameter | Examples for Possible Impact |
|---|---|
| PH | Corrosion of equipment, damage to biota, changes in biochemical processes.[a,b] |
| Electrical conductivity | Reduction of plant productivity, possible changes in soil properties.[b] |
| $Cl^-$ | Accumulation in soil may adversely affect plants.[b,c] |
| Sodium adsorption ratio | Exacerbate soil erosion, change soil hydraulic conductivity, and reduce plant growth.[b,d] |
| Boron | Accumulation in soil may be toxic to plants.[c,e] |
| Phosphorus | May cause eutrophication if excess concentrations reach surface water, might induce bioclogging of equipment.[b] |
| Nitrates | May contaminate groundwater and surface water.[b] |
| Surfactants | Accumulation may change soil hydraulic conductivity, plant toxicity.[f,g] |
| Oil and grease | Accumulation may change soil hydraulic conductivity.[h] |
| Xenobiotics | Toxicity to biota.[i,a] |

*Source:* Based on Maimon, A. et al., *Environ. Sci. Technol.*, 44(9), 3213, 2010.
[a] Eriksson et al. (2002).
[b] ANZECC (2000).
[c] Gross et al. (2005).
[d] Qian and Mecham (2005).
[e] Gross et al. (2007).
[f] Wiel-Shafran et al. (2006).
[g] Abu-Zreig et al. (2003).
[h] Travis et al. (2008).
[i] Eriksson (2002).

(Weil-Shafran et al., 2006). Furthermore, oils and fats in concentrations of up to 250 mg/kg lowered soil sorptivity by up to 60%, in an almost linear fashion (Travis et al., 2008). These data can be used to start creating a database that will include and update the PNECs of greywater's main pollutants.

#### 4.4.3.6 Risk Characterization

The comparison of PEC and PNEC is the foundation of environmental risk characterization and the cases where PEC > PNEC should be examined with extra care. Special emphasis should be placed on the biodegradation capacity of the substance and on the effects of its breakdown products. Biodegradable substances should be treated differently from nonbiodegradable substances, such as metals. A helpful distinction is also between substances that are degradable by biological treatment and those that have harmless breakdown products. The latter should be separated from more stable substances and those whose degradation products are hazardous. Other materials that merit special attention are substances, such as surfactants, that damage soil properties and increase runoff or percolation into groundwater by preferential flow, thus reducing the effectiveness of barriers and possibly increasing the health and environmental risk.

## 4.5 MEANS TO ATTAIN THE GOALS

At each stage of the project (Figure 4.2), various measures can be employed to reduce risks posed by greywater use. Risk reduction can be performed by combining three main methods: reducing risk factors at the origin by selecting certain water sources or preventing the entry of specific substances, treatment and/or disinfection that increases the water quality, and the use of barriers that reduce exposure to the hazards and hence the risk. More on the means for risk reduction can be found in Chapter 5.

When choosing from the means available, minimizing risk should not be the only consideration. The three goals for reuse of greywater should be taken into account: resource savings, public and environmental health, and economic feasibility. For this reason, the preferred means will not necessarily be that which will ensure minimal risks, but rather that which will attain maximum savings of resources while meeting the goals of safety and maintaining economic feasibility.

## 4.6 CRITICAL CONTROL POINTS

The last element of risk management focuses on desired risk level maintenance. To ensure the safety of the system without the need for extensive testing of each phase of the project, critical control points (CCPs) should be located and monitored periodically to assess the integrity of the entire system. A CCP is defined as a point, stage, or process in a system in which control is critical to prevent or halt a threat or lower it to a tolerable level (NSW, 2008). To locate the CCP, a decision-making chart can be used as shown in Figure 4.11.

## 4.7 SUMMARY

The popularity of greywater use in developed countries is growing. For example, the Australian Bureau of Statistics cites that approximately 55% of all households in Australia use greywater as part of their water sources (ABS, 2007). Furthermore, in the 493 articles published on greywater from 1977 until the end of 2012 (Scopus database), no negative epidemiological sanitary effects of greywater use have been reported. On the one hand, this may indicate that greywater use caused no significant outbreak of any disease. On the other hand, it could be that minor effects are not reported or simply do not penetrate the public consciousness. Therefore, extensive research on the subject is essential and can shed more light on the public health effects of using greywater. At the same time, the whole range of potential exposure scenarios should be explored further to improve the accuracy of the existing risk assessment models as much as possible.

Our literature review on the subject of environmental risks suggests that there are large gaps in information and that frameworks for evaluation and comparison of various risks are lacking. Despite the gaps, it can nevertheless be argued that the use of untreated greywater may cause an adverse change in soil properties and damage vegetation, which calls for greywater to be treated before use. Further research should be conducted on the subject to determine the optimal way to address such environmental risks.

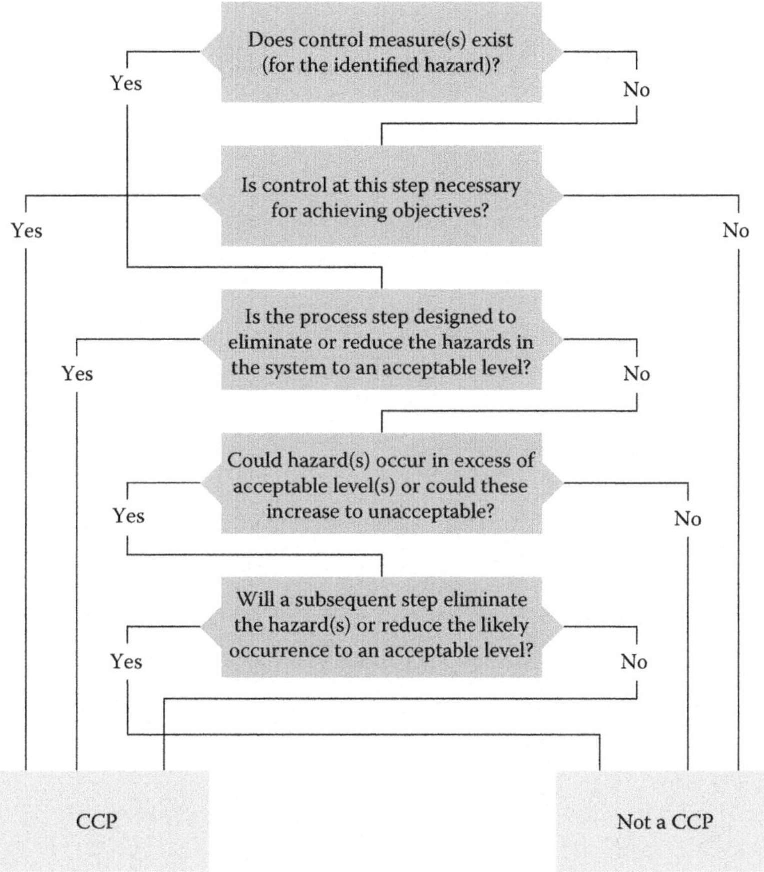


**FIGURE 4.11** Decision-making chart to locate CCPs. (According to NSW, *Management of Private Recycled Water Schemes*, New South Wales, Australia, Water for life, 2008.)

Frameworks for risk management are particularly relevant in areas of water scarcity. For example, in Israel, there is a public debate taking place between decision makers and various stakeholders on the safety of using this water resource. The framework presented in this chapter may help to inform decisions on future policy. The methods other countries have used to cope with the risks presented in this chapter, as well as the mechanisms that were adopted to reduce health and environmental risks, are presented in the next chapter on policy.

# 5 Policy and Legislation

*Written in collaboration with Prof. Alon Tal**

## 5.1 INTRODUCTION

The use of greywater is as old as modern plumbing, but the recognition of its value has grown in recent years. As the water shortage around the world has worsened, water prices have increased, quotas have been cut, and households and buildings have begun demanding hydrological independence. States, communities, and many households have started to realize the potential environmental benefit of reusing household wastewater; they can alleviate dwindling water balances and promote policies accordingly. This has led to the appearance of a new subcategory in water legislation, designed to allow the public to take initiative and adopt greywater systems while minimizing its damage to public health and the environment. Meeting these two needs and finding the appropriate normative balance are the main challenges facing greywater legislation.

Part of the recent popularity of greywater use is due to the growing culture of water conservation, which gives rise to the expectation that water reuse systems will be integrated in new and existing buildings. For example, the Leadership in Energy and Environmental Design (LEED) is a set of voluntary environmental standards for construction accepted worldwide. The LEED program established a standardized rating system of points for *green buildings* that reward water conservation as well as the reuse of greywater (USGBC, 2008). Though it does not constitute official law, such an international standard affects the tendencies of the construction industry, consumers, and the real estate market. Perhaps such voluntary standards are even more effective than regular legislative instruments in encouraging the public to install and use greywater systems.

While greywater systems have become more and more common in many countries, regulators are only beginning to formulate an appropriate response. Compared with the comprehensive legislation regulating wastewater treatment developed in many Western countries in the 1970s (Westman, 1972), efforts to address the prevalence of greywater systems, and the health and environmental risks associated with them, are relatively new. There is a great variety in laws dealing with greywater. The most common ones are attempts to limit the types of water that can be used in greywater systems, typically banning kitchen water containing relatively high levels of oils, organic matter, and bacteria. Some authorities, such as the state of Queensland in Australia or the state of California in the United States, consider greywater storage a health hazard and prohibit it by law (Radcliff, 2004). There is a wide range of

* The Ecology Department, Blaustein Institutes for Desert Research, Ben Gurion University of the Negev.

standards for maximum permissible concentrations for various environmental pollutants (Maimon et al., 2010).

Many authorities are still concerned about the so-called greywater experiment. In many countries, laws and regulations dealing with greywater still directly prohibit its use (Prathapar et al., 2005) or simply do not distinguish between the necessary treatment for greywater and the more stringent treatment levels required for municipal wastewater reuse (Allen et al., 2010). Nevertheless, most normative frameworks that currently influence greywater programs have a cooperative underpinning and are motivated by a desire to avoid negative consequences. Nonetheless, more attention to implementation strategies and complementary educational initiatives is called for.

Many U.S. states are environmental leaders and pioneers in environmental standardizations. The United States has served as a kind of laboratory to examine various greywater systems and differing approaches. The result, however, is a diverse menu of sometimes contradictory regulation strategies. This is also true in Australia, and recently several European countries have begun addressing the normative aspects of greywater use.

On every continent, it seems that there is an inverse relationship between the strictness of regulations that monitor the use of greywater and the level of water shortage. For example, on the east coast of the United States, where generally water shortage is not severe, the main emphasis is on possible negative health effects of greywater. However, in the arid West Coast, there is greater emphasis on its benefits. Another example, Cyprus suffers from a severe perennial shortage of water and has recently launched a program that subsidizes families with about 3000 USD per household to encourage watering gardens and toilet flushing with greywater (Tufvesson, 2009).

Many authorities around the world now recognize greywater as an untapped resource reflecting market failure. Like many other environmental laws, some of the legislation dealing with greywater is designed to optimize the use of natural resources by society. For example, in Tokyo, greywater recycling is now required in every building larger than 30,000 $m^2$ that has a potential daily consumption of more than 100 $m^3$/day of nonpotable water. Similarly, the municipality of Sant Cugat del Vallès, a suburb of Barcelona, began a process 6 years ago that requires the use of greywater systems in apartment buildings (Domènech and Saurí, 2010).

This chapter describes the approaches that have evolved around the regulation of greywater in legislative entities throughout the world. Enforcement problems are common to all of them and will be discussed in Sections 5.2 and 5.3.2, followed by an analysis of greywater policy (standards and regulations) presented in two parts: legislation pertaining to use of greywater (Section 5.3.1) and legislation designed for water treatment systems (Section 5.3.3). Finally, a case study will address the debate surrounding Israel's greywater bill (Section 5.5). Greywater guidelines, regulations, and laws have a common goal: to ensure safe use of a valuable source of water. Not surprisingly, the dramatic disparities between experts' perceptions of the relative risks of greywater use are also reflected in the legal field worldwide. Regardless, all parties involved have to recognize the need to deal more effectively with the problems of implementation and enforcement, which are still the biggest weaknesses in regulating greywater use.

## 5.2 COMPARATIVE REVIEW OF GREYWATER LEGISLATION: AN ATTEMPT TO RECONCILE DIFFERING REGULATORY MESSAGES

### 5.2.1 U.S. Experience

While the U.S. Congress enacted strict legislation to regulate many water-related issues and pollutants, greywater issues remain under the discretion of state legislatures. According to an estimate published recently, only about 30 of the 50 states have a law concerning the subject (Bahman, 2010), and the contents of the different legislations vary. For example, until a few years ago, all use of greywater was banned in North Carolina. A public campaign led to a moderate policy change allowing the use of greywater to water nonedible plants, but for extensive uses, full wastewater treatment is needed (Sturgis, 2008).

It is perhaps not surprising that legislative efforts to regulate the use of greywater began at the local level, in a semiarid California university town. In 1989, the town of Santa Barbara decided to legalize greywater use; when the city council discovered that the state of California did not prohibit the use of greywater, it released a local order to promote its use and regulation (Santa Barbara, 1989). The *Supervisors' Council* (city council) decided to define greywater separately from *blackwater* released from toilets. The order prohibited the use of water from kitchen sinks or laundry that could be contaminated by diapers. The exclusion of water from kitchen sinks, diaper laundry, and dishwashers is now common around the world. The regulation in Santa Barbara was among the first formal normative frameworks for greywater management in the world. The initiative helped create a domino effect in neighboring cities in California. Within 2 years, five other cities followed the example and issued local orders on the use of greywater.

Even though California cities were leading greywater progress, municipal legislative initiatives contradicted the international *Uniform Plumbing Code* (UPC) guidelines adopted by the state of California in 1992. As the UPC was a conservative model, it limited the reuse of greywater to mini–leach fields, an approach commonly used for releasing wastewater from cesspits.

The narrow application approved by the UPC code for the greywater was not what the initiators intended, so they quickly contacted the California legislature, arguably the most advanced U.S. state in environmental protection. In February 1992, the California legislature approved bill No. 3518, which added an entire chapter titled "greywater systems for single-family homes" to the state's water law. This regulation ordered the Water Resources Department to draft a formal legal code on greywater by July 1993.* The intention of this legislation was to improve the interface between the construction industry and the regulatory institutions supervising it.

This was a period of great momentum in the greywater field throughout the U.S.'s arid southwest. Over the years, the UPC began to adapt to the new reality of greywater use, but it still prohibited the recycling of kitchen wastewater in all greywater

* The subject is regulated today within the framework of the California Law: Water Code Section 14877.2.

systems. In September 1992, the annual convention of the International Association of Plumbing and Mechanical Officials (IAMPO) was held in Alaska. Seventeen western U.S. states joined the movement led by California in supporting emerging greywater legislation by turning to the UPC. The proceeding was proposed and quickly approved, before the promoters of health and environment issues had a real opportunity to consider its implications (OasisDesign, 2009). The code was not especially extensive; for example, it allowed the reuse of greywater only for subsurface garden irrigation in residential houses. Although it created an opening for the *legalization* of greywater reuse, the technology's continued proliferation still remained dependent on the initiative of states and local authorities.

A short time passed before the emergence of a *movement against greywater*, which called for the restriction or complete cancellation of its reuse mainly for health reasons. By this time, however, there were also commercial companies with significant economic interest in promoting the use of greywater. One of the amendments to the California Law was initiated by ReWater Systems, Inc., specializing in building greywater systems.* It called for a statutory extension of the use of greywater, to include greywater irrigation systems in condominiums (i.e., multiple family homes) as well as commercial and public buildings. In addition, the law addressed the need to ensure the supply of potable water for filters and supplemental irrigation. The expanded law was approved in 1997 and remained largely valid to this day, although a few amendments have been made (CBSC, 2009):

> All graywater systems shall be designed to allow the user to direct the flow to either the irrigation or disposal field or the building sewer. The means of changing the direction of the graywater shall be clearly labeled and readily accessible to the user.
>
> Water used to wash diapers or similarly soiled or infectious garments or other prohibited contents shall be diverted by the user to the building sewer.
>
> Graywater shall not be used in spray irrigation, allowed to pond or runoff and shall not be discharged directly into or reach any storm sewer system or any surface body of water.
>
> Human contact with graywater or the soil irrigated by graywater shall be minimized and avoided, except as required to maintain the graywater system. The discharge point of any graywater irrigation or disposal field shall be covered by at least (2) inches (51 mm) of mulch, rock, or soil, or a solid shield to minimize the possibility of human contact.
>
> Graywater shall not be used to irrigate root crops or edible parts of food crops that touch the soil.

During the 1990s, other U.S. states began to seriously consider the benefits of reusing greywater and quickly outpaced California in their efforts. Arizona has arguably surpassed California in having the most permissive legislation in the field. Its law of 2001 divided greywater systems into three categories or *levels* based on flow volume (ADEQ, 2001):

*Level 1:* Systems with a flow of less than 400 gallons (1.51 $m^3$) per day, meeting the 13 principles of best management practice (BMP)

---

* www.rewater.com.

*Level 2*: Systems with a flow of less than 400 gallon/day, which do not meet the licensing conditions of Level 1

*Level 3*: Large systems with a flow rate of over 3000 gallons (11.36 $m^3$) per day

The Level 1 system is the most common in households throughout Arizona. They can be operated under a general permit. Citizens are not required to inform the authorities that they are operating a greywater system, but they are expected to meet the 13 principles of BMP. The principles are intended to narrow the emission of—and human exposure to—hazardous substances in greywater. For example, they restrict irrigation by sprinklers and require that the storage tanks are covered.

Systems belonging to Level 2 do not meet, or are unable to meet, all 13 principles. For this reason, their use involves an approval by the Environmental Protection Agency (EPA) of Arizona and an individual permit to operate the system. This permit is valid for 5 years and then has to be renewed. The larger systems (Level 3) are required to operate in accordance with the terms of the individual permit from the state EPA (Allen et al., 2010).

It should be noted that the law enacted in California in 2009 probably reflected awareness of the method of dividing greywater systems into levels and is also based on such divisions, but only between two types:

*The laundry system:* A "greywater system that uses only a single washing machine in a one- or two-family residence"

Complex systems: Greywater systems that collect from any household appliance (excluding the toilet) or other water source

In the state of California, the first type of system can be used without obtaining construction permit. It is clear that the vast majority of greywater systems are not limited to treating water from washing machines. Thus, theoretically, almost any greywater system in California needs an official permit, and thus, the state's attempt to display *flexibility* is symbolic (CBSC, 2009).

In an example of pragmatism, Arizona grants permits to family systems automatically. The state's willingness to do so stemmed from recognition of the new household irrigation reality that had taken hold of the state. By 1999, about 13% of the residents of Tucson were already operating greywater systems. The new legislation was thus an attempt to prioritize those activities that carry excessive risk and to reduce nuisance as much as possible. The assumption was that the relative risk of operating greywater systems is low and that an attempt to enforce more stringent standards was unrealistic. Systems that treat larger volumes of water such as those in a large building may create a more significant problem, and therefore it was decided that regulation and enforcement should focus on them. The state recognized that kitchen sinks seem to pose the highest health risk, and the water they generated should be excluded from the definition of greywater. However, it seems they were not successful in enforcing this exclusion (Allen et al., 2010).

Within 2 years, New Mexico joined the greywater trend emulating the Arizona approach, as did Texas a short time later. Nevada also joined the movement, adopting a similar approach to the law in Arizona. More recently, the state of Washington developed a system that is based on a three-tier risk assessment (Washington-State,

2011). According to a new law, only when the water originates from the kitchen or from sinks is there need for treatment (Levels 1 and 2).

In Montana, the course of legislation was very different. Instead of relying on the more *permissive* approach of its neighbors to the west, the Legislature approved bill No. 259 in 2007, which seems to be influenced by some of the most stringent programs in Australia. The law may not be extremely popular, and greywater activists demanded that the state show more flexibility and establish constructed wetlands as part of its efforts to cope with its water crisis (Pedersen et al., 2007). The most lenient legislation is that of Wyoming, where approval is not necessary when using greywater for irrigation in an amount of less than 7.5 $m^3$/day (unlimited for a single household). In this case, the responsibility for compliance falls on the user.

In 2011, after 4 years of preparation, the National Sanitation Foundation (NSF) of the United States published standards and regulations for greywater treatment systems and its reuse (NSF, 2011). The organization operates as a *standards institute* responsible for publishing standards such as ISO 14000 and ISO 9000 and monitoring them. The standards issued by the NSF are voluntary (i.e., not legally binding). In light of the various legislative bodies in many states and countries in North America, it is not yet clear how these standards will affect actual legislation in the United States or the world.

### 5.2.2 Australian Experience

Australia is the most arid continent on Earth, and its extensive economic and agricultural activities have led to widespread pollution of its scarce freshwater sources (Connel, 1974). For this reason, greywater appeared to be an attractive alternative source to freshwater, especially given the high proportion of citizens who had gardens in a country characterized by urban sprawl. Greywater was officially promoted as an alternative to traditional water management in Australia, but even before this occurred, many studies were conducted on the subject. For example, in 1994, a review was published of the international experience in the field, especially in the United States and Japan. The Australian report confirmed that greywater may "contain fecal bacteria at concentrations high enough to pose a health risk due to the possible presence of pathogenic microorganisms." In the face of such health risks, the report recommended that caution be taken in reusing greywater. The federal government brought this to the attention of the state governments to bring awareness to these public health concerns. Proponents argued that appropriate guidelines could lead to a safe program that would lead to *significant water saving* (Jeppesen, 1996). According to their calculations, using appropriate technologies to recover greywater to water lawns and gardens would lead to savings of 30%–50% of potable water used in households in Australia.

The Australian government was convinced by the arguments of greywater advocates. Individual regulations were left to be defined at the state level, and the federal government settled for publishing a series of recommendations and guidelines regarding appropriate regulations (EPHC, 2008). Here, again, as can be expected from decisions about risk management, local climatic conditions influenced policy and subsequent legislation. There is a clear correlation between the stringency of the greywater standards in a state's laws and the region's level of aridity. Thus, in the dry southern parts of Australia, the use of greywater receives a greater encouragement and the standards are less stringent. The state of New South Wales (NSW) decentralized its decisions even further and certified the local authorities to establish a framework for approving the installation of greywater treatment technologies (NSW, 2006). According to state law, there are no restrictions on the use of untreated greywater for subsurface irrigation. The state of Victoria, too, has a permit system based on local regulators. In the capital, Canberra, no prior permission is required for greywater systems, but an inspection by the local municipal authorities is required after the installation of the system has been completed. That is, in Canberra, most of the responsibility for the system's integrity rests on the users, and they must ensure that the system they install meets the required standards (ACT-Health, 2007). In this respect, Canberra is unique in Australia, while in other states, such as the Northern Territory and Western Australia, most of the responsibility falls on the responsible authority. The user has to obtain a specific installation approval according to the place and the system specification and also approval from the appropriate agency.

In contrast, the humid Tasmania does not require a permit to irrigate gardens and lawns with greywater manually using a bucket. However, a permanent greywater system needs a special approval for piping, and greywater has to be treated before use (HCC, 2011). Additionally, on-site water reuse is prohibited where it is possible to connect to a central sewerage system (Tas-Environment, 2002). In this, the legislation in Tasmania is reminiscent of that in countries such as Israel, which prefer central treatment of the entire wastewater, with all its advantages and disadvantages, over local reuse. However, unlike Israel, Tasmania issued guidelines on using greywater for direct irrigation by hose or bucket, without treatment or storage. The guidelines state that this use of greywater is not at all recommended, but is not forbidden (HCC, 2011). The state of Western Australia, which requires an approval from the central government for such systems, also has a tighter system of permits.

Australia's federal government is actively committed to promoting the adoption of greywater systems. As part of the new national policy, the government is currently offering 500 AUD reimbursement for the installation of greywater systems. However, the power still remains in the hands of the state governments, some of which are not of the same opinion as the federal government. The new Australian guidelines on greywater reuse still have the status of a recommendation.

Although it is logical that different decisions on public policy arise out of different local climatic realities, that does not make it easy to deal with. Professional journals in the trade and industry field define the lack of harmony in standards and regulatory protocols, arising from the differences in legislation dealing with greywater in different states in Australia, as a "bureaucratic nightmare for both manufacturers and consumers" (Tufvesson, 2009).

### 5.2.3 European Experience

Although large parts of Europe do not suffer from significant water shortage, for the past 20 years, a binding directive of the European Union on the reuse of water requires that "treated effluent will be used in all appropriate circumstances." However, this binding directive does not present clear expectations on how and when the member states will translate the directive into action by installing greywater systems (EU, 1991). Thus far, no official all-European regulations on greywater have been published, although some have been in the preparation stage for quite some time. In its official publications, the EU encourages water reuse, with an awareness of the possible hazards as a result of salinity and the like. Recently, the European Science Institute website (EU, 2010) published a Jordanian research study confirming that treated greywater can potentially provide additional water sources without causing ongoing environmental damage (Al-Hamaideh and Bino, 2010).

The member states of the EU have displayed varying degrees of enthusiasm over greywater reuse, although this has not been translated into detailed legal instruments. For example, the British government issued two standards related to greywater systems (BS8525-1, BS8525-2; BSI, 2010) in 2010. Germany is among the continent's leaders in this field, and an important building in Berlin had already installed a greywater system in 1989. However, since then, no new standards have appeared, and until very recently, local legislation has not recognized the separation between *grey* and *black* wastewater (Nolde, 2005).

## 5.3 SETTING STANDARDS AND STRATEGIES TO MANAGE RISK INVOLVED IN GREYWATER

This part of the chapter examines the standardization of greywater regulation while emphasizing the different approaches to reducing the various risks involved in its exploitation. This is an essential scientific task. In most cases, the purpose of the greywater reuse policy is to balance three different goals that are not always consistent: conservation of water resources and their effective utilization, protection of public health (risk reduction), and maintenance of a healthy environment. To create an effective policy, the opportunities and risks involved in water reuse in general and in greywater reuse in particular have to be mapped, and the possible ways to minimize risk have to be examined.

Three methods are used to reduce the level of risk:

1. Prevention at source (e.g., removal of toxic substances that may harm the environment or negatively affect the treatment process)
2. Treatment to remove pollutants (e.g., biodegradation and disinfection)
3. Reduction of exposure (e.g., subsurface drip irrigation and banning the use of greywater for irrigating vegetables)

The regulations for reuse of greywater typically combine these three approaches and adjust the type of treatment to the water quality. For example, when using untreated greywater, it is prohibited to use kitchen wastewater or diaper laundry water (prevention at source), and its use is limited to subsurface irrigation of ornamental vegetation only (exposure reduction). However, when there is a high level of treatment, which includes disinfection, all sources of greywater can be used for many purposes. Figure 5.1 shows schematically the various ways to reduce the environmental and sanitary risks.

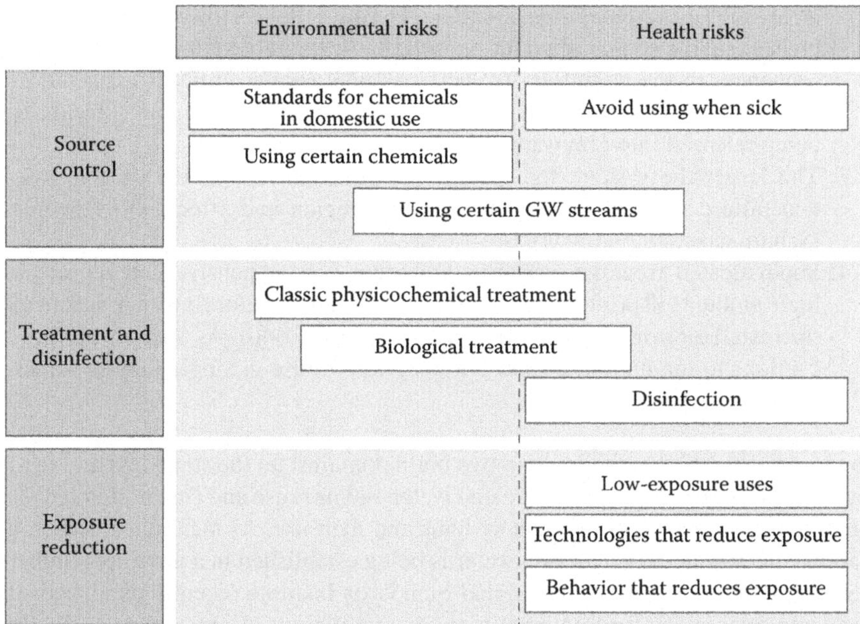


**FIGURE 5.1** The different ways to reduce environmental and sanitary risks of greywater use.

One of the basic ways to balance the various objectives of legislation is an essential separation between a small-scale reuse cycle that includes a single household and wider reuse cycles. This separation is very important due to several factors:

1. The probability of there being a sick person in one of the houses grows as the system services more homes (Dixon et al., 1999b). In a domestic system with a reuse cycle of a single house, excluding kitchen wastewater that may contain pathogens from food (and not from household members), pathogens will appear in the water mainly if one family member is sick or is carrying these pathogens. When one of the household members carries pathogens, there are many ways by which the other household members may be infected, including (though not primarily) through the greywater system. However, in a system that collects the greywater and distributes it among multiple households that probably have no other physical contact, the likelihood of the system being the vector of disease transmission is increased (Ashbolt et al., 2004). Accordingly, the water quality standard in this case should be more strict.
2. If multiple families are using the same greywater system, it is practically impossible to know which family contributes substances that may damage the treatment process or water reuse options. This causes a decrease in the personal responsibility of the *donors* for the water's quality and may result

a *tragedy of the commons* situation (Hardin, 1968). However, in a single household, the source of pollution is the same household that bears the consequences of the pollution. As such, personal responsibility is expected to be higher, and therefore there is a reduced risk of unsuitable substances being released into greywater.

3. The larger the system, the greater the potential risk in the case of system failure since it spreads over a larger region and affects more people (Schumacher, 1973).
4. Sophisticated treatment systems are often very expensive and require a high amount of professional maintenance and therefore are not practical for installation in single-family homes (Little, 2000). As such, other precautions to maintain the safety of greywater reuse in single-family homes should be taken.

Standardization can be divided into two main domains: (a) the standards and regulations regarding the quality of the treated water and its reuse and (b) the standards and regulations of greywater treatment systems and their use. As mentioned earlier, legislation concerning the use of greywater is being established in a growing number of U.S. states, and the American National Standards Institute recently published standards and regulations for treatment systems and the use of greywater (NSF, 2011). Similar standards and regulations can be found in the United Kingdom, Canada, and other countries around the world as well. The most detailed legislation, both in terms of water treatment/use and system standards, can be found in Australia. The legislation in Australian states relies on a risk assessment published in 2006 (NRMMC and EPHC, 2006). Different approaches to the challenge of regulating greywater, reflecting different values and different emphases, are usually divided along the three aforementioned policy objectives.

## 5.3.1 Greywater Management Strategy

### 5.3.1.1 Prevention at Source

The most basic way to guarantee reduced risk pertaining to greywater reuse is to prevent pollution from the greywater itself through reduction at the source (i.e., a careful examination that will determine which sources are allowed to be used as greywater and which are not). Across all countries, greywater is defined as wastewater not including effluents from toilets or urinals, but there is controversy about some other streams. For example, kitchen wastewater may have an adverse effect on soil, vegetation, and treatment systems and may also increase the sanitary risk, due to the high organic load, high fat content, and likely presence of pathogens from food (Ottoson and Stenstrom, 2003). Accordingly, many argue that it should be ascribed to the blackwater category (Lowe et al., 2007) (see Chapter 1). In contrast, others claim that removing the stream generated by the kitchen from the greywater leaves houses equipped with dry toilets (such as composting toilets) with a problem, because they do not produce blackwater at all. In their opinion, kitchen water should be permitted to contribute to greywater. In reality, not a single case of illness resulting from using kitchen water as part of domestic greywater has been reported thus far, which may indicate that a conservative approach on this point is not necessary (Carpenter, 2009). As middle ground, many countries recommend using kitchen water only when there is an appropriate treatment system (see Table 5.1).

To reduce the sanitary risk, various states expressly prohibit the use of water containing feces or urine, such as water used to wash dirty diapers, especially when

**TABLE 5.1**
**Greywater Definition in Several Countries**

| Greywater Is Household Wastewater that Does not Contain Toilet Wastewater and Does not Contain | List of Governing Bodies that Address One of These Three Prohibitions in Their Regulations |
|---|---|
| Kitchen wastewater | Idaho, Arizona, Texas, Utah, Nevada, New Mexico, California, Victoria (UT), Washington, South Australia (recommendation, UT), Canberra (recommendation), NSW (UT), Western Australia, Northern Territory (UT), Tasmania (UT), NSF, WHO recommendations, United Kingdom (recommendation) |
| Wastewater containing feces or urine such as diaper laundry | Idaho, Arizona (UT), Texas, Utah, New Mexico, California, Washington, Wyoming, Victoria (UT), Canberra (recommendation), NSW (UT), Western Australia (UT), South Australia (UT), Tasmania (UT), United Kingdom, WHO recommendations |
| Hazardous or toxic substances that can affect the treatment process or use options | Arizona, Texas, Utah, California, Washington, Wyoming, NSW (UT), Western Australia, South Australia (UT), Tasmania (UT), WHO recommendations, Israel[a] |

*Sources:* Roesner, L. et al., Long term effects of landscape irrigation using household graywater—Literature review and synthesis, Prepared for WERF, published with SDA, 2006; ADEQ, Direct Reuse of Reclaimed Water, Administrative Code, Title 18, Chapter 9, Article 7, Arizona Department of Environmental Quality, Phoenix, AZ, 2001; CDWR, Graywater Systems, California Administration Code, State of California, Department of Water Resources, Division of Planning and Local Assistance, 1997; CBSC, Chapter 16A, "Nonpotable Water Reuse Systems" as appearing in the 2007 California Plumbing Code, Title 24, Part 5, Chapter 16A, Part 1, 2009; VICEPA, Reuse Options for Household Wastewater, Environmental Protection Agency, Victoria, Australia, 2006; VICEPA, Code of Practice—Onsite Wastewater Management, Environmental Protection Agency, Victoria, Australia, 2008; SADH, Application for Alternative On-Site Wastewater/Waste Control system Installation, Department of Health, Environmental Health Service, South Australia, 2004; SADH, Installation of Permanent Onsite Domestic Greywater Systems, Department of Health, South Australia, 2006; SADH, Manual Bucketing and Temporary Diversion of Greywater, Department of Health, South Australia, 2007; ACT-Health, Greywater Use Guidelines for Residential Properties in Canberra, Australian Capital Territory: Department of Health, 2007; NSW-DEUS, NSW Guidelines for Greywater Reuse in Sewered Single Household Residential Premises, Department of Energy, Utilities and Sustainability, New South Wales, Australia, 2007; WHO, Overview of Greywater Management Health Considerations, Amman, Jordan, 2006b; Horovitz (2009).

*Note:* UT, untreated.

[a] Bill.

the reuse is practiced without treatment. Another recommendation aimed at reducing the sanitary risk by reduction at source is not to use greywater when a family member is ill. This recommendation can be found in several states in Australia, for example, in New South Wales (NSW-DEUS, 2007). To reduce the environmental risk and the possibility of damaging treatment quality, some states prohibit streaming of hazardous or toxic materials (such as hair color, paints, thinners, and greases) into greywater collection systems.

Some countries also recommend using environmentally friendly detergents or those containing low concentrations of phosphorus and sodium (NSW-DEUS, 2007), and some provide a list or website of recommended products (Prillwitz and Farwell, 1995). In Israel, due to the widespread use of effluent for agricultural irrigation, there is a standard that explicitly defines the maximum allowed concentration of boron, sodium, and chloride in washing powders and one that defines the maximum allowed concentration of boron in dishwasher powder (IS 438 and 1417, respectively). It is interesting to note that, despite the anticipated benefits, there are no empirical studies documenting the advantages of products considered *ecological* or *green* over conventional detergents for use in irrigation without treatment. This issue deserves further research, however, as a thorough study on the impact of different detergents on soil and vegetation may yet justify creating a standard for household chemicals permitted in greywater. Besides the definition of greywater in legislation, most recommendations for source reduction concern behavior and are not binding.

#### 5.3.1.2 Treatment and Disinfection to Reduce Risk

Proper treatment of greywater can reduce environmental and sanitary risks considerably while still allowing uses that involve relatively high exposure. As demonstrated in Chapter 2, there are various methods for treating greywater that produce water of varying quality. Treatment without disinfection usually decreases the sanitary risk by reducing the number of pathogens (through the creation of unsuitable conditions such as aerobic environments, competition and predation by microorganisms, and filtration). This also avoids fallout from chemical disinfectants such as chlorine, which can have a negative environmental impact and should be applied with care (Hattersly, 2000). However, without the addition of disinfection, the reduction of sanitary risk is usually limited (see Chapter 4).

Many countries permit the use of untreated greywater to a very limited extent and only for uses in which there is a low exposure level. As the level of water treatment rises, so does the number of water use options (Table 5.2). In Israel, the use of greywater is allowed only after intensive treatment, which is under the most rigorous quality control, and only in nonresidential buildings (Goldbereger, 2008). There are two primary legislative methods to ensure the system's treatment efficiency: (1) creating a standard for treatment units to which manufactured treatment units have to comply (in order to get official certification) or (2) instituting a specific authorization for systems that are built on-site and conducting a periodic integrity test of such systems.

#### 5.3.1.3 Exposure Reduction

Reduction in exposure is expected to reduce the risk level considerably, and, therefore, the poorer the water quality, the more barriers are needed to reduce human

**TABLE 5.2**
**Permitted Uses of Treated and Untreated Water Divided by Countries**

| | Prohibiting Countries | Permitting Governing Bodies | Possible Uses When Permitted* |
|---|---|---|---|
| Untreated greywater** | Israel*** | Arizona,[b,a] Texas,[a] New Mexico,[a] California,[c,a] Victoria,[d] South Australia,[e] Canberra,[f] New South Wales,[g] WHO recommendations[h] | Watering with a bucket, hose, or a subsurface system of an ornamental garden or trees (without water storage for more than 24 h). In certain places, toilet flushing is also permitted. |
| Treated greywater | — | Arizona,[b,a] Texas,[a] New Mexico,[a] California,[c,a] Victoria,[d] South Australia,[e] Canberra,[f] New South Wales,[g] WHO recommendations,[h] Israel*** | Surface irrigation, sprinkling, toilet flushing, laundry, car washing, sidewalk washing. |

* General review of common uses; not all the indicated uses are permitted in all the countries mentioned.

** Water that passed primary filtration only is considered untreated.

*** As part of the bill (not yet approved).

(a) Roesner et al. (2006), (b) ADEQ (2001), (c) CDWR (1997) and CBSC (2009), (d) VICEPA (2006) and VICEPA (2008), (e) SADH (2004), SADH (2006) and SADH (2007), (f) ACT-Health (2007), (g) NSW-DEUS (2007), (h) WHO (2006a), (i) Horovitz (2009).

exposure. Poor quality water can come from a more polluted source (such as the kitchen, diaper laundry), from low levels of treatment like filtration, or from water with no treatment at all.

It is possible to reduce human exposure through technological means such as subsurface irrigation, systems that require minimum maintenance, and moisture gauges in the soil that stop irrigation when the soil is saturated. In addition, it is possible to reduce exposure through behavioral factors: use of protective devices like gloves during work that requires contact with greywater, irrigating at times when there are no people in the area (e.g., at night), preventing water from accumulating on ground, and so on. In practice, monitoring the use of such protective measures is of course impossible.

In most countries, the use of untreated greywater for irrigation is allowed only for irrigation with a bucket or hose and/or for subsurface irrigation of ornamental plants or trees. Another common use of untreated greywater is in conjunction with a filtration and diversion device for subsurface irrigation. This application reduces the exposure to a minimum by collecting greywater and streaming it underground to a different area each time. Even when the greywater is treated, care should be taken to reduce human exposure. The highest risk lies in inadvertent cross-connections with adjacent drinking water systems. These cross-connections are significant points of failure, because the intrusion of greywater (even if treated at a high level) into a drinking water system will cause very high levels of exposure and thus high risk.

**TABLE 5.3**
**Common Barriers to Prevent Greywater Exposure**

| | |
|---|---|
| Greywater flow out of the plot boundaries | Prohibited |
| Intrusion of greywater into water bodies | Prohibited |
| Accumulation of water above ground (surface impoundment) | Prohibited |
| Contact with humans | Prevented |
| Irrigation by sprinkling | Limited |
| Irrigation of vegetables that are eaten raw | Prohibited |
| Overflow of excess water to the sewer | Required |
| Marking the piping as nonpotable water | Required |
| Location out of flooding or drainage areas | Required |
| Minimal distance from other facilities | Required |
| Minimal distance from groundwater water level | Required |

It should be emphasized that there are hundreds of thousands of small greywater systems at the single household level and no cases reported in the literature that link the use of greywater and disease outbreaks.

A list of the most common exposure barriers is presented in Table 5.3.

### 5.3.2 Enforcement and Supervision

Law enforcement of multiple decentralized contamination sources creates an enormous challenge for inspection teams addressing any environmental issue. The problem has been documented extensively in relation to water pollution from nonpoint sources (Rosenthal, 1990). It simply is not possible to have a policeman's presence felt at the site of all potential violations.

It may be especially difficult to justify the allocation of resources and governmental manpower for ensuring compliance with greywater programs because of the relatively small environmental nuisance associated with them. Worst-case scenarios of exposure to irrigation water from washing machines cannot compete in the field of risk assessment against hazards caused by toxic substances, radiation, chronic air pollution, or pesticides. This reality is reflected in enforcement policy. California is an example of a preemptive enforcement strategy that encourages systems to be designed properly in the first place. Hence, the following language appears in its new law:

> No construction permit for any graywater system shall be issued until a plot plan with appropriate data satisfactory to the Enforcing Agency has been submitted and approved. When there is insufficient lot area or inappropriate soil conditions to prevent the ponding or runoff of the graywater, as determined by the Enforcing Agency, no graywater system shall be allowed. (CBSC, 2009)

Although it is possible that behind the increased stringency of the regulations is healthy environmental logic, legislation should be judged also in consideration of its actual implementation potential. Thus, California regulators would be well advised to consider, first of all, the extensive violations perpetrated in the days when the standards were less stringent. Indeed, a recent research estimates that only 0.01% of the greywater systems in California comply with the legal permits (Bahman, 2010). As a result in 2012, the State of California abolished the requirement to obtain approval for domestic greywater systems as long as they meet the guidelines specified in the law (CBSC, 2012).

Once greywater systems become very common, no country will have enough manpower for inspectors to ensure the proper operation of each individual system. Instead, the best enforcement strategy will rely on awareness and, if possible, on deterrence. Most families do not have the technical qualifications required for redesigning their plumbing system. Given the dependence of most households on professional and commercial experts to redesign their homes, regulatory supervision and training of relevant professionals may be a more effective approach than supervising single families. For example, this was done to control the pollution of groundwater by nitrogen in Central Platte, Nebraska, where enforcement programs were successfully directed at the fertilizers' selling agencies and agents (Tal, 1998; Exner et al., 2010). It seems that the most effective action by which authorities can reduce human exposure to a minimum—without sabotaging the development of the greywater infrastructure—is to institute clear standards for installing greywater systems up front and heavy fines for disregarding them.

Of course, it is also possible to adopt conventional supervision tactics of enforcement in the style of *command and control*, especially if significant risks are identified. It is possible to create deterrence through an extensive advertising campaign accompanied by surprise inspection visits in homes that have greywater systems and the issuing of penalties to lawbreakers. However, it is very likely that such an approach would create a lack of political sympathy and ultimately dissuade the public from installing greywater systems. This would run counter to the ultimate goal of appropriately regulating greywater.

The real policy challenge is sharing information with the millions of people around the world who have already installed a greywater system, but failed to take into account the environmental precautions that are currently considered important. Stricter regulations and insistence on individual permits turn thousands of well-intentioned citizens into *lawbreakers* and only slightly affect their actions in practice. Grandfathering existing systems and creating a voluntary system of supervision and assistance (without the threat of punishment) may be a more effective strategy to improve greywater systems that pose a health or hydrological hazard.

Enforcement policy for the most part reflects usage tiers, and each tier entails a different regulation and enforcement method. In Australia, the tiers are ordered according to the treatment or use permitted. When the level of water treatment is lower, the permitted use of water is more limited, and therefore there is less need for certification, inspection, and so on. The first tier is using untreated greywater directly from the washing machine or shower, without storage, with a hose or bucket. This use is very limited, and it is assumed that watering by bucket or hose is done using small quantities of fresh greywater. It is only a temporary measure and is used mainly in times of water scarcity. For these reasons, it presents a low risk as compared to fixed systems that use larger volumes of water, in which faults may create increased risk. At this level, enforcement is problematic.

The second tier is using a facility that filters (coarse filter) and diverts the water to subsurface irrigation without storage. Here, too, the water quality is low (only primary treatment), and the exposure is low (subsurface irrigation), but the difference is that long-term use is often involved. In most states in Australia, the filtering and diversion facility has to have a certification. Its installation, which requires a change in the home plumbing system, should be performed by a licensed professional and reported to the authorities. The third tier is a treatment system. The permitted uses of water exiting the treatment system are many (e.g., surface irrigation, toilet flushing, car washing), creating higher exposure to the reused greywater. Accordingly, the risk level is considered higher, and hence the need for more certificates. The system and the quality of treated greywater it produces have to be preapproved. In addition, the installation of the system must be overseen by the authorities and be performed by a licensed professional.

In the United States, the usage steps are determined for the most part by the capacity of the system (volume/day). In Arizona, in order to facilitate and encourage the use of greywater, a blanket approval has been granted to all single-family house systems (up to 400 gallon/day, about 1500 L/day), which meet certain conditions. Any larger system (or one that is not for use by a single family on their own plot) or one that does not meet the conditions of the general approval needs specific approvals (ADEQ, 2001). Table 5.4 shows different approaches to supervising greywater systems.

Thus, it appears that different countries chose different ways to deal with the challenges posed by the reuse of greywater and that local circumstances such as climatic conditions and the extent of water shortage dictate the level of enforcement and supervision.

### 5.3.3 Standardization and Technical Aspects of Greywater Treatment System and Greywater Reuse

The following review presents three examples of standards of systems for greywater treatment. The United Kingdom and the United States have a key standard. In the United States, the key standard is in addition to the local standards of each state, while in Australia, each state has its own standards and there is no national standard. Therefore, the local standard of New South Wales (hereafter, NSW) is shown here as an example from Australia (NSW, 2008). The different standards are distinguished from each other by the definition of the type of treatable greywater and in their division into quality tiers and types of systems. In the British (BS8525-2) and NSW standard, the quality of the treated greywater that the systems are required to produce changes according to type of use, while in the U.S. standard (NSF/ANSI 350), the division is by type of user: single household or residential/commercial building. However, in division into types of systems, the British standard is exceptional in that it divides the system according to a specific element (having a storage tank) and not by the system volume as in the American and NSW standard. What is common to all three standards is the emphasis on the required quality of the treated greywater

**TABLE 5.4**

**Approvals Required for the Use of Greywater According to Usage Steps: Examples from Australia and the United States**

| | Australia | | | USA | |
|---|---|---|---|---|---|
| | **New South Wales**[a] | **Queensland**[b] | **Victoria**[c] | **California**[d,e] | **Arizona**[e,f] |
| *First tier* | Irrigation using a bucket or a hose | | | Washing machine or single connection | Up to 1514 L/day |
| User type | | Not defined | | Single or two households | Single household |
| System approval | | Not required | | Not required | Not required |
| Installation approval | | | Not required | | |
| *Second tier* | | Diversion facility | | Up to 950 L/day | Over 1514 L/day or not complying with the conditions of the general approval |
| User type | Not defined | Not defined | Single household | Single or two households | More than a single household |
| System approval | | | Required | | |
| Installation approval | Not required | Required | Not required | Required | |
| Volume limitation | Not defined | Up to 3000 L/day | Not required | Up to 950 L/day | Up to 11,356 L/day |
| *Third tier* | | Treatment system | | Over 950 L/day | Over 11,356 L/day |
| User type | Not defined | All consumers up to 50,000 L/day | Single household | Not defined | |
| System approval | | | Required | | |
| Installation approval | | | Required | | |
| Volume limitation | Between 720–900 L/day | Up to 50,000 L/day | Up to 5000 L/day | Not defined | |

*Notes:* (a) NSWDE (2007), (b) QDLG (2003) and QDIP (2007), (c) VEPA (2006) and VEPA (2008), (d) CDWR (1997), CBSC (2009), (f) Roesner et al. (2006), (f) ADEQ (2001).

**TABLE 5.5**
**General Definitions of Greywater, Division into Quality Steps, Types of Treatment Systems, and Types of Use (Britain, United States, and Australia)**

| | BS 8525-2 Britain | NSF United States | NSW Australia |
|---|---|---|---|
| General | Addresses household greywater treatment systems of up to 10,000 L/day, produced or installed by a single manufacturer and tested as a single unit | Greywater treatment systems of up to 5678 L/day in volume, in residential and in commercial buildings and in public laundries | Greywater treatment systems of up to 720–900 L/day |
| Greywater definition in the standard | Domestic sewage generated by hand basins, showers, and washing machines | Laundry and washing wastewater, excluding urinals, bidets, kitchen sinks, and dishwashers | Domestic wastewater excluding toilets and urinals |
| Division into water quality steps | Water designed for use without sprinkling; water designed for use with sprinklers | Class R, single household; Class C, residential and commercial buildings | Subsurface irrigation; *surface* irrigation; toilet flushing and laundry |
| Division into types of systems | Systems with storage tanks before gravity conveyance; systems with a pump for direct use from the treatment system | Systems that treat less than 1514 L/day; systems that treat the volume of 1514–5678 L/day | System for 8 people (720 L/day); system for 9 people (810 L/day); system for 10 people (900 L/day) |

*Sources:* BSI, *Greywater Systems BS 8525-1&2:2010*, British Standards Institute, 2010; NSF, *NSF/ANSI 350–2011, Onsite Residential and Commercial Water Reuse Treatment Systems,* 2011; NSW-DEUS, *NSW Guidelines for Greywater Reuse in Sewered Single Household Residential Premises*, Department of Energy, Utilities and Sustainability, New South Wales, Australia, 2007.

(at the outlet of the treatment system), and not on a certain method or treatment technology (Table 5.5).

### 5.3.3.1 Construction

While there is no preferred treatment technology in any of the standards, there are technical requirements that a system must meet. These requirements usually address the safety of the system and determine the baseline that all systems should meet. There are requirements common to the three standards: alert devices in case of error, the possibility of overflow/diversion of treated and untreated greywater to the sewer, and installing a license plate that contains basic information on the system. In the U.S. and NSW standards, the durability of electrical and mechanical components is considered, and there is even consideration given to noise intensity limits of the system. The British standard chose not to address these issues, but did tackle other issues not considered in the previous two: the minimum flow rate required and the existence of an electronic control panel (Table 5.6).

**TABLE 5.6**
**Construction Requirements of the Greywater Treatment Systems (Britain, United States, and Australia)**

| | BS 8525-2 Britain | NSF United States | NSW Australia |
|---|---|---|---|
| Materials | — | Should not show signs of structural change during operation, without bumps and dents, except where required. | Materials and system parts must meet specific product standards and be resistant under normal operating conditions. |
| Noise | — | Not exceeding 60 dbA. | Maximum noise should comply with relevant regulations. |
| Mechanical elements | — | Mechanical elements should not require an ongoing regular maintenance by the user. | Should be suitable for continuous and intermittent work, require low maintenance and be protected from harsh environmental conditions. |
| Electric parts | — | Should contain protection devices, such as fuses. | Should be suitable for continuous and intermittent work and meet specific product standards. |
| Access points | — | The system should have access points that allow visual inspection and removal of mechanical or electric parts, periodic cleaning and/or replacement of parts, sampling option, removal of residual material, and removal of treated or untreated water to sewage.<br>Access points should be protected against access by anyone unauthorized. | Safe and easy access should be possible for routine maintenance and repairs. |

*(Continued)*

**TABLE 5.6 (*Continued*)**
**Construction Requirements of the Greywater Treatment Systems (Britain, United States, and Australia)**

| | BS 8525-2 Britain | NSF United States | NSW Australia |
|---|---|---|---|
| Flow rate | >0.4 L/s. | — | |
| Volume | >80 L. | — | 720 L < V < 900 L |
| Water tightness | Without leaks. | Without leaks. | |
| Overflow | Overflow of 10% above the maximum flow. | — | Overflow or diversion of water to sewer should be enabled in case of failure or at user's command. |
| Control panel | At least one control panel should be installed if system includes electronic equipment. | — | |
| Alarm | Alarm should be triggered in case of power failure and when supply of backup water is interrupted. | When there is a failure in a critical part of treatment or in water flow, system should create visual and auditory signals heard and seen from a certain distance. Alarm should also be triggered in case of power failure.<br>Systems in commercial buildings should include alarm that sends a message through electronic communication. | Auditory and visual alarm should be triggered in case of a mechanical or electric failure in treatment, disinfection, or pumping system. |

(*Continued*)

**TABLE 5.6 (*Continued*)**
**Construction Requirements of the Greywater Treatment Systems (Britain, United States, and Australia)**

| | BS 8525-2 Britain | NSF United States | NSW Australia |
|---|---|---|---|
| Emptying treated greywater into the sewer | When treated greywater can be stored for up to 30 days, means required to release treated greywater into the sewer in case of nonuse or malfunction have to be included (even when there is no power supply). | Plans should be devised regarding outlet in case of failure or excess water. | If there is no need for treated greywater (e.g., in case of rainfall), greywater should be diverted to sewer. |
| Failsafe system in case of power failure | If there is a connection to tap water as backup, it must be operated automatically in the event of power failure. | — | — |
| Failsafe system, in case of malfunction of the disinfection system | When disinfection system is not working, water should be diverted to sewer and tap water should be pumped as backup. | — | — |
| System marking | Plaque on system should specify the number and date of British standard, type of greywater, category of treated water, daily hydraulic flow, maximum flow rate, maximum volume, details about disinfection process, maximum storage period, power supply requirements, and year of production. | Plaque on system should specify the manufacturer's name and address, model number, serial number, daily hydraulic capacity of system, restrictions on type of water entering the system (washing, laundry, or both), and classification system as for single household (R) or residential and commercial buildings (C). | Plaque on the system should specify the manufacturer's name, system's model, and year and month of production. |

#### 5.3.3.2 Water Quality Requirements

In the three different national standards, the nature of use dictates the water quality requirements of the treatment systems (Table 5.7). The British and NSW standards divide the uses according to the expected level of exposure. In the United Kingdom, the main division is into two: use by sprinklers (e.g., high-pressure rinsing, sprinkle irrigation, car washing) and uses that are nonsprinkling. The latter use is further divided into three types: garden watering, toilet flushing, and use in a washing machine. In NSW, the division is into subsurface irrigation, *surface* irrigation, and toilet flushing. As compared to these standards, the American standard offers a distinction between single household and residential/commercial building. In both cases cited in the U.S. standard, the treated greywater is intended for unrestricted irrigation and flushing toilets.

The three national standards address the microbial quality of water by using indicator bacteria for fecal contamination. Only the standard of NSW lists the option of subsurface irrigation and allows the use of low-microbial-quality greywater, for this purpose. This option opens up the possibility of using greywater for limited irrigation in a single household. Such a greywater application is permitted in the United Kingdom and in a number of states in the United States, but is not included in these standards. Water used for surface irrigation must not contain more than 30 thermotolerant coliforms/100 mL in NSW, while in the American standard, the maximum is 14 *E. coli*/100 mL in a single household and 2.2 *E. coli*/100 mL when a larger system is involved. In the United Kingdom, the breakdown is different; for garden irrigation, the treated greywater may be at a maximum of 25 *E. coli*/100 mL, but for sprinkler irrigation, the treated greywater should meet the highest standard (0 *E. coli*/100 mL). The most stringent requirement according to the NSW standard is less than 10 thermotolerant coliforms/100 mL (toilet flushing and laundry). The most stringent requirement in the American standard is 2.2 *E. coli*/100 mL (used for unlimited irrigation and toilet flushing in residential and commercial buildings). The most stringent demand in the British standard is 0 *E. coli*/100 mL (for use in sprinklers and washing machine), and there is an additional distinct microbial index in the British standard, for intestinal enterococci.

The standards of these three nations also address chlorine concentration after disinfection. All of the standards note that it is not necessary to use chlorine as a disinfectant but, if used, its concentration should stay in the specified range. For example, in the United Kingdom, the concentration of residual chlorine after disinfection is limited to a relatively low level when the treated greywater is intended for garden irrigation, due to the environmental damage that a high concentration of chlorine may cause.

Besides the microbial quality of the water, the chemical quality also plays an important role. According to the U.S. and NSW standards, treated greywater may not surpass maximum threshold concentrations of $BOD_5$ and total suspended solids (TSS). In NSW, when the greywater is used for irrigation, it must meet the standard of 20 mg/L $BOD_5$ and 30 mg/L TSS. This is as compared to greywater used for flushing toilets and laundry, which have to meet a more stringent standard of 10 mg/L $BOD_5$ and TSS. In the United States, the standard is uniform, and the treated

**TABLE 5.7**
**Quality Requirements of Treated Greywater (Britain, United States, and Australia)**

| | BS 8525-2 Britain | | | | NSF United States | | NSW Australia | | |
|---|---|---|---|---|---|---|---|---|---|
| | Sprinkler use | Nonsprinkler use | | | Single household R (average tests) | | Subsurface irrigation (90% of tests) | Surface irrigation (90% of tests) | Toilet flushing and laundry (90% of tests) |
| | Pressure wash, sprinkler, car wash | Toilet flushing | Garden watering | Washing machines | Toilet flushing, unlimited watering (except for edible parts of plants) | | | | |
| Thermotolerant coliforms/100 mL | — | — | — | — | — | | — | <30 | <10 |
| *E. coli*/100 mL | Not detected | 25 | 25 | Not detected | 14 | 2.2 | — | — | — |
| Intestinal enterococci/100 mL | Not detected | 10 | 10 | Not detected | — | — | — | — | — |
| Turbidity NTU | 10> | 10> | Irrelevant | 10> | 5 | 2 | — | — | — |
| pH | 5–9.5 | 5–9.5 | 5–9.5 | 5–9.5 | 6–9 | 6–9 | — | — | — |
| Residual chlorine mg/L[a] | 2> | 2> | 0.5> | 2> | 0.5<Cl<2.5 | 0.5<Cl<2.5 | | 0.2<Cl<2 | 0.5<Cl<2 |
| Residual bromine mg/L[a] | 0> | 5> | 0 | 5> | — | — | — | — | — |
| BOD mg/L | — | — | — | — | 10 | 10 | <20 | <20 | <10 |
| TSS mg/L | — | — | — | — | 10 | 10 | <30 | <30 | <10 |
| Odor | — | — | — | — | Not offensive | Not offensive | — | — | — |
| Foam and fatty layer | — | — | — | — | Not found | Not found | — | — | — |

[a] If this disinfectant was used.

greywater should be of a quality of below 10 mg/L $BOD_5$ and TSS for both of these categories of use. In addition to $BOD_5$ and TSS tests, U.S. systems must produce greywater meeting turbidity, pH, and odor requirements. In contrast, the indicators selected to indicate the chemical quality of the water in the United Kingdom are pH and turbidity only.

#### 5.3.3.3 Validation and Verification

There is a significant difference between national standards when it comes to the test duration required for system characterization and approval (Table 5.8). While 6 months of system testing are required in the United States and in NSW, 5 days are sufficient in the United Kingdom (or two rounds of system use). In contrast to the United Kingdom and the United States where system testing is conducted using synthetic greywater of known and predetermined composition, in NSW, the system is tested in the field using actual locally produced greywater. Only in the U.S. standard is the system required to meet different pressure tests, simulating common conditions such as vacation periods that create water shortage or alternatively an overload of water or cleaning agents.

#### 5.3.3.4 Operation and Monitoring

All standards indicate that an operating manual and user information should be attached to a system, as well as technical brochures designed for maintenance personnel. In the United States and NSW, the standard also specifies a minimum time period for which the system must be under a service contract and under warranty (Table 5.9).

## 5.4 INTERNATIONAL REGULATIONS VERSUS RISK ASSESSMENT

Examination of the regulations reveals that the indices for determining water quality are very similar in different countries, but the standards and regulations are very different. The reasons for this variance seem to be the risk perception versus the possible benefits in each country; however, this subject must be further explored. According to this theory (as already noted), it can be expected that countries that cope with water scarcity will tend to adopt a less conservative approach that permits a variety of uses with fewer restrictions. Accordingly, we would expect that the legislation in Israel would be lenient and allow extensive use of greywater, but the reality is actually the opposite. This could be explained by the fact that, unlike other countries where centralized water reuse is not common, greywater reuse in Israel is also perceived as a threat to the supply of recycled wastewater effluent for agriculture. Reusing water for irrigation has been a common and growing practice led by the government for over four decades. In addition, Israel has a historic tendency to solve problems in a centralized (vs. decentralized) manner. When these tendencies are taken into account, the strict legislation in Israel makes more sense.

As noted in Chapter 4, only a few studies on the level of sanitary and environmental risks associated with the reuse of greywater, and effective means to reduce them, have been carried out. Moreover, all of the studies were theoretical and used models,

**TABLE 5.8**
**Validation and Verification of Greywater Reuse Systems (Britain, United States, and Australia)**

| | BS 8525-2 Britain | NSF United States | NSW Australia |
|---|---|---|---|
| Test period | Five days or two rounds of system use (whichever is longer) | Twenty-six weeks | Twenty-six weeks. |
| Tests | To verify hydraulic characteristics and system's functioning using tap water: accepted flow rate and volume, watertightness, overflow, automatic emptying system, failsafe systems, water quality tests | Cleaning day test, entering 7 days' water amount in 3 days; electric/mechanical failure test, electricity will be stopped and the system will stop working for 48 h; vacation test, no entry of freshwater for 8 days; water utilization test, water will have 1.4 times the normal concentration for 1 week and reduction of 40% of daily water volume; detergent test, system will receive water containing a high concentration of trisodium phosphate (TSP) for 4.5 weeks. | System should be installed in *real* site, and use greywater produced on-site; suitable places to install experimental system are private homes, students' dormitories, trailer park areas, and business districts with wastewater of domestic character; approval for installation of experimental system should be obtained. |
| Types of water to be tested | Two types of synthetic greywater: to test systems that treat rinsing and washing water and to test systems that treat rinsing, washing, and laundry water | Three types of synthetic greywater: washing water, laundry water, and washing and laundry water. | — |
| Sampling | Before treatment; after treatment; after storage for maximum time period | Untreated water, twice a week; treated water, twice a week except for weeks without treatment (power failure test and vacation test); in weeks following each test, three samplings should be made instead of two. | Untreated water, every 12 days; treated water, every 6 days. |

**TABLE 5.9**
**Operation and Monitoring of Greywater Reuse Systems (Britain, United States, and Australia)**

| | BS 8525-2 Britain | NSF United States | NSW Australia |
|---|---|---|---|
| User information | Instruction manual regarding operation and actions to be taken during failure | Extensive manual that includes the following: type of system, type of water and substances that can be introduced into system, types of recommended use, comprehensive operation and maintenance instructions, actions to be taken during failure, troubleshooting, and manufacturer's service commitments and contact information. | User manual that includes the following: type of system, type of water and substances that can be introduced into system, type of advised uses, comprehensive operation and maintenance instructions, actions to be taken during failure, troubleshooting, and manufacturer's service commitments and contact information. |
| Technician information | Detailed operation manual that includes sufficient information for professionals about installation, operation, and safe dismantling of the system | Comprehensive instructions of installation and maintenance of system, which include the following: actions during failure, current maintenance, sample collection, and training service technicians and their regular updating. | Comprehensive instructions of installation and maintenance of system; periodic report on the system's status. |
| Service contract and warranty | — | System shall be supplied with service contract that includes examination and treatment if needed at least four times per year and emergency service within 24 h in case of failure. | All parts (except pumps and motors) should be guaranteed for at least 15 years of service; all mechanical and electronic parts, at least 5 years of service and minimum 1-year warranty; entire system should have minimum 3-year warranty; during warranty period, all parts and labor should be supplied without payment.<br>Periodic service according to manufacturer's instructions is user's responsibility and at his or her expense. |

and there are no epidemiological studies that examined whether a population that uses greywater is at a higher risk for certain diseases than a population that does not reuse greywater. The lack of acceptable scientific evidence leaves plenty of room for variance in legislation based on the *prospect versus risk* perception of policy makers in each state.

Another bias that is also related to the absence of specific research-based regulations for greywater use is found in setting quality standards based on domestic wastewater regulations. For example, the use of fecal indicators for microbial quality can be misleading since it is expected to be significantly lower in greywater. In addition, it is possible that turbidity and organic matter (found in domestic wastewater regulations) are less relevant than the concentration of surfactants, sodium and boron content that are high in greywater but are not regulated.

In most countries, when water is used for restricted irrigation in single-family homes, there is no microbial standard at all. However, in larger systems or in unrestricted irrigation, strict regulations prevail. The few existing regulations address fecal coliforms (*E. coli* in particular) to reflect fecal contamination that indicates the microbial quality of the greywater. In Israel, the regulations that permit irrigation near residential buildings are the same as those for unrestricted irrigation in cities. Accordingly, the limit is set at the median of <1 fecal coliforms in 100 mL and a maximum of 14 fecal coliforms in 100 mL (Halperin and Aloni, 2003). According to the Rotavirus risk assessment presented in Chapter 4, it can be concluded that a concentration limit of approximately 100 fecal coliforms in 100 mL will provide an adequate protection to users' health at a moderate level of exposure. Hence, the regulations in Israel are more stringent by at least two orders of magnitude (from an infection risk of 1/1,000 individuals to an infection risk of lower than 1/100,000) thus disinfection is required.

Another important tier in the precautionary principle is separation in legislation between reuse in the single-family home and in a broader circle. The sanitary risk of using greywater in a private garden is lower than the risk of using it in a public space or using greywater collected from a number of houses, so on-site use is not treated the same as public use. In most of the countries examined, there is indeed a separation in legislation between on-site reuse (single family) and broader circles of reuse that include a number of houses or families.

In general, it seems that the regulations for reuse of greywater provide a good protection against the spread of disease by making the first essential separation between the cycle of domestic use and the cycle of wider use, as well as by defining use tiers. In the Australian method, use tiers are determined by the greywater quality: the higher the quality, the more uses (and the greater the exposure) permitted. For public health protection, this seems to be the best approach. This method achieves health protection while still allowing a variety of reuse options according to the amount of water and the financial capacity to treat it optimally. However, it may not be necessary at the single home level.

Despite the growing evidence of environmental risk from studies, there is almost no reference to environmental risk in the regulations on the reuse of greywater. Even in places where a comprehensive environmental risk assessment of the reuse of greywater was conducted, such as in Australia (NRMMC and EPHC, 2006), one of the criteria for defining problematic pollution is the extent of its effect. This criterion is problematic when assessing the risk of greywater because in most cases, the impact is local; it does not cross the boundaries of the private property. Therefore, such an analysis ignores the fact that injury to many small areas may accumulate and cause wider environmental damage. In addition, if a particular behavior may negatively affect the vegetation or the soil, especially when the effects are felt only after an extended period of time, the public must be informed in advance to avoid the latent consequences.

As Israel is a world leader in using municipal wastewater effluent for irrigation, experience and research-based knowledge on the impact of effluent on soils and crops have accumulated in the country. Lessons can be drawn about the reuse of greywater from Israel's experience. We see that regulatory intervention on environmental issues can be on several levels: one assessing the quality of the treatment that the greywater undergoes in relatively large systems and the other according to the pollutant removal capacity. In addition, it is important to identify environmentally sensitive areas where legislation should be more stringent, such as areas where the groundwater table is particularly high or areas close to an important water source. Finally, the rating of chemicals according to toxicity, biodegradability, and biological accumulation (e.g., for household detergents and cosmetics) may contribute to a reduction at source of environmental risks, accompanied by a recommendation on the use of substances that pass a certain rating.

## 5.5 LEGISLATION IN ISRAEL

Israel is among the leading countries in the effective management of water resources and has the highest rate in the world of reusing municipal sewage (Tal, 2006). Greywater systems have actually been in use in Israel since the late 1980s. Nevertheless, government offices were hostile to the spread of these technologies in the beginning, especially in households, and initially intervened to prevent their adoption. The use of greywater is in fact only allowed to a limited degree in buildings, such as clubs and soccer fields, where there is organized maintenance by professional personnel. In addition, treatment of greywater is required to bring it to a very high quality that will allow unrestricted irrigation in cities (Goldbereger, 2008), and any system installed at these locations needs an individual permit. In single-family homes, greywater reuse is forbidden entirely, but there is no enforcement of this regulation. The lack of enforcement together with aggressive advertising about saving water and the introduction of a drought tax, created a confusing situation: on one hand, the Israeli public is pushed to save water and on the other hand greywater reuse which seems as an attractive alternative is banned.

The Ministry of Health traditionally stands at the forefront of the *opposition* to greywater use in Israel. The ministry's position was recently summarized in a review on legislation dealing with greywater, which was prepared for the Knesset (Ronen, 2009). In the following are the Ministry's reservations (as stated by Eng. David Weinberg, from the Ministry of Health, 2012):

## THE OPINION OF THE MINISTRY OF HEALTH ON GREYWATER USE IN PRIVATE HOMES

Pathogens (bacteria, viruses, protozoa and parasites) found in fecal secretions cause infectious disease in humans. More than two hundred different diseases are known to be transmitted by water. Pathogens are the most common health risk associated with drinking water. Disease infection may be caused while drinking and even through contact with water contaminated with these microorganisms.

Transmission of disease sources from sewage to humans can occur in a number of situations:

- A person's physical contact with wastewater.
- Transfer by water: the smallest amount of sewage, effluent, or water of insufficient quality that enters the water system is enough to pollute large quantities of water, resulting in infectious disease outbreaks.
- Transfer by food: vegetables and fruits exposed to direct contact with wastewater and/or effluent or water of insufficient quality.
- Transfer by various animals that come into contact with sewage, effluent, or water of substandard quality.
- By flying insects that come into contact with sewage, effluent or water of insufficient quality, and then with people and/or crops.

In the past, the frequency of infection and transmission of infectious diseases through contaminated water was very high. Improvements made over the years in the methods and manner of wastewater treatment, in the distancing and conveyance of wastewater away from the vicinity of humans, and the advancements in protecting water sources and the treatment methods of the supplied water have lowered the number of incidents of water-borne infectious diseases in the Western world.

Greywater is defined as sanitary wastewater with no contact with liquids from toilet stalls. In recent years, there is an increased public demand to promote solutions to reuse greywater in urban uses such as for flushing toilets and watering gardens.

Greywater may contain high concentrations of pathogenic microorganisms (disease generators) that may endanger the health of the public who come in contact with it. This contact may become possible during failures in treatment and reuse systems, contact with water used for irrigation, cross connection of greywater with potable water systems, and more.

Therefore, the use of greywater requires the construction and operation of advanced treatment facilities, with efficient controls. Moreover, the construction of greywater conveyance systems that are completely separated from potable water systems with effective controls to prevent cross-connections between the greywater system and the potable water system should be established.

The solution for wastewater that is preferred from the sanitary and environmental aspect is its distancing from the public and treating and reusing it in such a way that public health, water resources, the environment, soil, and crops will not be harmed. The State of Israel invests heavily in these solutions, and currently about 80% of the wastewater are reused in agricultural irrigation, industry, and public gardening (an unparalleled scale in the world).

The Ministry of Health is aware of the issue and therefore published in June 2008 conditions for the reuse of greywater for gardening and flushing toilets in places that will not endanger public health. These conditions relate to, among other things, the location of the facility in the building area, location of reuse areas, the level of treatment required, and various engineering measures required in the facility, as well as facility ownership and responsibility for its operation. The purpose of these conditions is to reduce the sanitary risk.

The guidelines allow greywater reuse in facilities operated, maintained and controlled by local authorities and in licensable businesses where there is appropriate maintenance staff. In these places, effective monitoring of the facilities and reuse systems can be performed.

The guidelines do not permit greywater treatment and reuse facilities to be built and operated in residential buildings, whether single-family houses or high-rise buildings. The reasons for this are:

1. The guidelines require that greywater treatment will be set to a quality defined as "excellent." Facilities are required to reduce fecal *E. coli* bacteria 10,000 times or more, and this requires advanced technology, such as biological treatment, granular depth filtration with flocculation, and chlorination with sufficient contact time. Such facilities require great knowledge for their operation and maintenance and regular monitoring. In the absence of proper control and maintenance, the facilities will not function at the required level.
2. There is no real possibility of controlling the level of maintenance of the tens of thousands of domestic installations to be constructed. In the absence of control, the maintenance level of the facilities may deteriorate. Poor maintenance can cause the dispersal of greywater pollutants into the environment, allows the public to come in contact with the pollutants, and pollutes the environment. Additional risk to the public due to the reasons listed above is neither desirable nor acceptable.
3. It is impossible to control the changes in water systems that the flat owners might make in the buildings in which greywater is reused for toilet flushing will be approved. If systems in which greywater are conveyed will be connected to systems that supply drinking water,

penetration of polluted water into drinking water will occur, which can cause extensive morbidity, both in the building where the cross-connection was performed and in nearby buildings.

4. Treatment facilities in residential buildings have the potential to create environmental and safety hazards, especially in light of the experience in Israel and abroad regarding the decline in the level of maintenance over time.
5. The use of greywater at the domestic level will reduce the amount of water for agriculture and may cause damage to agriculture in Israel, expressed in various negative environmental aspects. It should be noted that the use of greywater for watering private gardens does not constitute a saving for the water sector in Israel.
6. Separation of greywater and its large scale reuse may affect the transport of wastewater and the wastewater treatment systems, both by changing the composition of the wastewater (it will become more concentrated and its transport properties will be affected) and the fluctuation in the flows conveyed to the municipal treatment plants (in the winter the intensity of irrigation in private gardening will decrease and thus the excess greywater be conveyed to the municipal treatment plant).

According to these guidelines, it is possible in principle to approve a greywater treatment facility of a residential neighborhood for irrigation of public landscaping, provided that the collection, treatment, and effluent reuse will be done by the local authority.

Even at remote sites, such as individual farms, where the problem of access to maintenance and control is often increased, the guidelines do not allow for the construction of facilities for greywater reuse. It is important to make sure that even at these sites the solutions for wastewater will be safe in terms of sanitation.

In summary, the sewage solutions presented for the approval of the Ministry of Health are examined first in terms of protecting public health and only then from economic and savings aspects. Sewage solutions in Israel should be chosen with reference to the characteristic factors including population density, the complexity of infrastructure, systems maintenance, the current high level of wastewater reuse, protection of agriculture, and more.

The last event of *E. coli* morbidity in Germany illustrates the risks that may be created by environmental pollution and the need to avoid them.

Sincerely,

David Weinberg
National Planning and Treated-Effluent Engineer

Due to these reservations, the Israel Ministry of Health continues to restrict the use of greywater systems and allows their operation only in projects under the responsibility of the local authority or in commercial businesses operating under a business license. The operation of greywater systems is forbidden in single-family homes, apartment buildings, kindergartens, and state schools, as well as in food and pharmaceutical plants. If a commercial facility is interested in adopting a greywater system, it must pass a strict licensing process. For example, the license application has to be accompanied by an engineering opinion given by an expert on the effects of the greywater unit on the sewer transport system. The greywater has to be treated to an *excellent* level. This means that no measurable concentration of fecal coliforms will be found in the treated greywater, the turbidity should be low and it should be odor free, and monitoring should be very frequent. Moreover, a survey has to be taken of the effects that any installed system has on the environment, including kitchen sinks, dishwashers, and washing machines.

The Ministry of Health also requires that the treatment facility and reuse system will be under the responsibility of an organization with proven experience in the maintenance of similar systems. This organization will act in accordance with a treatment plan that includes a means to maintain sanitary conditions and a monitoring program of the facility operation. A backup system of freshwater must be installed, as well as a way to deal with cases of malfunctions or surpluses in which a discharge of greywater to the sewer system is called for. The greywater plan is also required to contain detailed measures to ensure that greywater for flushing toilets will not be in contact with potable water systems. Well-intentioned citizens who consider installing a greywater system need high motivation to find their way through the tangle of bureaucratic requirements.

Israel's Water Authority is an independent governmental agency that works alongside the country's Ministry of Health. It is responsible for ensuring Israel's water supply and for the development of water sources. The authority's position regarding greywater stems mainly from the high rate of wastewater effluent reuse in Israel: arguments in favor of greywater accepted throughout the world that position greywater favorably—as another potential source of water—do not apply in Israel. This is because nearly 80% of the wastewater in Israel is already reused, so allowing the population to use greywater in private gardens would simply divert it from reuse in agriculture (Shani, 2008). From a regulatory perspective, it is much easier to monitor the quality of treated wastewater in several dozens of central wastewater treatment plants around the country than in hundreds of thousands of households. However, recently (2011), the Water Authority recognized that in some areas of the country, there is an effluent surplus today, so it is willing to consider setting up systems to privately reuse the excess greywater in these regions (The Israeli Water Authority, 2012).

The Israeli government's historical lack of interest in promoting greywater is not consistent with the feelings of the public. Like citizens of other countries, a growing segment of the Israeli public is simply to ignore the ban and to construct greywater systems without permission from the government. Environmental groups are aware of the contribution of greywater to the water resources of other countries and so have joined the movement by promoting legislation that could change the situation.

One of the organizations where this issue is a priority is *Shomera for a Better Environment*. Their primary motivation is saving water and, according to the organization (which relies on the work of Friedler (2008)), if "by 2025, 30% of the household sector will recycle its greywater, water savings are expected to equal the water consumption of a medium-sized city with about 300,000 residents." This organization began promoting experimental initiatives and policy changes in the field, in partnership with a variety of organizations under the responsibility of the local authority and with industrial organizations (Allen et al., 2010). It is only a matter of time before these movements will be reflected in legislation. The greywater policy recommendations of such environmental groups are presented in the following.

## RECOMMENDATIONS ON GREYWATER POLICY BY THE ENVIRONMENTAL GROUPS (GERMAZ ET AL., 2010)

### Regulation by Legislation of the Possibility of Greywater Reuse

Government legislation on greywater reuse will allow for a controlled water market, where a private enterprise can provide alternative water sources in the urban sector, abiding by clear recycling rules.

The law must balance the encouraging of recycled greywater use, and prevention of hazards to public health and the environment.

The law should include mechanisms for approving facilities that considers the needs of the water sector and public health.

Greywater reuse standards should be established under realistic set of operational conditions.

### Formulate a Government Program to Encourage Greywater Recycling

A government program should be initiated to encourage greywater reuse, as part of efforts to balance the national water sector on the basis of enhanced sustainability.

Such a program can save the water sector considerably by replacing some desalinated water, and avoiding some of the desalination environmental costs.

The encouragement program for greywater reuse will allow the introduction of greywater treatment and reuse systems into the Israeli water sector. Within two decades, greywater sources will be able to occupy an important place in the national water profile and help prevent damage to the water supply for agriculture.

### Establish a Mechanism for Monitoring and Control

A control mechanism for greywater reuse facilities should be promoted.

The track of prior approval of technologies and services should to be pursued, as well as the issuance of a *certification* for technologies that will allow the installation of standard systems without the need for a detailed inspection of each house.

The certification will entail a rigorous qualification process for professional installers. Those who comply with the requirements will become certified professionals permitted to install greywater treatment and reuse systems.

Following the installation of the systems, a control mechanism should be put in place for their operation and periodic quality reporting. (The reporting subjects of interest are displayed in the table below).

The regulations should be defined and implemented so that the companies operating in the market bear the responsibility for the proper operation of the systems and the prevention of hazards. This will protect public health without relying heavily on the Ministry of Health. This will follow a similar principal to that used in licensing elevators or electrical and gas installations, which transfers the responsibility from government offices to the commercial company, and to the examiner who is authorized by the Ministry.

| Subject | Details |
|---|---|
| **Regulations of greywater recycling facilities** | Define the required quality of greywater reuse systems |
| **Administrative regulations** | Set rules for installation, operation, maintenance and control of greywater reuse systems |
| **Certification of professionals** | Define requirements for installation and maintenance professionals, and establish training and certification program |

## Promoting Recycling of Greywater in Single Family Homes

Dealing with many small systems functioning without professional operation and maintenance makes it difficult for management and supervision to prevent a public health and environmental risks.

The risk posed by greywater facilities can be markedly reduced by the implementation of appropriate greywater regulations that rely on the commercial companies and certified examiner.

It is therefore recommended that a maintenance contract be required also in single-family homes, as well as periodic sampling of the reclaimed water, to reconcile the need for such systems in rural development and the possible risk they pose.

## Water Reuse in Apartment Buildings (E.g., Condominium Complex) and Commercial Construction

Water reuse in licensed businesses allows an entity to be appointed that will be responsible for the functioning and operation of the greywater reuse system. Therefore, in apartment buildings, greywater reuse could be allowed only in buildings that are served by a contracted professional maintenance company. Priority should be given to promoting greywater use in such domains, since the water saving potential in multi-family buildings is highest.

**Promoting Research and Development in the Field of Greywater**

Investment into further research on greywater is recommended, particularly that which can help foresee problems and design preemptive solutions in urban water recycling. There are several topics ripe for examination (some of which have already been explored): dedicated products and services for this domain; the effect of greywater reuse on sewage flows and solid movement in sewer systems; surveys and analysis of disease-causing pathogens and epidemiology; and full functioning over time of the treatment and reuse systems.

In 2008, Parliament member Esterina Tartman suggested a private bill on the recycling of greywater (Tartman, 2008), but since she was not reelected in the 2009 elections, her legislative initiative was not continued. Instead, a coalition of eight Parliament members headed by former journalist Nitzan Horowitz, representing a wide range of political parties and tendencies, suggested a new and expanded bill in 2009. After making many amendments to the original bill, the government agreed to support the initiative (GWI, 2010). The multiple compromises and the stamp of the Ministry of Health are manifest in the bill draft. The bill boasts a lengthy title—"Protection of the Environment Bill (rational use of natural resources – allowing greywater reuse), 2009" (Horowitz, 2009)—but in fact is relatively brief. It is just three pages long and contains only eight sections.

The introduction of the law reads:

> "The small amounts of rains recently, and the depletion of water resources in Israel requires taking measures to conserve water. One of the most efficient and immediate conservation methods available to the water system in Israel is reuse of greywater" which are produced year round through household water supply. It is proposed to allow the public, pursuant to certain limitations, to have reuse of these water, and to save and reduce drinking water consumption. The limitations in the proposed law are designed to both protect the health of the public that uses this water and the fauna and flora that are near the water system. Greywater is a water source that can be used subject to Environmental and hygiene constraints. It is relatively easy to use, technically convenient and supported by clear economic advantages. Greywater reuse is consistent with "Agenda 21" and government decision…."

Article 2 of the law offers a broad definition of *greywater*, which includes "household wastewater from showers, washing machines, washbasins, bathtubs, bathrooms, condensation water from air conditioners, and runoff gathered in gutters. It does not include water which is used in the kitchen and dishwashers, wastewater from humans or animals, industrial wastewater or wastewater polluted with hazardous substances."

The law opens by permitting residences and guesthouses to install greywater reuse systems, but poses a number of conditions (Section 3):

1. No use will be made of the greywater for reuse except for toilet flushing, watering non-edible plants, in ornamental ponds, fountains, and waterfalls.

2. Treated greywater will be used only in the residence or guesthouse premises where the greywater was generated.
3. A sign will be posted where treated greywater is being used, indicating that the water is non-potable.
4. The treated greywater will be used for irrigation only through subsurface drip irrigation.
5. Direct contact between humans and animals and reused greywater will be prevented.
6. Treated greywater shall not be reused for irrigation in protection zones of potable water wells.

In addition, the Israeli law imposes a number of specific design requirements for the irrigation system. These include that the storage tank be completely sealed and emptied every 12 h and that it should be connected (by a one-way valve) to the sewage system to allow the emptying of the storage tank to the sewage when needed. It also requires that the system include a solids filter to prevent clogging of the system components and that it be located at a distance of at least 25 m from any water source. Finally, it should not be installed in areas prone to flooding (Section 5.5).

Regarding enforcement, the proposed Israeli law stipulates that the system operation undergo annual examination by an inspector authorized by the Minister of Industry, Trade and Labor. The integrity check should include taking a composite sample of minimum 8 h at the outlet of the storage tank. This sample should be analyzed for the following pollutants, the concentrations of which should be below a maximum concentration: fecal coliforms, TSS, coliforms, residual chlorine, and turbidity (Section 5.6). The law finishes by granting the health minister the power to set additional requirements to ensure that the greywater is differentiated from the potable water and sewage systems. As a whole, this is probably the most complicated series of requirements of all greywater laws currently in effect worldwide.

The complexity did not curb the enthusiasm of the civil society in Israel. According to an evaluation by the *Friends of the Earth Middle East*, households will be able to reuse 40% of their water (Mazuri, 2011). Other high assessments quoted in the press referred to a saving of 164 MCM/year following a full implementation of the law, although it is not clear what assumptions formed the basis of this calculation. Other organizations presented a more modest scenario, expecting a possible additional 50 MCM/year (Israel Union for Environmental Defense, 2009).

However, it is unclear whether the public will have the required patience and financial resources to meet the high demands made by the new comprehensive legislation. Even if the law passes, the message received by the public will be far from that of independence and savings as envisioned by the parliamentary coalition that initiated the original law. Instead, the public will face an expensive, complicated, and potentially disheartening option. Even strong compliance efforts may end up inadequate, leaving the citizens exposed to penalties for environmental violations, despite the good intentions that motivated them to pay for a greywater system.

The Standards Institution of Israel (SII) began preparing construction standards for installing greywater treatment and reuse systems. The standards will be part of engineering requirements that construction companies and professional gardeners who install greywater systems will need to meet. The focus on installation and on the professional community may be a much more promising strategy for the proposed law, compared to placing the responsibility of complying squarely on the shoulders of the general public. In the following is a description of the work of the committee preparing this Israeli standard for greywater at the Standards Institute, written by Eng. Rami Halperin, the chairman of the experts committee appointed by the SII (Standard 6147, to date still not confirmed):

## EXPERTS COMMITTEE FOR DRAFTING AN ISRAELI STANDARD ON GREYWATER SYSTEMS (RAMI HALPERIN EXPERT COMMITTEE)

In September 2010, the Technical Committee 923 (for installation instructions on sanitation, water and sewage) appointed two expert committees to jointly draft an Israeli standard on greywater systems: Expert Committee 92311 (for greywater—design and manufacturing) and Expert Committee 92312 (for greywater—installation and maintenance).The members of the two committees concluded that it would be detrimental to separate the parts of the standard for which each was individually responsible, so it was agreed that the two would work together in preparing the standard. Accordingly, all members of both committees were invited to joint meetings.

### The Format of the Standard Preparation

The Committee's first hearing focused on the methodology for preparing the standard. One option was to adopt an existing standard of another country, and make the necessary amendments to adapt it to the local conditions. This would allow the preparation of the standard to be accelerated. The second option was to prepare a new Israeli standard, which would necessarily extend the preparation time. The committee decided on a compromise: to prepare a new Israeli standard based on the framework of an existing standard. The committee examined the standards of the United Kingdom, Canada, Arizona and Australia, and decided to draft the standard based on the British standard's framework.

### The Principles of the Committee's Work

The Committee agreed on two principles as the basis for its work, according to which the proposal for the standard would be prepared. The first principle was to ensure public health regarding collection, treatment, and reuse of greywater. The need for this principle was obvious, and it was designed to prevent risks of morbidity resulting from the operation of greywater systems. The second principle was to prepare a standard with "fixed" requirements for structure and system performance, which would not depend on "ad hoc" decisions of any supervisor or functionary. This principle is critical for designers and manufacturers of greywater systems because the designers can know the system requirements in advance. This allows them to assure that it can perform all tasks imposed upon it without "danger" of additional or different requirements being imposed after the system is operational.

The existence of "fixed" criteria for the design and construction of greywater systems makes possible the construction of a series of systems (primarily treatment facilities). It also allows for an entire series of "off the shelf" systems to be approved in bulk, instead of manufacturing and approving systems one at a time.

### The Basis for the Committee's Work

In recent years, the use of greywater has increased all over the world, and different countries have set standards and guidelines to regulate the treatment and reuse of greywater. As studies on the subject accumulate, the Committee members felt that they should not itself explore the subject, but rely on the scientific/professional knowledge already existing in Israel and around the world. They looked to comparable regulations or standards set forth in developed countries. The scientific knowledge existing in Israel and abroad was introduced in the Committee meetings by representatives from research institutions in Israel; the Technion University and Ben-Gurion University of the Negev. Epidemiological knowledge was also presented by the managers of the Central Laboratory of the Ministry of Health in Jerusalem. Standards and guidelines from around the world studied by the Committee members included the British, Canadian, Arizona, and Australian Standards, and guidelines of the World Health Organization and the American National Sanitation Foundation (NSF) standard.

### Areas of the Standard Reference

The standard being prepared on greywater systems will address the following areas:

Sources of greywater suitable for reuse;
Possible uses of treated greywater;
The expected quantities of greywater and its expected consumption in the different uses;
Greywater collection systems;
Treatment and storage facilities of greywater;
The required quality of treated water with respect to its use/exploitation;
Preparation of a risk management plan as part of the system design;
Inspection and approval of the greywater treatment facilities; and
Ongoing control of the quality of the treated water.

**Significant Disagreements with the Ministry of Health**

During the Committee's deliberations, significant disagreements emerged between some of the representatives of the Ministry of Health and the rest of the Committee. The disagreements revolved around which existing requirements with which the standard must comply, and on the professional requirements (including the required quality of the treated greywater and control requirements) to be included in the standard.

One of the underlying factors in the disagreement was that the Ministry of Health representatives felt that their office was primarily responsible for public health, and thus the standard should comply with the existing requirements of the Ministry of Health relating to greywater systems. The majority of the Committee members disagreed, claiming that the standard should be based on scientific/professional data, taking the positions of the Ministry of Health into account but not necessarily adhering to them.

An important implication of these disagreements lies in how the standard would be implemented. The Ministry of Health strongly opposed the application of the standard on greywater systems in single-family homes, while the professional information and the reference of the existing standards in the world maintain that the risk in single family homes is smaller than that in residential buildings.

The difference of opinions on this issue was brought before the Central Committee on Water at the Standards Institute, which decided that the standard should consider greywater systems in all types of buildings, including single family homes.

## 5.6 CONCLUSIONS

There appear to be two conflicting motivations behind greywater legislation. On the one hand, many advocates of increased water conservation and efficiency see this as an unusual opportunity to engage the public in a highly effective program to save freshwater resources. Yet, there are many who are alarmed at what they deem to be a growing risk to public health and hydrological contamination. Like all environmental legislations that seek to protect the commons from individual actions, they see greywater laws as the best way to protect common resources and prevent unnecessary disease. These two perspectives are not diametrically opposed and could conceivably complement each other. But the record to date suggests that greywater laws have not yet found the optimal interaction between them.

The Israeli case is a strong example of the significant compromises made by supporters of greywater reuse when addressing the objections of an opposing government office. The resulting (proposed) legislation sets so many limitations and creates bureaucratic obstacles that it is doubtful whether the general public will be willing to pay for a greywater system that brings with it so much inconvenience (i.e., requires a valid certificate, requires a chemical and biological treatment facility in the yard of each home). A much more likely scenario is that after passing the law, it will become another example of a *symbolic legislation* in which the regulators express their intentions and aspirations, and the general public continues for the most part with business as usual (for more on symbolic legislation, see Dwyer (1990)). It is doubtful whether such a model is desirable for an effective environmental legislation.

One theoretical way to reconcile these two motivations is through a liberal paradigm of risk management. Here, risks that are considered negligible will be allowed in light of the environmental gain of saving water, similarly to the approach taken in Arizona. The general public interested in installing a modest system for the lawn and garden will enjoy great latitude in operating greywater systems. This is much easier—and potentially more fruitful—than using conservative legal approaches to public health risks. A family with common sense will not be eager to make an extensive investment in the monitoring and maintenance of a greywater system based on concerns about possible public health risks and the quality of groundwater. It is clear to them that the exposure to fecal coliforms by children who play on the lawns used by their pets is much greater than the risks inherent in water coming from the bath or even from the kitchen sink.

It should be remembered that legislation has to be an important part of a public campaign for safer use of greywater. The campaign should start by sending an educational message to the public, along with an uncompromising regulatory message to contractors and to the companies that install these systems. The integration of safe greywater infrastructure in construction standards, as local authorities began implementing in Japan and Spain, is also a desirable direction. If greywater reuse is indeed an economically efficient step, imitating incentive programs for existing structures like those in Cyprus and Australia should be considered. It would be similar to the subsidies granted for energy-saving renovation, a measure that is becoming more and more common (Bar-Ilan et al., 2010).

Ultimately, for countries that do not have a substantial infrastructure for recycling sewage, greywater is a significant resource that can be developed. The ecological damage and potential pollution of alternative sources, such as desalination or sewage recycling, are generally much higher. It is important for legislators to remember that greywater is not the problem but part of the solution. Finally, greywater laws should not ignore the limitations of legislative action. For example, without public support, it would be completely impractical to try to influence the behavior of millions of homeowners. Greywater legislation will reach peak efficiency when it is used only as a normative element within a broader strategy of water management and conservation.

## 5.A APPENDIX: SUMMARY TABLES OF LEGISLATION IN AUSTRALIA, BRITAIN, AND THREE U.S. STATES

### Australia

*Mode of regulation*:

*WM (WaterMark) is the certification of a product approved according to the plumbing code of Australia (PCA).

**Australia**

*Mode of regulation:*

*Greywater definition*: Greywater shall never include toilet wastewater.

*Use definitions*:

*Technical definitions*:

| | **New South Wales** | **Western Australia** | **South Australia** | **Queensland** | **Victoria** | **Tasmania** | **Australian Capital Territory** | **Northern Territory** |
|---|---|---|---|---|---|---|---|---|
| Agency responsible for systems' certification | Department of Health | Department of Health | Department of Health | Planning and infrastructure dept.; systems above 50,000 L/day, Water and Nature Dpt. | EPA | Department of Justice | No agency in charge of greywater | Department of Health |
| Agency responsible for installation approval | Local authority | Local authority | Relevant authority | Council approval | Local council and water authority | Local council | | Inside building control area, Department of Planning Infrastructure; outside, Department of Health and Community Services |
| *First tier* | | | Untreated (watering with bucket or hose) | | | | Diversion of untreated water | |
| Consumer type | Single household | Single household | | | | Single household | | Single household |

(*Continued*)

| | New South Wales | Western Australia | South Australia | Queensland | Victoria | Tasmania | Australian Capital Territory | Northern Territory |
|---|---|---|---|---|---|---|---|---|
| System approval required | | | | | | | WM | ✓ |
| Installation approval required | | | | | | | | ✓ |
| Quantity limitation | Small quantity | Small quantity | Small quantity | Small quantity | | Small quantity | | |
| *Second tier* | | | Filtration and diversion facility | | | | | |
| Consumer type | Single household | Single household | Single household | | | Single household | | |
| System approval required | WM | WM+ Health Dpt. | ✓ | WM | ✓ | Domestic recycling of all types of wastewater forbidden in places connected to the central sewage system | | |
| Installation approval required | | ✓ | ✓ | ✓ | ✓ | | | |
| Quantity limitation | | | | Up to 3000 L/day | | | | |
| *Third tier* | | | Treatment facility | | | | Treatment facility | |
| Consumer type | Single household | Single household/ apartment building/ business building | Single household | Every consumer up to 50,000 L/day | Single household | | | Single household |
| System approval required | ✓ | ✓ | ✓ | ✓ | ✓ | | | ✓ |
| Installation approval required | ✓ | ✓ | ✓ | ✓ | ✓ | | | ✓ |
| Quantity limitation | Minimal capacity, 8 people; maximal, 10 people | | | Up to 50,000 L/day | | | | |

| Greywater Is Defined as Household Wastewater That Does Not Include | NSW | Western Australia | South Australia | Queensland | Victoria | Tasmania | Australian Capital Territory | Northern Territory |
|---|---|---|---|---|---|---|---|---|
| *First tier* | Without treatment (watering with bucket and/or hose) | | | | | | Diversion of untreated water | |
| Wastewater from the kitchen | Forbidden | Forbidden | Not recommended | Not mentioned | Not mentioned | Forbidden | Not recommended | Forbidden |
| Wastewater containing feces or urine (diaper laundry) | Forbidden | Forbidden | Forbidden | Not mentioned | Not mentioned | Forbidden | Not recommended | Not mentioned |
| Wastewater that contains dangerous substances, toxic materials, oil, or grease | Forbidden | Forbidden | Forbidden | Not mentioned | Not mentioned | Forbidden | Not mentioned | Not mentioned |
| *Second tier* | Filtration and diversion facility | | | | | | | |
| Wastewater from the kitchen | Forbidden | Forbidden | Not recommended | Not mentioned | Forbidden | The use of greywater in regions connected to central sewage is forbidden. In other areas, the regulations refer to all wastewater. | | |
| Wastewater containing feces or urine (diaper laundry) | Forbidden | Forbidden | Not mentioned | Not mentioned | Forbidden | | | |
| Wastewater that contains dangerous substances, toxic materials, oil, or grease | Forbidden | Forbidden | Not mentioned | Not mentioned | Not mentioned | | | |

(*Continued*)

| Greywater Is Defined as Household Wastewater That Does Not Include | NSW | Western Australia | South Australia | Queensland | Victoria | Tasmania | Australian Capital Territory | Northern Territory |
|---|---|---|---|---|---|---|---|---|
| *Third tier* | | | Treatment facility | | | | Treatment facility | |
| Wastewater from the kitchen | Not mentioned | Forbidden | Not mentioned | Not mentioned | There are no special regulations for greywater. The regulations are for a treatment facility of household wastewater. | | Not recommended | Not mentioned |
| Wastewater containing feces or urine (diaper laundry) | Not mentioned | Not mentioned | Not mentioned | Not mentioned | | | Not recommended | Not mentioned |
| Wastewater that contains dangerous substances, toxic materials, oil, or grease | Forbidden | Forbidden | Not mentioned | Not mentioned | | | Not mentioned | Not mentioned |

| | NSW | Western Australia | South Australia | Queensland | Victoria | Tasmania | Australian Capital Territory | Northern Territory |
|---|---|---|---|---|---|---|---|---|
| *First tier* | Without treatment (watering with bucket and/or hose) | | | | | | Diversion of untreated water | |
| Allowed type of use | Direct watering of private garden without storage. | Direct watering of private garden, toilet flushing, laundry, without storage. | Direct watering of private garden without storage. | Direct watering of private garden without storage. | | Direct watering of private garden without storage; not recommended. | Private garden watering; water is not stored for more than 24 h and is used directly for watering. | Direct watering of private garden without storage. |
| *Second tier* | Filtration and diversion facility | | | | | | | |
| Allowed type of use | Subsurface irrigation watering of private garden without storage | Subsurface watering (depth of at least 10 cm) of private garden without storage | Subsurface watering of private garden without storage | Subsurface watering of private garden without storage | Subsurface watering of private garden without storage | | | |
| *Third tier* | Treatment facility | | | | | | Treatment facility | |
| | | From a list of approved facilities | Approved facility that treats water according to the required level in the *national Guidelines for Water Recycling* document | (A)<br>20 mg/L BOD<br>30 mg/L SS<br>30 cfu thermotolerant coliforms/100 mL<br>(B)<br>10 mg/L BOD<br>10 mg/L SS<br>10 cfu thermotolerant coliforms/100 mL | | | (A)<br>20 mg/L BOD<br>30 mg/L SS<br>(B)<br>20 mg/L BOD<br>30 mg/L SS<br>10 cfu thermotolerant coliforms/100 mL | |

*(Continued)*

| | NSW | Western Australia | South Australia | Queensland | Victoria | Tasmania | Australian Capital Territory | Northern Territory |
|---|---|---|---|---|---|---|---|---|
| Permitted uses | Toilet flushing, washing machines, watering private gardens | (A) Without disinfection—subsurface watering (B) After disinfection—surface watering, toilet flushing, laundry with cold water | According to the level of treatment—subsurface watering, unlimited watering, toilet flushing, laundry and car washing | (A) Surface watering (B) Toilet flushing, laundry with cold water, car washing, walls and pavement washing, and sprinkler watering | Watering | | (A) Watering private gardens at a depth of 100–300 mm (B) Toilet flushing, washing machine, washing cars and watering private gardens; surface watering + sprinkling | Watering, toilet flushing, or laundry |
| Waterlogging outside the plot prohibited | ✓ | ✓ | ✓ | ✓ | ✓ | | ✓ | |
| Waterlogging to surface water sources prohibited | ✓ | ✓ | ✓ | ✓ | ✓ | | ✓ | |
| Ponding prohibited | ✓ | ✓ | ✓ | ✓ | ✓ | | ✓ | ✓ |
| Avoiding contact with humans | ✓ | ✓ | | ✓ | ✓ | | ✓ | ✓ |
| Preventing public nuisance | ✓ | ✓ | ✓ | ✓ | ✓ | | ✓ | |

(*Continued*)

| | NSW | Western Australia | South Australia | Queensland | Victoria | Tasmania | Australian Capital Territory | Northern Territory |
|---|---|---|---|---|---|---|---|---|
| Watering with sprinklers prohibited/subsurface watering mandatory | According to level of treatment | According to level of treatment | According to level of treatment | According to level of treatment | | | According to level of treatment | |
| Watering vegetables prohibited | ✓ | ✓ | ✓ | ✓ | ✓ | ✓ | ✓ | ✓ |
| Hazardous materials prohibited | ✓ | ✓ | ✓ | ✓ | | ✓ | | |
| Watering during rainfall prohibited | ✓ | ✓ | | | ✓ | | ✓ | |
| Overflow to sewage | ✓ | ✓ | ✓ | ✓ | ✓ | | ✓ | ✓ |
| Minimal tank size | ✓ | In systems that treat all greywater, a settling tank of at least 1820 L in volume is mandatory, or a form of aerobic treatment. | The treatment facility should hold a minimum of 600 L/day. | In the treatment facility, a tank size sufficient for retention time of at least 36 h should be installed, according to average daily supply. | ✓ Depends on size of the plot and location. | | | |

*(Continued)*

| | NSW | Western Australia | South Australia | Queensland | Victoria | Tasmania | Australian Capital Territory | Northern Territory |
|---|---|---|---|---|---|---|---|---|
| Marking water as nondrinkable | ✓ | ✓ | | | ✓ | | ✓ | |
| Filtration | | | ✓ | | | | | |
| Limiting the water pressure in the irrigation system | | | | | | | | |
| Location away from the drainage and flooding route | | | ✓ | ✓ | ✓ | | | |
| Setting a minimal distance for treatment facility from other structures and facilities | ✓ | ✓ | ✓ | ✓ | ✓ | | ✓ | |
| Distance from groundwater | | ✓ | ✓ | ✓ | ✓ | | Forbidden above drinking water aquifer | |

**Britain**

*Mode of regulation*:

*Greywater definition*: Greywater shall never include toilet wastewater.

*Use definition*:

*Technical definitions*:

| | |
|---|---|
| Responsible agency | The main responsible agency is the environment agency. |
| *Treatment Level A* | Direct use of water with bucket or hose. |
| Consumer type | |
| System approval required | |
| Installation approval required | |
| Amount limitation | Small amount. |
| *Treatment Level B* | Filtration. |
| Consumer type | |
| System approval required | The materials used (pipes, connectors, etc.) should be approved by the WRAS. |
| Installation approval required | The local authority should be informed (water authority, environmental health, and public health). |
| Amount limitation | |
| *Treatment Level C* | |
| Consumer type | |
| System approval required | The materials used (pipes, connectors, etc.) should be approved by the WRAS. |
| Installation approval required | The local authority should be informed (water authority, environmental health, and public health). |
| Amount limitation | |

| Greywater is defined as wastewater that does not include | |
|---|---|
| *Treatment Level A* | Direct use |
| Wastewater from the kitchen | Not recommended |
| Wastewater containing feces or urine | |
| Wastewater that contains dangerous substances, toxic materials, oil, or grease | |
| *Treatment Level B* | Use not by sprinklers |
| Wastewater from the kitchen | Not recommended |
| Wastewater containing feces or urine | |
| Wastewater that contains dangerous substances, toxic materials, oil, or grease | |
| *Treatment Level C* | Use with sprinklers |
| Wastewater from the kitchen | Not recommended |
| Wastewater containing feces or urine | ✓ |
| Wastewater that contains dangerous substances, toxic materials, oil, or grease | |

| | |
|---|---|
| *Treatment Level A* | Direct use with bucket or hose |
| Type of use permitted | Direct watering without storage |
| *Treatment Level B* | Use without sprinkling<br>100 enterococci/100 mL, 250 *E. coli*/100 mL, 1000 total coliforms/100 mL residual chlorine lower than 0.5 mg/L (watering© and 2 mg/L toilet flushing) |
| Type of use permitted | Garden watering and toilet flushing |
| *Treatment Level C* | Use with sprinklers<br>10 total coliforms/100 mL, 10 legionella 100/mL, 0 enterococci/100 mL, 0 *E. coli*/100 mL<br>Residual chlorine lower than 2 mg/L |
| Type of use permitted | Watering by sprinklers, pressure rinsing, car washing, and laundry |
| Waterlogging outside plot prohibited | Pollution prohibition |
| Waterlogging into surface water sources prohibited | Pollution prohibition |
| Ponding prohibited | |
| Avoiding contact with humans | |
| Preventing public nuisance | ✓ |
| Watering with sprinklers prohibited/subsurface watering mandatory | ✓ Depending on the water quality |
| Watering vegetables prohibited | Not recommended |
| Hazardous materials prohibited | |
| Watering during rainfall prohibited | |
| Overflow to sewage | ✓ |
| Minimal tank size | |
| Tank lid | ✓ |
| Installing a device to prevent backflow of treated water into the drinking water system, which is mandatory | ✓ |
| Marking water as nondrinkable | ✓ |
| Filtration | |
| Limiting the water pressure in the irrigation system | |
| Location away from the drainage and flooding route | |
| Setting a minimal distance for treatment facility from other structures and facilities | |
| Distance from groundwater | |

**USA**

*Mode of regulation*:

*Greywater definition*: Greywater shall never include toilet wastewater.

*Use definition*:

*Technical definitions*:

| | Arizona | California | Washington | Wyoming |
|---|---|---|---|---|
| Responsible agency | Department of Environmental Quality | Department of Housing and Community | Department of Health | Department of Environmental Quality |
| *Treatment Level A* | Up to 1514 L/day | | Untreated | Untreated |
| Consumer type | Single household | One or two households | Single household | |
| System approval required | | | | |
| Installation approval required | | | | |
| Amount limitation | Up to 1541 L/day | Washing machine or single connection | Up to 227 L/day | 7570 L/day |
| *Treatment Level B* | | | Untreated | |
| Consumer type | More than a single household | Single or two households | Residence/nonresidential building | |
| System approval required | ✓ | ✓ | ✓ | |
| Installation approval required | ✓ | ✓ | ✓ | |
| Amount limitation | Between 1,514 and 11,357 L/day | Up to 947 L/day | 1300 L/day | |
| *Treatment Level C* | | | Treatment system | |
| Consumer type | | | (A) More than 947 L/day (B) Tertiary level treatment system | Residence buildings/offices/commerce/environmentally sensitive areas |
| System approval required | ✓ | ✓ | ✓ | |
| Installation approval required | ✓ | ✓ | ✓ | |
| Amount limitation | Above 11,357 L/day | | 1320 L/day | |

| Greywater Is Defined as Wastewater that Does not Include | Arizona | California | Washington | Wyoming |
|---|---|---|---|---|
| *Treatment Level A* | | | Untreated | Untreated |
| Wastewater from the kitchen | ✓ | ✓ | ✓ | |
| Wastewater containing feces or urine | | ✓ | ✓ | ✓ |
| Wastewater that contains dangerous substances, toxic materials, oil, or grease | ✓ | ✓ | ✓ | ✓ |
| *Treatment Level B* | According to the regulations of traditional wastewater treatment | | Untreated | |
| Wastewater from the kitchen | | | ✓ | |
| Wastewater containing feces or urine | | | ✓ | |
| Wastewater which contains dangerous substances, toxic materials, oil, or grease | | | ✓ | |
| *Treatment Level C* | According to the regulations of traditional wastewater treatment | | Treatment system | |
| Wastewater from the kitchen | | | | |
| Wastewater containing feces or urine | | | ✓ | |
| Wastewater that contains dangerous substances, toxic materials, oil, or grease | | | ✓ | |

| | Arizona | California | Washington | Wyoming |
|---|---|---|---|---|
| *Treatment Level A* | | | Untreated | |
| Type of use permitted | Watering by flooding or dripping | Subsurface irrigation | Subsurface irrigation in a single household | Irrigation |
| *Treatment Level B* | | More than washing machine or single connection | Untreated | |
| Type of use permitted | Depends on the level of treatment | Subsurface irrigation | Subsurface irrigation in residence buildings/ offices/commercial buildings | |
| *Treatment Level C* | | Tertiary treatment system | | |
| Type of use permitted | | Use in and out of the house | Subsurface irrigation in residence buildings/ offices/commercial buildings/ environmentally sensitive areas | |
| Waterlogging out of the plot prohibited | ✓ | ✓ | ✓ | ✓ (without a written approval from the neighbors) |
| Waterlogging into surface water sources prohibited | ✓ | ✓ | ✓ | ✓ |
| Ponding prohibited | ✓ | ✓ | ✓ | Minimized |
| Avoiding contact with humans | ✓ | ✓ | ✓ | ✓ |
| Preventing public nuisance | | | Preventing odor nuisance | |
| Watering with sprinklers prohibited/ subsurface watering mandatory | Spraying forbidden | ✓ | ✓ | |
| Watering vegetables prohibited | ✓ | ✓ | ✓ (It is forbidden to water edible plant parts.) | ✓✓ |
| Hazardous materials prohibited | ✓ | ✓ | ✓ | ✓ |
| Watering during rainfall prohibited | | | | |

| | Arizona | California | Washington | Wyoming |
|---|---|---|---|---|
| Overflow to sewage | ✓ | ✓ | ✓ | ✓ or to secondary treatment system |
| Minimal tank size | | | | |
| Tank lid | ✓ | | | |
| Marking water as nondrinkable | ✓ | ✓ | ✓ | |
| Limiting the water pressure in the irrigation system | | | ✓ (in systems to be used in single households) | |
| Location away from the drainage and flooding route | ✓ | | | ✓ |
| Setting a minimal distance for a purification facility from other structures and facilities | | ✓ | ✓ | |
| Distance from groundwater | ✓ | ✓ | ✓ | Should not affect groundwater |

# 6 Perceptions and Attitudes toward Greywater Recycling
## *A Review*

*Adi Inbar**

## 6.1 THE IMPORTANCE OF PUBLIC'S ATTITUDES AND PERCEPTIONS REGARDING GREYWATER RECYCLING

On July 9, 2006, the residents of Toowoomba in Queensland, southeastern Australia, were asked to decide whether they would allow or oppose the flow of treated sewage into one of the city's open drinking reservoirs. Federal and local government agencies put forward this recycling plan in response to extended droughts, a steady increase in population size, and the demand for a potable water supply. Public consent was deemed necessary because, although the technology was completely safe, it was not in the government's purview to force citizens to use it.

The vote was the culmination of an extensive public debate over the proposed plan, and it won nationwide attention. Both parties used scientific evidence to prove the good or, alternatively, the poor quality of the treated water and waged an extensive public battle. The final count showed that 62% of the voters rejected the plan. This, in turn, led to an alternative program to address the needs of urban water, which included the construction of a pipeline for transporting water at much higher costs. A major argument of the opponents who waged their campaign under the slogan "Think before you agree to drink" was, in the absence of a more professional term, "the yuck factor." This refers to the negative emotional response that can be triggered by the idea of drinking sewage, even when it is treated at a high level. Subsequent studies suggest that the voting pattern was also (and perhaps mainly) determined by the lack of direct public involvement in the decision-making processes, making it more susceptible to external influence by interested parties seeking political gain (Hurlimann and Dolnicar, 2010).

Predominantly, this case indicates the importance of perceptions and attitudes among the public regarding the success of water recycling projects, including greywater recycling. Evidence from around the world suggests that a negative public perception is a major barrier to the implementation and expansion of various water

* Social Studies Unit, Blaustein Institutes for Desert Research, Ben Gurion University of the Negev.

recycling technologies and the usage of the treated water they produce, even in cases where scientific, engineering, and economic authorities offer full support. Accordingly, this chapter focuses on the following topics: the importance of the public's perceptions and attitudes with respect to their actions regarding greywater recycling; mapping of the key factors involved in the shaping of these perceptions and attitudes, based on global experience and research on the subject in recent decades; an extended review of the research evidence in Israel; and finally, a possible outline for future research in the Israeli arena.

The term *perceptions* refers to the meaning individuals give their experiences. It also represents the cultural and social world of those individuals, which is usually deeply related to how they understand their national, familial, and technological context (among other things). Perceptions tend to be abstract and unconscious in nature, an example being a general statement such as "water conservation is important," which many might agree with. This is different from *attitude*, which is usually described in the literature as a limited and conscious translation of the general perception of a given issue and is more strongly linked to the individual's tendency toward a particular behavior. For example, the sentence "I support the recycling of shower water for watering the garden" is a clear and detailed attitude, and research indicates that this attitude will predict more strongly the individual's willingness to act accordingly and install a facility for recycling shower water in his house than the general aforementioned perception.

The important point for our purposes is that *perception*, *attitude*, and *behavior* mean different things, and one cannot automatically deduce the existence of one from the other. Thus, many may support the general idea of saving water, but only some of them will feel comfortable with greywater recycling in a private household or consciously adopt a determinate attitude toward it. An even smaller group will eventually install greywater recycling devices in their homes. This disparity is due to the differences in the levels of concretization; as something becomes more concrete (from perception to attitude to behavior), the number of factors relevant to one's decision about it increases, leading to a higher probability of reaching a different decision. Moreover, a perception, attitude, or behavior may change if the available data—new information, technical changes, changes in cost, sharing in decision making, etc.—change. It is a dynamic system, and as such we cannot argue in advance what the widespread public perceptions will be regarding any water recycling project.

In addition, one should remember that the word *public* embodies different social groups. It is possible that in the same society, in the same *public*, different attitudes toward the question of recycling greywater will exist. This can result from variability in factors such as geographic location, age, number of family members sharing the household, previous experiences with the technology in question, or economic status, as we will explore in further detail later in this chapter. The important point here is the understanding that when we seek to study perceptions and attitudes of the public on the subject, and from there attempt to predict their behavior, we should try to understand the complex and stratified contexts in which the factors leading to a perception or an attitude are shaped. This will vary by societal group, so it is important to acknowledge that *technology*, *knowledge*, and *society* shape each other reciprocally and symmetrically.

The vast majority of existing research on public perceptions in the field of water recycling deals with the reclamation of treated sewage for various purposes. Only in recent years has public and research attention turned to cases of greywater recycling. This recent change is due to the expansion of greywater reuse in light of the increasing demand for water as well as the damage done to available water resources, a process described extensively in this book.

The great importance of public perceptions and attitudes in the case of greywater recycling is due mainly to the nature of the technology itself. In most cases, a small-scale facility is involved, which serves a relatively small number of households. This size creates a central dependency between the quality of the treatment facility operation and the activity patterns of individual people. The activity patterns are relevant in two directions: for greywater production (the number of showers a day, the type of detergents used and their concentration, whether a dishwasher is used or dishes are washed by hand, etc.), and for consumption of the treated product (number of toilets, garden size, number of vehicles, and the like). Additionally, in some cases, the operation of the recycling facility itself, or basic control of its proper functioning, is under the responsibility of the very residents using and contributing greywater, so the question of their training and their understanding of the operation arises as well. In contrast, the operation of a central water treatment system that copes with large volumes of water is controlled exclusively by professional personnel who are supervised. Thus, in domestic greywater treatment systems, the activities of individuals—whose behavior varies according to their perceptions and understanding of the importance of the subject—are highly important.

The transition from national and regional water treatment centers to localized greywater recycling technologies has significant political implications. The decentralization of water recycling facilities is in sharp contrast to the common practice in many of the developed countries in the modern era, where water treatment is centrally managed and supervised by government agencies only. Today, central regulation is often justified through the discourse of national water needs as well as public health. This means that the introduction of decentralized greywater recycling has a direct impact on the level of public involvement in the national agenda and decisions regarding these issues. As such, the public's perceptions, attitudes, and behaviors are crucial when it comes to the matter of localized greywater recycling, not only in terms of the uptake and distribution of the technology but also—and perhaps even more importantly—in terms of the national implications of such actions.

Against this backdrop, we turn now to a review of research evidence from several countries around the globe.

## 6.2 KEY FACTORS SHAPING ATTITUDES AND PERCEPTIONS REGARDING GREYWATER RECYCLING: LITERATURE REVIEW FROM AROUND THE WORLD

This section maps the key factors that shape perceptions and attitudes regarding the topic of greywater recycling. As noted, existing literature on this subject is relatively meager as compared to that on the recycling of sewage water. As such, it includes mainly exploratory studies conducted in populations with little or no experience with greywater recycling. Still, these initial studies allow us to frame current knowledge on the topic and offer the main points of focus when seeking to learn about the subject.

Aside from the preliminary nature of research in this field, and perhaps due to it, current studies of public perceptions and attitudes regarding greywater recycling are characterized by high internal diversity. First, as may have been made obvious throughout this book, the term *greywater* has a broad definition based on the principle of exclusion, whereby greywater is the total domestic wastewater that is *not* the product of toilet wastewater. In practice, greywater facilities may recycle various types of wastewater, usually pending on the decision of the homeowner which to recycle: from washing machine, showers, kitchen wastewater, and so on. The variety of practices represents not only different technical capabilities and limitations and differences in knowledge but possibly different attitudes and perceptions about which water is appropriate to or worthy of being recycled and which is not. This may result in the willingness to engage in one type of recycling but not another.

Second, different households in different national and geographic contexts use greywater for different purposes. Some use treated greywater only for underground drip irrigation, while others use it to wash their cars or flush toilets. This variance in experience with recycled products could reflect different systematic constraints, as well as different perceptions regarding the quality of possible recycling product.

Third, currently, there is no one single dominant technology for greywater recycling, and each of the studies presented in Sections 6.2.1 through 6.2.4 is based on data for different technologies. In some cases, a garden hose or a bucket is used for carrying water directly from the washing machine to the garden without any treatment, while in other cases, a treatment and recycling system that is based on physical or biological treatment or a combination of both is used. This means that people views are likely to be affected by their different experience of greywater reuse, which poses a difficulty in characterizing *greywater recycling* as one single cohesive phenomenon.

Fourth, and directly related to the point made earlier, technologies for greywater recycling are in a process of constant change, aimed at reducing costs, simplifying treatment, and improving product quality. This suggests that the data presented in the following studies is wanting, and does not represent the current full spectrum of technologies available for users or the current performance potential of the technology.

Finally, the timing of the study should be noted, as well as the population to which it refers. A research study focusing on a particular future project may yield different results than a study conducted in a population that already has experience with the technology, which in turn may yield additional different results from a study examining only general attitudes.

These five examples of possible internal differences—in raw materials that could be recycled, patterns of use of the recycled water, type of recycling technology, performance of a given technology over time, timing of research, and the study population—make it difficult to compare the findings of the different studies presented. However, at the same time, these studies were chosen so that the reader may see diverse research models as well as diverse data, representative of the current status of a given domain, which is in constant motion and variance. As such, the different case studies serve as important sources of information to be considered in total, while paying attention to the differences between them.

In this light, a number of studies conducted in different countries—United Kingdom, Spain, Oman, and Australia—will be presented in Sections 6.2.1 through 6.2.4, exemplifying different national and cultural contexts, different research approaches (from exploratory research into populations with no experience with greywater recycling to a community that has been recycling for almost 3 years), etc.

### 6.2.1 Britain

Jeffrey and Jefferson (2003) report on an exploratory study conducted in the general British population, examining public attitudes toward using water recycling technologies for toilet flushing. In particular, the research sought novel data about socioeconomic aspects that shape perceptions among the British public on this issue, what water issues matter to the public, and their relative importance to that public compared to other public issues.

Another topic in question for this research was the language that the British public uses to discuss water issues in general and water recycling in particular. This last point may seem insignificant, but it is a critical dilemma when it comes to public opinion polls, where the wording of the questions may affect the nature of the answers. For example, consider the different associations regarding water quality that each of the following terms might carry with it and how the choice of one term over the other may impact the response an individual or a community might have to a greywater recycling technology or project: *recycled water, recovered water, purified water,* or *reclaimed wastewater.*

Due to the initial nature of this study, it was decided to sample the entire population of the United Kingdom, without focusing on a particular segment of the population or a given geographic location. The results are based on a representative sample of 300 respondents over the age of 18 who live in a house connected to the national water grid and who carry some responsibility for paying the household water bill. Here are some of the findings:

The initial finding, and perhaps the most important one, is that the vast majority of respondents supported the idea of using recycled water. Of the respondents, 88% said that they were willing to recycle the shower and bathroom water for flushing the toilet in their home, as long as it did not involve a health risk. This rate of agreement decreased when external water sources were offered as alternative to the recycling of only in-house greywater: greywater from a service area near the motorway (60%), the house next door (57%), restaurant (57%), hospital (50%), and the entire neighborhood (49%).

Secondly, the percentage of those agreeing to use recycled water, especially from external sources, was higher among residents in homes with a water meter installed (i.e., where water consumers were aware of the rate, amount, and cost of consumption of domestic water). Water meters are installed only in about 40% of the homes in the United Kingdom (OFWAT, 2010). Third, among the total respondents, the support rate for using recycled water was higher among those who stated that they have already been taking various actions to save water at home than among respondents who declared that they do not take such actions. In addition, the more aesthetic the properties of the treated water were in terms of color, turbidity, and sediments, regardless of its actual health risk, the higher the respondent's willingness to come into physical contact with it.

It is also interesting that no differences in the respondents' answers were found by gender or by age and that the issue of climate change was not perceived by the respondents as affecting the amount of rainfall in the United Kingdom and therefore was not seen as a relevant factor in their water conservation perceptions, attitudes, or practices.

The study also revealed another interesting finding. Respondents were asked to rate the organizations they trust the most for setting the standards of water quality and monitoring it. The organizations that won the highest degree of trust in this survey were the British Medical Association, Friends of the Earth, and the Environment Agency. However, when the statistical relationship was tested between the degree of confidence in the ability of these organizations to monitor the issue of recycling and the degree of support for the use of recycled water, it became clear that it is impossible to predict support based on trust. Thus, while the public is willing to put its trust in certain organizations for maintaining public health, it seems that where personal behavior is concerned (i.e., the decision of whether to be exposed to recycled water or not), British citizens trust their own judgment rather than that of an external organization.

In conclusion, this exploratory study allows us to look at the nature of a population that has not yet been widely and consciously exposed to the practical aspect of greywater recycling and the questions that arise from it. The questions presented in this research concerned a wide range of factors that can shape the attitude of the respondents, and many findings must be qualified (such as the large support rate to the idea of greywater recycling) mainly due to lack of exposure of the study population to greywater recycling in practice. Another interesting point that arose in the study was the lack of public trust in the decision making processes and recommendations of organizations. Instead, individuals tended to base their decisions on the subject of greywater reuse off of their own personal perceptions and calculations.

### 6.2.2 Spain

In 2002, the town Sant Cugat del Vallès in the metropolis of Barcelona was the first in Spain to decide on local regulations requiring the installation of greywater recycling systems. The regulations were set for new residential buildings with eight or more apartments and any other building that has showers consuming more than 400 cubic meters of water per year. Various technologies were implemented in the town for water treatment—biological and physical—and all have a common action model: recycling of shower water and bathwater for toilet flushing. By 2008, about 3800 apartment buildings were built in the town that included systems for greywater recycling.

Domènech and Saurí (2010) report a survey conducted in 2008 among 120 respondents. The survey was conducted due to continuous drought, restrictions on the use of drinking water for secondary purposes (not for drinking), and the growing number of residential buildings due to urbanization. Its goal was to examine public perceptions regarding greywater recycling and to determine the key factors that shape them, with the hope of providing tools to promote the use of this technology. The respondents were all residents of the town, and during the study period, they lived in homes that had active greywater recycling systems. The respondents' hands-on experience with the technology ranged between 3 months and 3 years. The survey examined the following: (1) the degree of acceptance of greywater recycling, (2) factors that shape the measured degree of acceptance, and (3) socioeconomic variables involved in shaping the observed social perceptions and attitudes on the subject.

The following findings were of particular interest.

Overall, the level of acceptance of greywater recycling systems was high, particularly at the beginning of system use. However, as the extent of use and experience with the system increased, opposition to reuse grew stronger. Additionally, a higher acceptance level was noted among tenants, as compared to property owners.

In terms of the factors that shape measured acceptance levels vis-à-vis the use of recycled greywater for toilet flushing, it was found that the main defining factor is the perception of the health risk involved. Thus, according to Domènech and Saurí, the high rate of support indicates that survey respondents did not perceive the use of recycling greywater in flushing toilets as incurring a high health risk. This might be explained, they suggest, by the fact that blue color is added to the recycled water, making it more aesthetically pleasing. This is an interesting finding worth thinking about: aesthetics can help imply a *healthy* recycled product and therefore can raise support. The second most important factor found in the survey was the management regime of the technology: how it is installed and used and the gap between its expected capacity and its actual function. Factors such as incorrect installation or an erroneous evaluation of domestic use patterns led to negative experiences and hence to a decrease in the acceptance level.

The third most important defining factor was the perception of cost; most users viewed greywater recycling systems as inexpensive. However, the authors noted that most users were not aware of the real costs of the systems. This demonstrates the importance that perceptions hold and how they may be formed regardless of available data.

The study found that the fourth most important factor is environmental awareness. Surveyed users held that a major benefit of greywater systems was the reduction of potable water use. A high level of environmental awareness was found among all respondents; these attitudes aligned with higher degrees of acceptance for greywater recycling systems.

It was also found that only 43% of the users who participated in the study were able to accurately describe the source of the recycled water used in their toilets, indicating a low level of knowledge about the system's operation. When asked, more than 50% of the total respondents felt that they did not receive comprehensive information relating to this new technology. Hence, it is very probable that the high support rates indicated in this survey would drop as users learn more about the systems that they are exposed to on a daily basis.

The study did not report findings regarding the effect of socioeconomic variables, since no statistical relation was found between those variables and the acceptance rate of the technology. The researchers attribute this to a statistical problem resulting from the absence of variance in the sample, since most of the respondents who participated were young couples having at least a bachelor's degree and high to middle income, living in their own home (rented or purchased) for the first time. The researchers mention that this is precisely the population that is believed to have a high tendency to accept technologies that help the environment. Although no statistical relation was found, high support rate is consistent with the characteristics of the study population. These population characteristics are also compatible with the fact that, despite complaints raised during the survey about the smell of the water, a high acceptance rate was still recorded.

Lastly, the authors argue in favor of the importance of the cultural–national–local context when assessing the factors that shape perceptions. This context is manifested in the sum experiences of the community and shapes their perceptions on various topics, including the necessity of water reuse. Thus, a society that suffers from water shortage would be more willing to implement strategies required for solving this problem. In this case, it was found that favorable perceptions, attitudes, and behaviors toward greywater recycling are compatible with the principles of the new *water culture* in Spain, and in Barcelona in particular.

Thus, the Spanish study portrays widespread perceptions in a society that has already had some experience with greywater recycling, even if for a relatively short period. This sort of experience may actually have a negative influence on the initial level of support, and its effects are even stronger among people who are compelled to use greywater recycling in the long term. This can be seen in the difference measured between renters and owners of flats where greywater recycling is in use. Therefore, the researchers claim that the interaction between the characteristics of the technology, its management, and the cultural–national–local context (in this case drought and water regulations) strongly impact public perception on the issue.

### 6.2.3 Oman

The Sultanate of Oman, on the shores of the Arabian Sea and the Gulf of Oman, has experienced both population growth and accelerated economic development in recent decades. As a result, total water consumption has expanded, causing the country to turn to diverse solutions including desalination and well drilling. However, these solutions do not suffice to meet the increasing demand for water, as evidenced by the decreasing groundwater levels and salinization of water in wells near the shoreline. In an attempt to test alternative water sources, the possibility of introducing greywater treatment technologies was also raised in Oman.

An article by Prathapar et al. from 2005 refers to an exploratory study and survey on the urban population of the Sultanate. The survey, conducted in 2004, examined the public's attitude toward the use of recycled greywater. It also included a review of the water quality and usage patterns. In the survey, 221 households were sampled, which comprised 1543 residents of the urban population in Oman. The population was heterogeneous in terms of lifestyle, culture, and income and was not exposed to greywater recycling technologies in practice, either before or during the study period. Several key findings are described in the following paragraphs.

Of the total respondents, 84% supported the idea of greywater treatment and its reuse, and 74% felt that installing a greywater recycling system would be economically beneficial. Eighty-two percent felt that treated greywater should be used for irrigation, 72% for flushing toilets, and 42% for car washing. However, 61% of all respondents felt that the use of treated greywater may harm the environment. In addition, 41% felt that the use of recycled greywater may pose a health risk. According to the researchers, the most surprising finding was the support of 16% for using treated greywater for drinking, which seemed relatively high. Those who opposed the use of treated greywater mentioned religion, health, groundwater pollution, and cost as the main reasons for their opposition.

Another survey mentioned in the same article was conducted among the student population of the Sultan Qaboos University in the capital Muscat. The survey found that 62.5% of the respondents supported the separation of greywater from blackwater, 58.5% said that they would use treated greywater for several household needs (not specified), and 43% were willing to pay more out of pocket to change the existing pipeline infrastructure to allow the use of greywater. The primary reasons mentioned as a basis for resisting the use of treated greywater included health (40%) and religion (37%).

Another study also exploring the Muscat public's attitudes on the subject was published by Jamrah et al. (2008). Here again, the study population had not yet been exposed to widespread implementation of greywater recycling technologies. The study sampled 169 households of university faculty members located on the campus of Sultan Kavos University, which comprised 1365 persons. The following is a partial list of the findings.

The survey found that 87% of all respondents supported the installation of new plumbing in their homes to collect greywater (7.7% object). It found that 76.3% support the use of treated greywater for irrigation, 66.3% for flushing toilets, and 53.3% for car washing. Those who opposed the use of treated greywater mentioned the five following main reasons for their opposition: safety (88.2%), religious considerations (60.1%), negative effect on the environment (52.6%), unreasonable economic cost (24.5%), and groundwater contamination (22.3%).

Overall, the data presented from studies in Oman indicated a cultural willingness to use treated greywater which is consistent with patterns presented in other countries. The more distanced the proposed use is from physical contact, the more acceptable it is to the public.

At the same time, the case of the Sultanate of Oman emphasizes the importance of understanding implications of local context on support for greywater reuse. In this case, in addition to the objections mentioned in other studies (i.e., the fear of a health risk), respondents also objected to the use of recycled greywater based on religious considerations. This reflects Muslim beliefs concerning purity and specifically, the central role of water in Muslim purification rituals.[*,†] Clearly, in this particular context, this is an important cultural fact, and the researchers offer a possible way of addressing it. They note a *fatwa* (a ruling according to Muslim law) set in 1978 that allows the use of recycled water, even for religious purposes, as long as the water is treated to the required level, and does not pose a health hazard. The *fatwa* also recommends enlisting *Imams* (having religious authority) in campaigns to change public opinion (Prathapar et al., 2005).

In conclusion, this case study illustrates how greywater recycling technologies, which are supposedly neutral technologies are valued differently in varying cultural contexts. Hence, any attempt to implement the technologies sweepingly across different cultures may encounter difficulties. Clearly, cultural differences should be identified and understood, and supporters of greywater recycling should act to identify the most appropriate solutions for a given society. They cannot assume that the *blind* transfer of a technology that thrives in one cultural and social context will be successful in another.

### 6.2.4 Australia

The last example of research on this subject is from Australia, the driest inhabited continent and perhaps the most obvious case in the field of the use of greywater recycling technologies. Most of Australia's population is dispersed around a system of freshwater rivers, and the pressure on them is growing. This is particularly true in the southeast of the continent, due to climate change, economic development, and population growth. In response, many local governments in Australia set regulations to limit water consumption in recent decades, as can be seen in the example presented at the beginning of this chapter. The Australian government started to examine alternatives to the existing water regime, including desalination, wastewater recycling, and the establishment of decentralized water recycling systems and began actively encouraging its population to install alternative water systems. As a result, Australia leads the world in terms of the scope of alternative water technology used in the municipal sector. This includes collecting rainwater and recycling greywater for domestic and community use. Australia also leads the public debate on practical uses and relevant technologies.

For example, in Victoria, Australia, a state faced with ongoing droughts in recent decades, limitations have been imposed on the use of freshwater and government campaigns, such as a refund for installing water-saving devices, have been established in order to encourage water conservation. At the same time, likely due to an increase in the collection of domestic wastewater for reuse, the local authorities recorded a considerable decline in the quantities of domestic wastewater transferred to the main sewer systems. In a 2008 study conducted in Victoria by

* Another example of the important role that purity ritual in Islam affects environmental behavior can be found in the work of Paz et al. (2012). It explores decisions made by Muslims in Israel regarding the use of compost in agricultural settings.

† For a detailed description of the religious Muslim arguments for and against water reuse, see Farooq and Ansari (1983).

Hurlimann (2011), the researcher tried to understand what alternative water sources were being used, to what extent they were used, and to fulfill what needs. In a representative sample, data was collected from 410 local households. The following are some key findings of the study.

First of all, research findings demonstrated that most households in Victoria use decentralized water sources, such as rainwater, recycled greywater, and untreated shower and bathwater. The researcher found that 74.8% of respondents indicated the use of these alternative water sources in regular or occasional irrigation of their private ornamental gardens, with laundry water (usually untreated) being the most common water source for this activity. In general, it seemed that users of alternative decentralized water sources designated certain water sources to fulfill particular needs and not others.

In addition, survey respondents listed the following factors as barriers to using alternative decentralized water sources: (1) One is being tenants and thus unabling to change the house infrastructure and install permanent devices to exploit alternative water, such as systems for greywater recycling. Many respondents suggested developing a set of incentives for homeowners to install such systems in properties for rent. (2) Maintenance costs of these water systems were often perceived as being too high. (3) The state of Victoria requires a qualified plumber to install rainwater harvesting tanks, which the survey respondents perceived as a limitation.

The Australian case serves us by demonstrating that despite clear government policy and incentives, as well as a long-standing local tradition of using alternative water sources, there is still a gap between *perception* (i.e., the general support of water recycling and reuse) and *behavior* (i.e., the actual installation and use of water recycling devices). This highlights the importance of conducting additional research into these questions each and every time. Though the rate of actual use of these technologies in Australia is indeed very high relative to the rest of the world, it is not as high as the acceptance rate that this population demonstrated in various surveys.

This study also reemphasizes the point that greywater reuse is context-sensitive, due to differing local and cultural perceptions of and behaviors regarding this topic. For example, in this case the majority of the population feels comfortable recycling laundry wastewater, but not dishwashing water. Additionally, the population preferred to use the reclaimed water for ornamental garden irrigation rather than, for example, flushing toilets. These preferences exemplify some of the important considerations necessary when introducing a new technology or a change in policy.

The Australian case illuminates once again the importance of homeownership versus tenancy. Not only do homeowners seem more sensitive to poor performance of the recycling devices as we have seen in the Spanish case, but in cases where they rent out the property to others (and therefore do not pay the usage bills), they have less of an incentive to install them in the first place.

### 6.2.5 Interim Summary

The water and ground division of the Commonwealth Scientific and Industrial Research Organisation (CSIRO), the Australian umbrella organization for research, conducted a worldwide literature review of the factors that shape public perception regarding water reuse (Po et al., 2003). The review relates to the subject of water recycling as a whole (not only greywater) and includes a list of factors that give the authors' explanation of the persistent disparity between the overwhelming rate of support and the much lower rate of actual wastewater recycling technology adoption and usage. This text, which lists 11 factors, will serve here as a general summary to the issue:

1. *Risk perceptions associated with the use of recycled water*: The main measured risk associated with the use of recycled water is the perception of health risk, evidence of which has accumulated in dozens of studies. In addition, it seems that fear of health risk increases when children are expected to come in contact with the recycled water. The authors of the report also note that a documented disparity exists between the risk perceptions of *experts* and *the public*. The public seems to tend toward a broader conceptualization of the risk in question, based on aspects such as uncertainty, anxiety, and the possibility of an extreme case occurring. Experts, on the other hand, tended to define risk according to the conceptualization of probabilities, where the possible extreme cases are much rarer. This gap is often a cause for public disputes regarding the benefits and hazards of water recycling projects.
2. *The "yuck factor"*: Though acknowledged in the literature since the 1970s and cited as major psychological barrier for using recycled water for various uses, there is still no complete body of work explaining its mechanisms or quantifying its particular impact on the issue. In general, it is suggested that the feeling of repulsion is fueled by the perception that the water is *dirty* and the unsubstantiated fear of infection following use.
3. *Source of recycled water*: There is conflicting evidence regarding people's preferences as to the source of recycled water. Several studies have found a preference for using water recycled from the household where the water was originally used, while other studies have found a preference for recycled water from the entire city or neighborhood. The explanation for this variance might be found in the perception of the quality of the water in each specific case, so that in some cases, people prefer to rely only on themselves, while in other cases, they would prefer external management of the recycling process.
4. *Purpose of use*: Feelings of repulsion and risk perception are also key in determining what is an acceptable use of the greywater. These factors influence both attitudes and behaviors regarding water reuse. Many studies, both quantitative and qualitative, indicate that the closer the physical contact between the users and the recycled water for a given use, the lower the support for that use.
5. *Trust in the authorities and scientific knowledge*: It is still unclear exactly how government and scientific authorities affect public support regarding water recycling. As different studies provide distinct evidence, these factors should be considered on a case-by-case basis.

6. *Attitudes toward the environment*: Data collected from various studies indicate that proenvironmental attitudes are associated with a positive perception of using recycled water. However, this association was not found as being the most important factor in determining wastewater reuse. Moreover, it seems that a *proenvironmental attitude* is too broad a definition. Instead, the target population's attitudes and perceptions on the subject of *water* should be clarified and distinguished from those on the *environment*. (For such an example, see the extended case of Israel in the following section.)
7. *The question of necessity*: There is a greater tendency to accept the use of recycled water in cases where water supply is limited.
8. *Locus of control*: The report also notes the importance of responsibility: who the public considers responsible for water shortages and the subsequent need to consider new water recycling schemes. For example, if the public is convinced that the shortage of drinking water is the product of mismanagement on behalf of the authorities and not necessarily due to the wasteful behavior of the community, the community may refuse to take action to ameliorate the situation. Similar recalcitrant behavior can also occur if the community feels that its efforts to recycle greywater do not have a significant impact on the shortage that they seek to solve.
9. *Environmental justice issues*: Objections to water recycling projects can also stem from public criticism of the distribution of water resources, from who bears the costs for the recycling project and who is eligible to receive its profits, whether certain groups in the population or economic sectors will be given preference, and in what ways. There is evidence that several projects around the world were stopped following such objections.
10. *The cost of recycled water*: In general, people expect to pay less to use recycled water since it is perceived as having a lower quality and since it is not as versatile (i.e., cannot be used for all activities such as drinking or showering). Studies show that the introduction of an economic incentive, such as savings or subsidies, is a highly effective way to encourage the use of recycled water.
11. *Sociodemographic factors*: Lastly, a comprehensive review of studies found there are no consistent results regarding the effect of sociodemographic characteristics such as gender, age, education, and income, on the degree of support for the use of recycled water.

Throughout the case studies described in this section, some interesting trends exist. First of all, it seems that the degree of acceptance of the subject of greywater recycling is high. This is not obvious and can perhaps be explained by the fact that all the countries presented here—the United Kingdom, Spain, Oman, and Australia—have experienced a significant decrease in the available drinking water resources for their populations. They are simultaneously experiencing an increase in population size and in water consumption for industry and other sectors. However, studies still identify a disparity between general support and personal attitudes regarding greywater recycling and the willingness to act on them. In addition to the preceding factors listed, the following is a summary of the additional factors from cases explored earlier in this section:

- *Cultural–national–local context*, such as droughts, water shortages, national culture, and religious beliefs
- *Personal knowledge and awareness* of the amount of water consumed, its cost, and the state of the water sector
- *Lifestyle* that supports the operations of recycling and saving
- *Aesthetic* property of recycled water (e.g., color, turbidity, sediment, odor)
- *Extent of actual experience* of the individual with recycled water and the quality of the experience
- *Rental versus ownership* of residential property leading to a disconnect between those who are required to invest financially in installing the system and those who enjoy its products
- *Management regime* (i.e., installation and functioning of the systems)
- *Personal economic cost* and the perception of that cost

The review presented so far provides useful background for those seeking to understand the perceptions and attitudes toward greywater recycling in general. The following section is dedicated to one extended case study: that of Israel. Israel has unique local characteristics, and it will serve to highlight in application the general factors reviewed so far.

## 6.3 EXTENDED CASE: ISRAEL

The current formal Israeli state policy on the subject of greywater recycling is clear, represented and managed by the Ministry of Health: water recycling, regardless of the origin or use of the water, has direct potential impact on public health and therefore must be centrally managed by the state and/or be under its direct daily and routine supervision. Thus, decentralized water recycling technologies and practices are treated with much suspicion. In 2003, the Israeli Health Ministry published a document detailing the quality of potable water required for various purposes in the urban sphere, setting criteria for minimum treatment and quality and control (Halperin and Aloni, 2003). In 2008, the ministry issued an additional series of regulations, this time specifically on the subject of sanitary conditions required for the recovery of greywater for toilet flushing and garden watering (Goldberger, 2008). This indicated the Ministry of Health's perceived need to regulate what seems to be a growing practice. The regulations state that "greywater may contain high concentrations of pathogenic microorganisms and endanger the public's health when coming in contact with it" and that each facility for greywater recycling must have the approval of the Ministry of Health. This approval is conditional on several factors, one of which being that a local authority or a licensed business will be responsible for the facility (Article 4) and on the facility being located far from the place of residence (Article 8). The ongoing result of this position is that since 1988, the Ministry of Health in Israel approves greywater recycling facilities to a very limited extent, mainly in public buildings such as sports facilities (Ronen, 2009).

Interestingly, it turns out that a growing number of systems are in practice installed in Israel, most of them without the knowledge of the Ministry of Health. Unofficial numbers, estimated by researchers and activists in the field, range from 5,000 (a conservative estimate) to 15,000 (high estimate) greywater recycling systems in the country. Today, the vast majority of these unofficial systems are installed in the courtyards of detached single-family homes in the countryside and suburbs. They are primarily used for recycling laundry and shower wastewater and for the irrigation of ornamental gardens. The expansion of the greywater recycling practice into the urban sphere in Israel is symbolized by the garden estate in the local council of Ganei Tikva, built several years ago in the Tel Aviv suburbs. The apartments in the neighborhood buildings were built with dual piping that leads the wastewater from the bath and showers to recycling facilities. It is then used for irrigation in the public gardens, and the building project was marketed to the public as *green*. However, the local council encountered many difficulties in its efforts to receive the approval of the Ministry of Health to operate the system, and the issue has remained a point of controversy for a long time, despite the fact that the buildings were already populated.

The issue of greywater recycling in Israel is currently debated in two main legal domains: (1) a regulation process within the Israel Standards Institute (Israeli Standard 5281), which deals with *green* construction in general and also refers to greywater recycling, and (2) a legislative process (which joins the legislation efforts of the last decade), which would allow recycling in a regulated manner in both public and private constructions. Together, the use, regulation, and legislative efforts compose an active community of individuals and organizations in Israel eager to promote the subject of greywater recycling. Acting as a whole, they receive a significant boost and enjoy an expanding public discussion.

This begs the following questions: Who exactly are the people who want to promote greywater recycling in Israel? Are they a representative sample of the entire population in Israel or a particular segment? And given the importance of cultural factors in the shaping, implementation, and dissemination of the use of greywater, what are the most significant factors that will affect the expansion of greywater recycling in the Israeli context? Answering these questions is essential, as we have seen, to any successful campaign that wishes to introduce greywater recycling technologies into a new social context.

Current evidence on the subject of public perceptions and attitudes toward greywater in Israel arose from a pioneering exploratory study that dealt with the Israeli public's attitudes toward the use of treated wastewater in the municipal context. In 2004, the research team of Friedler et al. conducted a survey in which 256 respondents from the urban population living in condominiums in the town of Haifa (considered the third largest city in Israel) took part. In the immediate environment of this population, no technologies for wastewater treatment were in use at the time, and wastewater was not reused at all. Since this was an exploratory research, the aim was to gather information about the following aspects: (1) the attitudes of the urban public in Israel regarding possible uses of treated wastewater, (2) the main issues that trouble this public regarding to these possible uses, and (3) the major sociodemographic characteristics of those who express opposition to the use of treated wastewater.

The authors of the survey chose to use the relatively blunt term *treated wastewater* (Hebrew) rather than other more delicate terms such as *recycled water* or *purified water*, so the survey findings should be considered accordingly. It could be assumed that a terminology that emphasizes the source of the water—*wastewater*—may produce stronger adverse reactions among the respondents, even subconsciously, than other terminology (i.e., the "yuck" factor mentioned earlier), although this assumption has not been empirically tested among the Israeli public.

The research team divided the various possible uses of treated wastewater presented to the respondents into three categories. The first was defined as *low contact*, in other words the use of treated wastewater for purposes that do not include direct contact between the user and the water itself. In particular, this refers to agricultural irrigation and recharge of aquifers for the purposes of agricultural irrigation. These uses have been common across Israel for over 30 years. The second category was called *medium contact* and included treated wastewater uses that can be applied in an urban context, but do not involve direct human contact (and are not implemented currently in Israel in a regulated manner or on large scale) such as fire hydrants, industry, construction, watering public and private ornamental gardens, and flushing toilets. The third category is *high-contact* applications including direct and intentional body contact, such as domestic laundry, swimming pools, and aquifer recharge for drinking. These uses are not seen today in Israel at all. The respondents were asked to indicate to what extent they supported each use on a scale ranging between *strongly oppose* (0) and *very supportive* (4).

When publishing the study results (Friedler et al. 2006), the researchers focused on analyzing the second use category, that of *medium contact*, because in their view this category was more likely to be applied on a large scale in Israel in the near future. They statistically examined whether a correlation existed between the various attitudes and beliefs of the respondents (such as beliefs about anticipated economic gain or about the authorities' capabilities to deal with the subject) and their support (or lack of support) for implementing the treated wastewater uses described in this category. The following presents their findings, analyzing them in light of additional research regarding the Israeli case. This will allow us to consider the importance of not only asking the right questions but also placing the answers in their right particular context.

### 6.3.1 Low Contact, High Support?

Findings from previous studies, including those mentioned earlier in this chapter, predict that as the treated wastewater is (perceived as) designated to uses that have higher physical contact with people, the degree of public support decreases for this use. Interestingly, the findings of the Israeli study seemingly contradicted this assumption, as the highest rate of support was awarded to uses that were defined by the researchers as holding potential *medium contact*, such as fire hydrants, air conditioning, gardening, and toilet flushing. On the scale of support, these usages were then followed by those defined as holding *low contact*, and the lowest rate of support was given—as expected—to usages belonging to the category of *high contact*.

According to the researchers, explanation for the unexpected finding regarding *medium contact* might be found by looking closer at how they constructed the category *low contact*. They included in it three usages of treated wastewater, based on previous research conducted in other countries. It received the following rates of support from survey participants: (1) watering orchards of fruits that have to be peeled before eating (such as citrus) (48.8%), (2) aquifer recharge for agricultural irrigation (61.7%), and (3) irrigate cotton and fodder crops (86%). Moreover, these three usages have been in common use in Israel for decades, and thus the research team had even less reason to consider the possibility of low support rates.

As we can see, the results do follow what we know about physical contact and support rates in one way: the lowest support rate is attributed the irrigation of edible fruit, a little higher to general irrigation in agriculture, and the highest support rate given to irrigation of cotton (a product used mainly for clothing or industry and is not eaten). However, the fact that these usages did not win the highest support rates (compared to usages such as irrigation of public ornamental gardens or fire hydrants) suggests the possible existence of an unpredictable intervening factor, which is here suggested to be local Israeli culture. Though these are three common usages, the relatively low support rate might suggest that in fact the majority of Israeli public, or at the very least its urban population, is unaware of the main water sources used for agricultural irrigation in Israel. This may also suggest a gap between how the public in Israel perceives possible health risks involved in these usages of treated wastewater and the way professionals in the fields of agriculture and health perceive them. It seems that this possible lack of awareness by the public and the disparity between the perceptions of the professionals and the general public may have extensive ramifications in future attempts to examine actual implementation of uses of other treated wastewater, in the city and in general. This research can be used as a platform to consider future policy in Israel.

### 6.3.2 Not Healthy, No Thanks

A negative correlation was found between the support for applying treated wastewater uses of the *medium-contact* category and the way the potential health effects from this application are perceived by survey participants. This implies that when the Israeli public sees a use of treated wastewater as having a higher health risk, it opposes its implementation more strongly. This finding is consistent with the finding presented in the previous section and with the literature in general that identifies the perception of health risk as a key factor in shaping public perceptions regarding the subject of water recycling.

### 6.3.3 Not in My Backyard

Another finding from this study consistent with other evidences from around the world is that public support for using treated wastewater in public areas is higher than for using this water in the private sphere. For example, the study found higher support for using treated wastewater to flush toilets in offices (86%) than in private houses (79%) and for watering public parks (90%) than private gardens (80%). It should be noted that the support rates for usages in the private sphere were quite high; however, the disparity observed between usages in the private and public spaces should be investigated to find out how it may represent actual attitudes toward future projects.

### 6.3.4 High Economic Gain Equals High Support

A strong positive correlation was found between the individual's belief in possible economic profit for himself or herself and the individual's support of treated wastewater plans of the *medium-contact* category. A moderate positive correlation was found between the individual's belief in financial gain for the municipal authority and his or her support of these usages.

The first finding regarding personal profit also appears in other studies, but the relationship found between the possibility of municipal profit and the individual's support for applying treated wastewater usages is not obvious; it may be a unique property of the Israeli society. The possible implications of these findings on future plans to implement greywater recycling, as well as on water recycling technologies in general in Israel, lead to some obvious questions. To begin, can there actually be economic profit from the use of greywater recycling technologies in Israel? If so, could this be used as an incentive to distribute the use of greywater recycling technology in Israel? And if personal gain may come at the communal–national expense, will it still have such high support?

These questions are still open. There are studies suggesting that under certain conditions, significant economic savings for consumers living in condominiums in Israel is possible (Friedler, 2008). There are also studies concluding that greywater recycling will have financial benefits on a national level (Adel et al., 2012; Chapter 7). However, there are also economic calculations at the national level, for the entire Israeli water sector (Becker et al., 2010). These look into considerations such as infrastructure and the costs of treating the remaining black wastewater. These show that greywater recycling might in fact be the most expensive option for dealing with the national water shortage in Israel. Hence, the question of consumer economic feasibility is not clear in the Israeli context, certainly not to consumers living in private homes. In addition, it seems that there is a conflict between the economic interests of the private home economy and that of the national economy. There may also be a conflict between the national interests and the interests of competitive organizations such as desalination plants and water corporations. Such conflicts may affect the extent of support from an Israeli consumer in investing in this kind of water recycling, and certainly more data should be collected to shed light on this significant issue.

### 6.3.5 Awareness Does Not Equal Support

No correlation was found between personal awareness of water and environmental problems in Israel and the rate of support for urban use of the *medium-contact* (category 2) treated wastewater. This finding is surprising since, as noted earlier in this chapter, studies from around the world claim a direct relationship between awareness of water and environmental problems and support for water recycling.

Friedler et al. (2006) suggest that the explanation for this unique finding may lie in the particulars of the Israeli context: the total Israeli public is highly exposed to the shortage of national water resources and is aware of it through extensive national campaigns on the subject. They are also sensitive to it because of regulations that limit the manner and extent of water use, a well-publicized increase in water prices (*drought tax*), and discussions regarding water sources in peace agreements with neighboring countries (as was the case when negotiating peace with Jordan). This means that there is no internal variance in the study population with regard to this variable (awareness of water shortage), and therefore this factor cannot explain the variance in the rate of support for different treated wastewater uses. This conclusion is supported by the work of Katz-Gerro (2009) who examined the social roots of environmental attitudes in Israel. Based on surveys conducted in 2000 and 2006, it was found that current environmental concerns receive generally wide support among the Israeli population, regardless of any social grouping. According to Katz-Gerro, the broad social base results from the fact that environmental dangers are becoming more pervasive and are more extensively covered by the media, leading to greater public awareness. However, the explanation of this finding suggested by Friedler and others (2006) is essentially statistical and is based on the absence of variance.

A different direction to look for a possible explanation that has not yet been tested in research might be found in the way that Friedler et al. constructed the category of *personal awareness of water and environmental issues in Israel*. In this research, the terms *awareness of water problems* and *awareness of environmental issues* were unified based on the thought that the two are complementary problems. However, perhaps more significantly in the case of Israel, a distinction between *awareness of water issues* and *awareness of environmental problems* can help us. Several researchers (e.g., De-Shalit, 1995; ibid) claim that environmental issues in Israel involve significant questions regarding national-economic development and national security, including those dealing with natural resources. Hence, public awareness in Israel, as well as its mobilization in the area of water savings, is anchored and embedded in more than just a question of environmental stability; foreign relations, political dependency, and Middle East conflict management also play a part. Evidence of this can be seen in the many discussions on national infrastructure projects, where a significant gap is felt between those who believe that they represent the interests of national security (embodied in the idea of the undesirability of resource dependence) and those who seek to represent the environmental interests. Sometimes, this gap results in an open struggle between the government offices themselves, most notably the ministry of infrastructure on the one side and the Ministry of Environment on the other.

In light of this hypothesis, it seems that despite the finding presented in the study, it has not yet been fully elucidated at this stage how the attitudes of the Israeli public on water and environmental issues affect their attitude regarding the use of treated wastewater in the Israeli-urban context. Further research on the subject should focus in particular on the actual perception of the issue of water among the Israeli public and suggest in what ways and to what extent the subject is shaped by the concept of national security as compared with environmental perceptions.

### 6.3.6 I Will Make My Own Decision, Thanks

Of the total respondents, only 32.5% expressed confidence in the ability of the municipal authority to manage treated wastewater facilities and to ensure their safe use. Another 16.8% had no opinion, and 50.7% expressed lack of confidence in the ability of the local authorities on the subject. No correlation was found between the degree of confidence in the management ability of municipal authorities and the extent of support for the implementation of usages of the *medium-contact* category.

Friedler et al. argue that this finding indicates that the Israeli public—which in general is quite skeptical of the authorities' capability in general and their ability to provide water specifically—believes that there is no connection between the authorities' capability and the application of uses of treated wastewater. In conjunction with the other studies presented earlier in the chapter, this may be a wider phenomenon not unique to Israeli society: individuals prefer to shape their own position on the basis of personal calculation, and not necessarily on the basis of the recommendation of public organizations.

Referring to the Israeli case, this finding raises fascinating possible implications, especially when it is presented in close proximity to another argument made in the same study: that a large part of the Israeli public is not aware of the extensive use of treated wastewater in agriculture in the country. It may well be that the public is not knowledgeable about the health and environmental dangers inherent in unprofessional management and monitoring of the various systems of water recycling. Combining these two findings suggests that any public discussion on the topic in Israel, a discussion that can be seen to be growing and expanding every day, will have to include both a process of exposure and investigation of the treatment and monitoring mechanisms of Israel's water recycling. It will also have to include a dialogue on the role and authority of state agencies and local government on this issue. Discussion of this kind, by nature, will inevitably be rooted in political questions of administrative and public responsibility alike.

### 6.3.7 Don't Know, but Support

The study found no correlation between awareness of the existence of a suitable technological solution for wastewater treatment and the support of merely applying uses of the *medium-contact* category. Closer examination reveals that the respondents disagreed regarding the very existence of such technology: 28.7% of the respondents believe in the existence of suitable technology, 43.5% do not know whether such technology exists, and 27.9% believe that suitable technology does not exist.

This is further cumulative evidence of what appeared to be superficial familiarity of the Israeli-urban public with the greywater technology subject. Apparently, in this case, most of the public does not understand the nature of the production process of recycled water and the quality control performed during this process.

### 6.3.8 Population X

In terms of social research, one of the more interesting findings of the study in question is precisely what was not found. In examining the possible links between the level of support by the respondents for applying the *medium-contact* category and six different sociodemographic background variables (education, income, gender, age, marital status, and age of children), no significant correlation was found. Thus, it is difficult to understand fully who supports the use of treated wastewater in Israel and who opposes it. As shown in Section 6.2, world literature on the subject is also divided and has difficulty in establishing a steady and uniform argument between the various case studies examined and the permanent effect of one of the socioeconomic background variables.

In contrast, another study conducted in Israel (based on a survey from 2000) found that education, awareness of implications, and environmental literacy are factors that do significantly shape attitudes and environmental behavior in Israel (Ne'eman-Abramovich and Katz-Gerro, 2007). The study also did not find an effect related to gender, ethnicity, or extent of religiousness, but some influence was traced to the individual's place of residence; residence in the suburbs or in a rural and secluded place predicts a more positive perception toward nature and a greater sense of environmental responsibility, as compared with residence in the city.

Clearly, as with other cases worldwide, in Israel, the influence of sociodemographic background variables on public perceptions of the subject of the environment and water is far from clear. Ne'eman-Abramovich's and Katz-Gerro's work might provide initial clues and offer a basis for further exploration.

### 6.3.9 Rationalization for Support

Finally, after completing the questionnaire, each respondent who indicated that he or she supports any of the uses of treated wastewater was asked to rank the reasons for his or her support out of four possibilities. These are presented here with their final ranking by the survey respondents, from most important to least: the most important reason was "reuse of treated wastewater will help the water sector"; the second in importance was the statement "reuse of treated wastewater will decrease the dependency on imported water," followed by the equally important, sharing third and fourth rankings, reasons that "reuse of treated wastewater will save the costs of infrastructure and will improve the economy" and "reuse of treated wastewater will improve the quality of the environment."

These findings raise some interesting points in regard to the Israeli case in particular and the literature on the topic in general. First, "helping the water sector" is a general statement, a fact that may explain its popularity among the respondents. The fact that "reducing dependence" ranked in second place clearly owes its popularity to the public debate in Israel: whether it is feasible to purchase water from foreign sources in the context of the complex political situation in the region. As suggested earlier, it is probable that in Israel attitudes toward natural resources policy, and water policy in particular, are directly related to attitudes on national security issues. This can be seen, for example, in the common narrative among the Israeli public and policy makers in which Israel is described as a "small country surrounded by enemies." One of its widespread implications is that Israel should make sure that it is not dependent on external sources for its needs and ultimately its existence.

Secondly, the fact that the most statistically significant reason for high support for a greywater recycling policy is that "reuse of treated wastewater will help the water sector" (more so than *contribution to the economy* and *environmental improvement*) suggests that this consideration is of greater importance than the other two according to the urban public perception in Israel. This makes further research on this factor all the more important.

Thirdly, the fact that the *economic* consideration and the *environmental* consideration carry the same weight for the respondents in this study raises interesting questions: In regard to water recycling, is the size of the population that finds the economic consideration important indeed equal to the size of the population that finds environmental considerations important? Does this mean that their weights are therefore equal? If the size and weight are not indeed equal, is it possible that within the Israeli context *economic* and *environmental* considerations are in fact two sides of the same coin?

It is important to emphasize that at this stage, these are all speculations; they have not yet been empirically tested in Israel. In addition, these postulations overlook the possibility that concern for the water economy may be deeply related to the respondents' fears of the effects that a struggling water economy could have on their own condition. This theory, if found correct, would cast the original findings in an entirely different light. In addition, it should be noted that no statistical comparison was made to test the importance of these reasons for support over other basic categories of tested attitudes that could be influencing the same belief system. These other spheres of influence could include personal economic profit or public health impact. Therefore, it is difficult to determine whether a concern for the water sector in general—or for Israel's political situation—is a more significant factor here than, for example, public health concerns.

### 6.3.10 Conclusion

In conclusion, the article discussed here in detail (Friedler et al., 2006) represents a significant milestone in the examination of Israeli public opinion on greywater use, although it was based only on a sample of urban population. The study shows that broad Israeli public support for using treated wastewater of the *medium-contact* level is possible. It also shows that the economic gain to the individual and to the city and public health risk perception are influential factors and that concern for the Israeli water economy is a major reason to support the use of treated wastewater. In addition, the study found no association between the degree of support for using treated wastewater and the sociodemographic characteristics examined, awareness of water and environmental issues, belief in the existence of a suitable treatment technology, or trust in the authorities.

These findings, as well as additional work carried out in Israel in recent years, indicate that the Israeli public is not fully aware of how wastewater is treated in Israel, what wastewater is used for today, and what are the dangers involved in improper treatment. Moreover, it has been suggested that the Israeli public distinguishes between environmental issues and issues related to water and natural resources and tends to decide on issues relating to water on the basis of collective-national considerations rather than personal or environmental considerations. Thus, it appears that in Israel, as in the case studies presented from around the world, the question of the national–cultural–local context is very important. In addition, the relatively high level of support (especially in light of the use of the term *treated wastewater*), the lack of clear sociodemographic predictors, and the emphasis on public health and economic gain cost are all in line with the findings from the world literature. They indicate that there may be common social forces influencing people on this issue in all western societies.

Lastly, as noted earlier, the work of Friedler et al. is pioneering. As such, it serves as a platform to delineate the future research needed in the field more than it provides final answers and resolves debates on the topic.

## 6.4 SUMMARY, CONCLUSIONS, AND OUTLINE FOR FUTURE RESEARCH

As can be seen in this chapter, global and Israeli research in the field of human views on greywater is still in its initial stages. However, it is clear that the increase in water consumption, the decrease in the amount of natural resources, and technological and scientific developments are all intertwined and are driving the field to expand. Along with this expansion, a clear consensus emerges among those who are dealing with the subject: It is impossible to understand the question of greywater recycling without understanding the different and complex ways in which this question is translated and understood in the social, cultural, and geopolitical local context. Understanding this context is the key to the future of greywater recycling in Israel and worldwide.

This chapter offered several possible perspectives from which to understand the nature of the relationship between the public and the subject. These ranged from broad social factors (macrofactors) such as religion, culture, and government policy to individual factors (microfactors) such as the number of children and ownership of an apartment. All of this was gleaned through questions about the functioning of the technologies themselves or experience gained with them.

In addition, it is clear that the Israeli case itself needs further study. Evidence from the study by Friedler et al. (2006) here presented in detail suggests that indeed Israel has an active community with an expansion potential in the field of greywater recycling. This is supported by the growth of public debate on the subject. However, together with the specific evidence emerging are equally important questions that have to be addressed by future research in this field. These include questions that are also helpful to anyone interested in this kind of research anywhere in the world. For example, do groups that promote legislation on the subject, or who participated in the survey in the study in question, represent the entire Israeli society? Does the actual acceptance among rural or suburban dwellers predict acceptance in the future and in the urban sector? Is the Israeli public aware of the extensive use of water recycling technologies in Israel in agricultural irrigation for producing food? What will happen if and when this issue becomes a broad and open public discussion, beyond its current sector demarcations? How will those who seek to advance desalination technologies respond to the attempt to expand the use of greywater recycling instead, an action that potentially has economic implications? How will the agricultural sector that is fed by sewage water respond to this? Does the position of the Health Ministry on the subject have extensive implications for the development of the sector and in what ways? Is the question of water resources indeed primarily linked to questions of national security in the Israeli case? If so, how is it reflected in the way different groups in Israel relate or may relate to the subject?

The central goal of this chapter has been achieved: The importance of the sociocultural dimension in the debate on greywater recycling has been revealed, and possible channels for progress in this relatively new area have been described. Answers to these questions and many others can be found by joint research by teams combining the social and natural sciences, as well as by studies that seek to compare different populations and countries as opposed to focusing only on a local case. The data presented in this chapter can serve as a basis.

## ACKNOWLEDGMENTS

I would like to thank the authors of the book, as well as Oshrat Hochman, Yaakov Grab, Tali Katz-Gerro, and Isaac Meir, for their useful notes during the writing of this chapter.

# 7 Technoeconomic Aspects of Greywater Reuse

## 7.1 INTRODUCTION

Population growth combined with a constant tendency toward urbanization, which is accompanied in many cases by an increase in specific water consumption, leads to a steady increase in water demand in the urban sector in many parts of the world. Today, many urban regions suffer from water shortages, even regions considered water-rich such as Japan and Europe. This compels the development of more distant (surface water) and deeper (groundwater) water sources, the construction of dams and long conveyance systems, and seawater desalination. Utilization of these new water sources usually results in high direct costs (i.e., construction, operation, and maintenance) and high indirect costs (i.e., external costs induced by an increase in negative environmental effects). Consequently, reducing total water consumption in the urban sector has become a goal for companies supplying water and for entities that outline water policy like central governments.

There are many measures that can be combined to reduce the total water consumption in the urban sector. These include improving the efficiency of water conveyance and distribution systems (i.e., decreasing real losses by reducing the physical loss of water), installing water savings facilities (e.g., water aerators, water-efficient washing machines), raising public awareness regarding water conservation, and introducing water reuse as a viable alternative water source.

For reusing water, on-site greywater reuse has a considerable potential for potable water savings. In developed countries, household water consumption stands on average at 149 LPD (liter/[person × day]) (excluding garden irrigation and other external uses), ranging from 95 to 224 LPD (standard deviation 31 LPD), which is 54 ± 11 $m^3$/[person × year]. Domestic consumption in these countries comprises 30%–70% of the total urban consumption.

The reuse of greywater for toilet flushing has the highest likelihood of being adopted in dense urban areas. This measure reduces domestic water consumption by 40–60 LPD (14.5–22 $m^3$/[person × year]). If it becomes widespread, this practice could reduce urban water consumption by 10%–25%. For example, by 2025 Israel's population is expected to reach about 10 million. If we assume a penetration rate of 30% (i.e., 30% of homes will have systems for greywater treatment and reuse), greywater reuse for toilet flushing in the household sector would lead to savings of about 50 MCM (million cubic meters) per year. This would save approximately 5% of the total anticipated urban water consumption in Israel in 2025, savings that equal the output of a medium seawater desalination plant. If the government encourages and promotes local reuse of this type, a penetration rate of approximately 30% over 20 years is feasible (Friedler, 2008).

As we have seen earlier in this book, greywater can be quite contaminated. Therefore, it may pose a health and environmental hazard or can be not aesthetically pleasing (Rose et al., 1991; Almeida et al., 1999; Dixon et al., 1999a; Nolde, 2000; Diaper et al., 2001; Ogoshi et al., 2001; Friedler, 2004). To address these challenges, effective and reliable systems for treatment, storage, and conveyance of greywater are needed. Many potential treatment systems are described in the literature on greywater (see Chapter 2 dealing with greywater treatment), which differ in their level of complexity and extent of treatment (Shin et al., 1998; Diaper et al., 2001; Hills et al., 2001; Jefferson et al., 2001; Ogoshi et al., 2001; Friedler et al., 2005).

Greywater treatment and reuse projects should be examined from a technoeconomic perspective to provide a complete picture of their implications. Widespread implementation of local greywater reuse will be possible only on the condition that reuse projects will be financially balanced—that the benefit yielded due to water savings will be at least equal the total cost incurred in the initial investment, maintenance, and operation of the systems. Therefore, the direct and real cost or benefit of the project can be estimated only after examining its technoeconomic aspects at the planning stage. An examination of these aspects allows the most appropriate alternative for treatment and conveyance to be chosen.

It should be mentioned that the daily flow of greywater *produced* within the house is higher than the existing demand for toilet flushing (Figure 1.1). Therefore, in cases where the primary use of greywater is for flushing toilets, it is possible and preferable to reuse only *light* greywater (the less polluted greywater streams, from bathroom sinks, baths, and showers), thus reducing both the cost of treatment and potential adverse effects (Friedler, 2004).

## 7.2 BASIC CONSIDERATIONS

When discussing the economic aspects of greywater treatment, there are two distinct entities that carry the cost of treatment and derive profit or benefit from greywater reuse (Table 7.1) (Friedler and Hadari, 2006):

1. The private/individual consumer—the apartment owner, family, or a group of tenants who live in an apartment building
2. The general public—the public, the authorities, the water corporation, and the central government

The individual consumer benefits from potable water savings—as expressed in the water bill—or from the possibility of enjoying a garden, even in arid areas. On the other hand, he or she carries the major burden of the cost of treatment: cost of construction, operation, and maintenance of the system. In contrast, the public does not bear the burden of the cost of greywater treatment, but enjoys the benefit resulting from greywater recycling in several ways. First, since the water consumption of the individual user decreases, the water costs for the entire public are reduced due to

**TABLE 7.1**
**Principle Cost–Benefit Analysis of Greywater Reuse**

| Benefit | Cost |
|---|---|
| *Individual consumer* | |
| Savings in water and sewage bills | Installing double piping: collecting greywater and blackwater separately, separate supply of potable water and of treated greywater. |
| Allowing for a garden in arid areas | Transporting treated greywater: energy and storage costs. |
| | Greywater treatment system: return on investment, operation and maintenance, control. |
| *General public* | |
| *Water sources*: cancelling or postponing development of new water sources | |
| *Water treatment*: energy savings, chemicals savings, option to postpone expansion facilities and in future build smaller new plants | |
| *Water supply systems*: energy savings, option to postpone expansion and in future install smaller diameter pipes | |
| *Sewage collection*: energy savings (pumping), piping expansion that could be postponed and new systems that could be of smaller diameters | *Sewage collection*: lower flows may require higher maintenance (higher probability of blockages). |
| *WWTP*: lower pollutant load, energy savings, potential for postponing enlargement of WWTP and new WWTP that could be built smaller | *WWTP*: higher pollutants concentrations (less dilution). |

*Source:* Based on Friedler, E. and Hadari, M., *Desalination*, 190(1–3), 221, 2006.

the reduced need for developing new water sources and the subsequent reduction in the costs of water production, treatment, and conveyance. Secondly, reducing water consumption reduces the amount of wastewater produced and consequently affects the sewage collection, transport, and treatment process. Reducing wastewater flows may lead in some cases to an increase in the rate of blockages in sewers. However, a large proportion of the sewer systems are close to the upper limit of their design capacity. As the penetration process of greywater reuse is expected to be long and gradual, it is likely that it will not lead to an increased rate of sewer blockages, but rather will be of benefit by postponing the need to upgrade or expand the existing wastewater conveyance systems. In addition to the benefits that can be quantitatively assessed, there are benefits that cannot be directly calculated, such as the possibility of developing private or public gardens in arid regions that makes them more *livable*.

The cost distribution model proposed below is just one of the possible models for the distribution of costs and benefits between the individual user and the public. This model is very extreme: the public does not bear any of the costs, but enjoys most benefits (except for a reduced water bill and sewage fees). As such, this model can serve as a case study for demonstration purposes because, if it proves economic feasibility, it is likely that less extreme models of the distribution of the financial burden will also be feasible. Therefore, this chapter focuses on examining the direct costs and benefits to the individual user (of the greywater). The basis for the analysis is a *typical* multiresidential apartment buildings in Israel. The data on these buildings were taken from publications of the Israel Central Bureau of Statistics (ICBS, 2010) and Ministry of Construction and Housing in which it is stated that the average size of a family is 3.4 persons, one family living in each apartment, four apartments in each floor, and the height of each floor is 3 m. In addition, it was assumed that the interest rate is 3.5% and the payback period is 15 years, which complies with the expected serviceable life of the main elements in the system.

## 7.3 CALCULATING ANNUAL COST OF GREYWATER SEPARATION AND TREATMENT

To compare the various alternatives of greywater treatment and reuse, they should be calculated using on a common economic base. There are several ways to calculate economic alternatives:

1. Net present value of the investment, operation, and maintenance costs (Tang et al., 2006; Chen and Wang, 2009; Nazer et al., 2010)
2. Financial life cycle impact assessment (Nazer et al., 2010)
3. Number of years required for investment return period (Friedler and Hadari, 2006)
4. Net annual cost (Friedler and Hadari, 2006)

In this chapter, the comparison will be made primarily in terms of the number of years required for the return of the investment and the overall annual cost (net annual cost).

### 7.3.1 Cost of Separating Greywater and Blackwater Streams

The cost of separating water streams in a building is simply the cost of the additional plumbing required for a dual system of greywater conveyance. The greywater conveyance system is divided into three main parts (Figure 7.1): (1) collection system of raw greywater from the apartments to the treatment facility, (2) conveyance system of treated greywater to a central storage tank (usually located on the roof), and (3) a distribution system of treated greywater from the storage tank to the toilet cisterns in each apartment (Friedler and Hadari, 2006).

The cost of separating streams in apartments of new multiple-family buildings (specifically when this is done during construction) should not be high because in high-rise buildings, the toilets and showers are located close to the infrastructure shaft of the building, so the extra plumbing needed for each apartment is minimal.

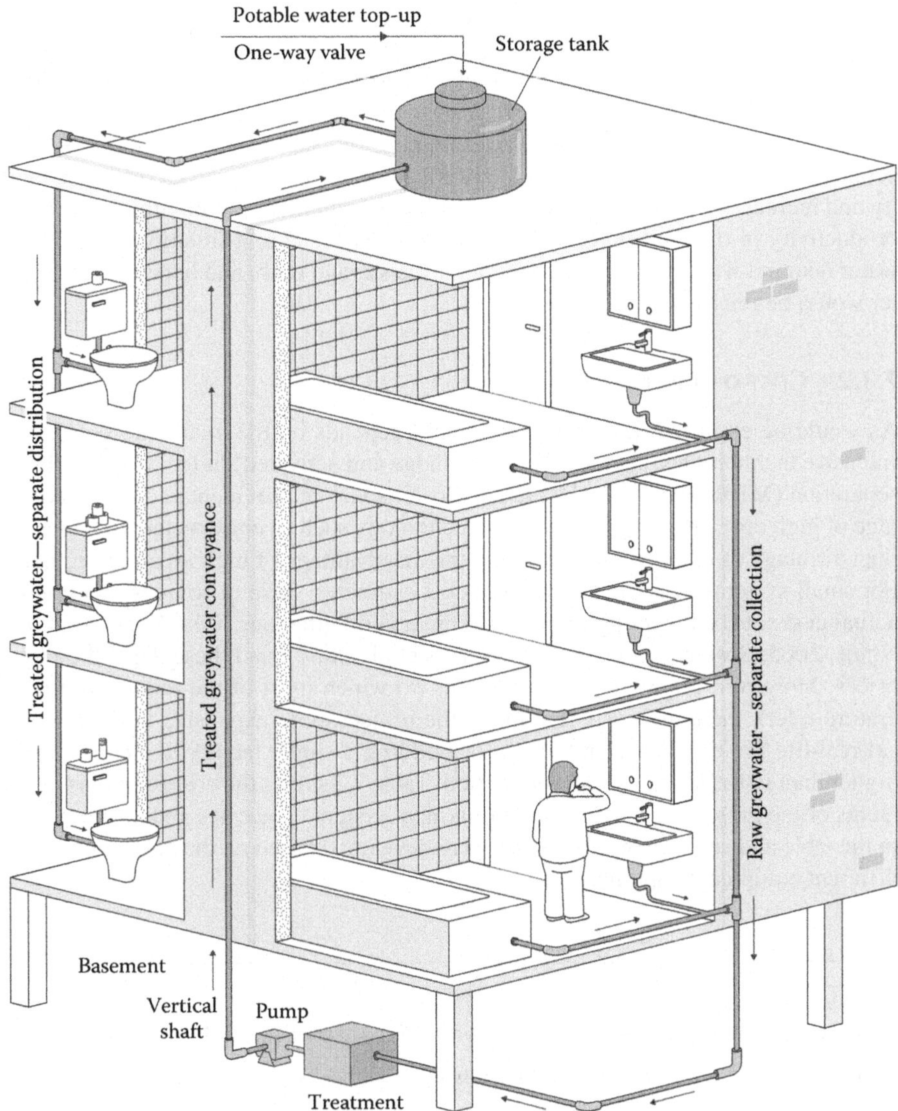


**FIGURE 7.1** Schematic description of collection, treatment, conveyance, and supply of greywater in a building of several floors. (Based on Friedler, E. and Hadari, M., *Desalination*, 190(1–3), 221, 2006.)

It was estimated that for the separation of streams, extra piping at a length of 5 m would be required for each apartment. An additional 9 m per floor (3 × 3 m) would be needed for collecting raw water for the treatment facility, transport of treated greywater to a storage tank, and conveyance of treated greywater from the storage tank to each floor (en route to the toilet cisterns). Another 15% should be added to this cost estimate for additional expenses such as fasteners, valves, and angles (Friedler and Hadari, 2006).

Several studies have addressed the challenge of finding the optimal volumes of a greywater collection tank and storage tank of the treated greywater (Fewkes and Ferris, 1982; Dixon et al., 1999a). The collecting and storage volumes have different economic effects: On the one hand, increasing the volume raises the cost of the facility and increases its footprint. On the other hand, a smaller volume limits the overall productivity of the facility. In other words, this will lead to a situation in which the water demand would exceed the volume of the storage tank, and hence less greywater would be reused.

### 7.3.2 Cost of Treatment

As would be expected, the cost of treatment depends on the treatment technology. Intensive technologies, such as activated sludge and activated sludge with membrane separation (MBR), have the advantage of lower space requirement and the disadvantage of high cost. In contrast, extensive technology such as constructed wetlands has the advantage of relatively low cost and the disadvantage of high area requirements. For small systems, it is difficult to find cost estimates. Several sources mention the actual costs of the recycling project (Brennan and Patterson, 2004; Lin et al., 2005; Nolde, 2005; Šostar-Turk et al., 2005; Gross et al., 2007), and these depend on local prices. Moreover, among research projects (to which most of the reports in the literature refer), there are costs that make the project more expensive than an identical real-life project. When producing or building a single facility, the cost is much higher than in the case of serial production. Table 7.2 shows the cost functions of elements of the greywater transport and treatment system. The costs and cost functions in the table and subsequent discussion were developed based on the costs reported by different equipment suppliers.

**TABLE 7.2**
**Cost Functions of Elements of Greywater Treatment in High-Rise Buildings**

| Item | Basis for Cost Calculation | Units | Cost Functions |
|---|---|---|---|
| Apartment piping | Length ($L$) | US$/m | 3 |
| Central piping | Length ($L$) | US$/m | 6 |
| Tank | Volume ($V$) | US$/$m^3$ | $C = 144 \cdot V^{0484}$ |
| Pump | Flow rate ($Q$) | US$/($m^3$·d) | $C = 594 \cdot Q^{0.0286}$ |
| MBR | Flow rate ($Q$) | US$/($m^3$·d) | $C = 18,853 + 17,954 \cdot \ln(Q)$ |
| RBC (Including a sedimentation basin) | Flow rate ($Q$) | US$/($m^3$·d) | $C = 3,590 \cdot Q^{0.6776}$ |
| Chlorination | Facility | US$/unit | $C = 1,670$ |

*Source:* Based on Friedler, E. and Hadari, M., *Desalination*, 190(1–3), 221, 2006.

### 7.3.3 Operation and Maintenance Costs

Operation and maintenance costs of a greywater system consist of the following components: energy used for treatment and conveyance, personnel, materials, maintenance (such as membrane maintenance), chemicals (e.g., chlorine for disinfection), replacement, and repair of wear and tear (e.g., pumps).

#### 7.3.3.1 Energy

The energy required for treatment depends, of course, on the chosen treatment technology. For example, the energy consumption for a rotating biological contactor (RBC) system was calculated as follows (Friedler and Hadari, 2006):

$$P = \frac{19.27 + 10.94 \cdot Q}{1000} \tag{7.1}$$

where
$P$ is the energy consumption (kWh/[$m^3 \times$ day])
$Q$ is the flow rate ($m^3$/day)

Based on data from manufacturers and the literature, the energy consumption of a membrane bioreactor (MBR) was estimated at 1.5 (kWh/$m^3_{\text{treated}}$) (Friedler and Hadari, 2006).

When the greywater is reused for toilet flushing, the water has to be pumped from the treatment facility to the storage tank, usually located on the roof of the building. The higher the building, the higher the height that must be overcome and the higher the head losses in the pipes. As such, the energy cost also increases. The head loss in the pipeline can be calculated using the Hazen–Williams equation:

$$\Delta H_f = 1.131 \times 10^9 \cdot \left[\frac{Q}{C_{H\text{-}W}}\right]^{1.852} \cdot D^{-4.87} \cdot L \tag{7.2}$$

where
$\Delta H_f$ is the head loss (m)
$C_{H\text{-}W}$ is the Hazen–Williams coefficient (–) (150 for plastic piping)
$Q$ is the flow rate ($m^3$/h)
$D$ is the pipe diameter (mm)
$L$ is the pipe length (m)

The head differentials that the pump must overcome are

$$\Delta H = \Delta Z + \Delta H_f \tag{7.3}$$

where $\Delta Z$ is the difference between the height of the treatment facility and the height of the upper storage tank (m).

The energy required for pumping is described by

$$P = \frac{\gamma \cdot Q \cdot \Delta H}{\eta} \tag{7.4}$$

where

$P$ is the output required from the pump (W)

$\gamma = \rho \cdot g$ (kg/[m$^2$ s$^2$]) ($\rho$ is the water density [kg/m$^3$], $g$ is the gravity acceleration [9.81 m/s$^2$])

$Q$ is the flow rate (m$^3$/s)

$\eta$ is the pump efficiency (–)

It should be noted that in high-rise buildings, potable water has to be pumped to the upper floors (because the pressure head in the municipal water supply network is not sufficient for this purpose). Therefore, in a tall building that reuses greywater, the greywater replaces some of the potable water and results in a decrease of energy required for pumping potable water to the roof of the building or the top floors (in addition to water savings).

#### 7.3.3.2 Labor

Labor is required for maintenance, monitoring, and repairs. It is estimated that for intensive systems, about 1 h of work per week is required to complete these tasks. For extensive systems this is about a quarter of an hour per week.

#### 7.3.3.3 Materials

Materials are required for flocculation (if performed) and disinfection (if chlorine disinfection is performed). The amount of flocculants required depends on the raw greywater quality, and the amount of chlorine required depends on the efficiency of the treatment process (before the disinfection stage). One study used a dose of flocculating agent ($FeCl_2$) of 40–50 mg/L in the treatment of raw light greywater as a pretreatment before membrane filtration (Friedler et al., 2008). This dose is on the same scale as the dosage required for wastewater effluent of good quality (Soffer et al., 2005) and much lower than the dose required for primary effluent of municipal wastewater (Abdessemed and Nezzal, 2002). There are those who indicate a low chlorine demand of about 3 mg/L to reach residual chlorine of 1 mg/L after 1 h of contact in treated greywater, treated with an RBC or MBR (Friedler et al., 2006, 2011). In contrast, others point to chlorine demand higher in one order of magnitude (45 mg/L) for greywater treated only at a basic level (March et al., 2004).

### 7.3.4 Comparison of the Cost Separation, Collecting, and Treatment of Greywater

Costs of various facilities for greywater treatment systems as reported in the literature are depicted in Figure 7.2. The systems examined offer varying levels of

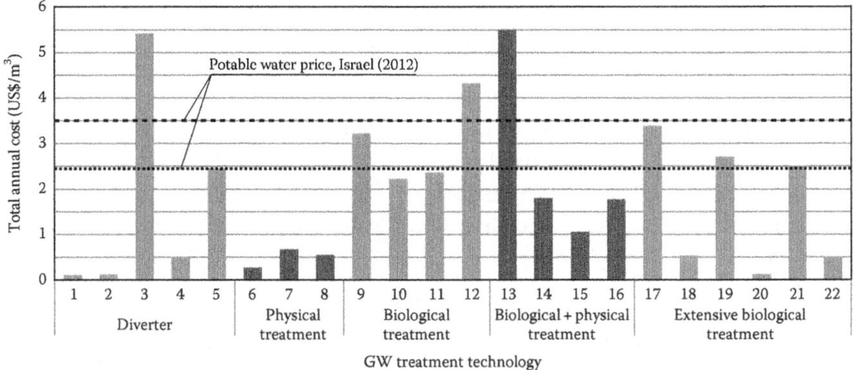


**FIGURE 7.2** Comparing the cost per cubic meter of treated water in different treatment technologies. *Sources of data*: (1) Laundry diverter (PlanetArk, 2007), (2) diverter + filter (PlanetArk, 2007), (3) diverter + tank + installation (PlanetArk, 2007), (4) diverter + tank + pump (PlanetArk, 2007), (5) diverter + filter + tank + pump (PlanetArk, 2007), (6) electro coagulation (Lin et al., 2005), (7) coagulation + sand filter + GAC (Šostar-Turk et al., 2005), (8) primary treatment (Brennan and Patterson, 2004), (9, 10) RBC + sedimentation + chlorination (Friedler and Hadari, 2006), (11) AquaCycle (Nolde, 2005), (12) secondary treatment (Brennan and Patterson, 2004), (13) MBR + chlorination (Friedler and Hadari, 2006), (14) biological oxidation + ultrafiltration (Lin et al., 2005), (15) biological oxidation + microfiltration + UV (Lin et al., 2005), (16) MBR (Šostar-Turk et al., 2005), (17) biological + chemical + physical (PlanetArk, 2007), (18) RVFCW (Gross et al., 2007), (19) RVFCW + filtration + UV (Gross and Friedler, unpublished data), (20) sedimentation + gravel and sand filtration + aeration + chlorination (Godfrey et al., 2009), and (21, 22) VB + UV (calculated in this chapter).

treatment, ranging from diverters (which actually constitute no treatment at all) to those with a combination of biological and physical treatment. The specific cost (in US$/m$^3$ treated greywater) in 14 out of 22 of the reported facilities (64%) is lower than the cost per m$^3$ of freshwater in Israel (2.4 US$/m$^3$, at the lower water tariff in 2012), and the specific cost in 19 out of 22 facilities (86%) is lower than the higher tariff for potable water (3.5 US$/m$^3$ in 2012). The specific cost of a cubic meter of treated greywater in the diverters' group is generally the lowest, as would be expected, but the water is reused without any treatment at all. Also in systems comprised of physical treatment only, the treatment cost is relatively low. However, separation facilities and physical treatment usually yield treated greywater of inadequate quality (Friedler and Alfiya, 2010). Facilities that incorporate biological treatment as part of the treatment train are usually more expensive, and sometimes, the cost of treated greywater is higher than the savings resulting from reducing the use of potable water. The specific cost of extensive biological treatment is usually low compared to the cost of intensive one. However, extensive treatment requires a larger area and is usually not suitable for dense urban areas (characterized by high-rise buildings). It is best for single-family houses in rural or suburban areas.

## 7.4 COMPARISON OF THREE BIOLOGICAL TREATMENT TECHNOLOGIES AS A CASE STUDY

Since there is a wide range of technologies and varieties of treatment systems, it is impossible to analyze each one of them separately. Therefore, in order to demonstrate the cost distribution, the total treatment cost of three different treatment systems was calculated to serve as a case study (Figure 7.3): (1) MBR, representing the latest generation of compact intensive wastewater treatment technology; (2) RBC,

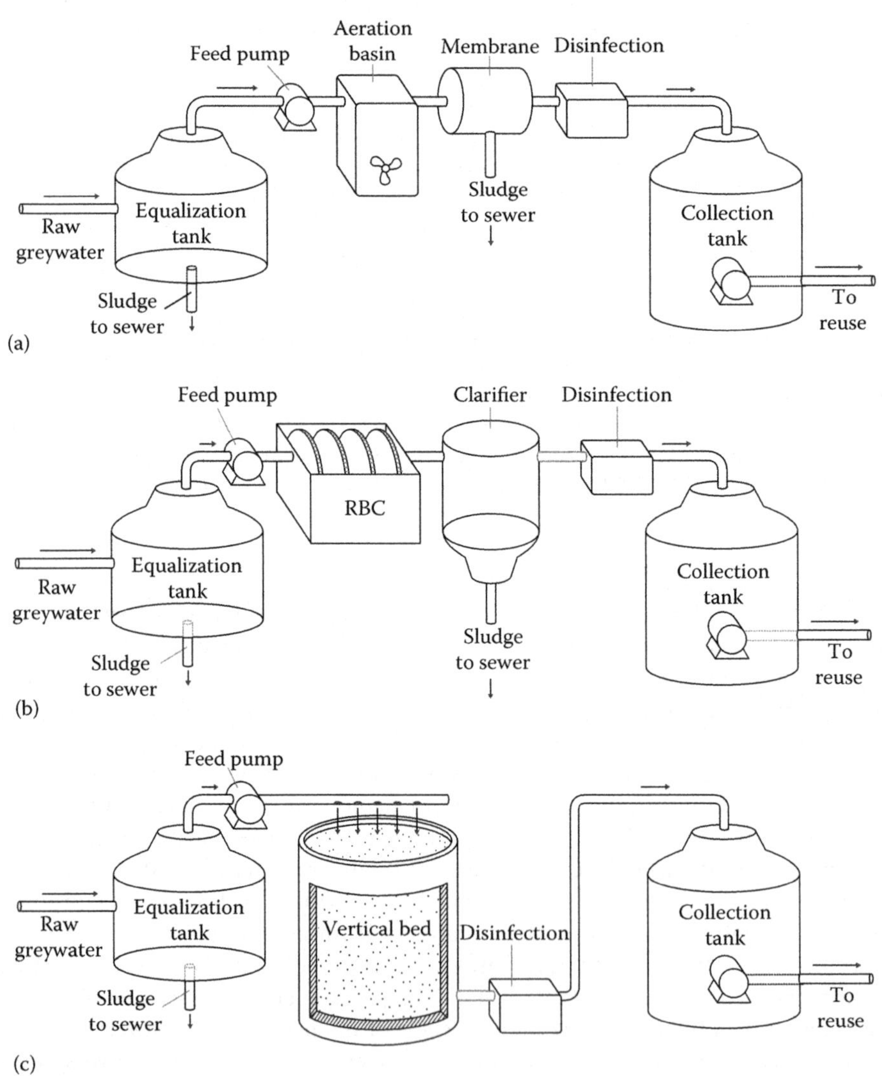


**FIGURE 7.3** Schematic of the treatment process through (a) MBR, (b) RBC, and (c) VB.

**TABLE 7.3**
**Economic Comparison of the Three Treatment Alternatives, Calculated for Treatment of 3.3 m³/day (Equivalent to Greywater Generated by 55 Persons)**

| | Equalization | Biological Treatment | | | Disinfection[a] | |
|---|---|---|---|---|---|---|
| **Criterion/System Component** | **Tank[a]** | **MBR** | **RBC** | **VB** | **UV** | **$CL_2$** |
| Required area ($m^2/m^3$) | 0.27 | 0.37 | 0.42 | 1.2 | — | 0.06 |
| Capital return (US$/$m^3$) | 0.015 | 3.67 | 0.98 | 0.28 | 0.06 | 0.12 |
| Energy consumption (US$/$m^3$) | — | 0.21 | 0.06 | 0.01 | 0.04 | — |
| Operation and maintenance (US$/$m^3$) | 0.08 | 1.89 | 1.27 | 0.30 | 0.12 | 0.12 |
| Total cost (US$/$m^3$) | 0.095 | 5.77 | 2.31 | 0.57 | 0.22 | 0.24 |

[a] Costs of equalization tank and disinfection are already included in the treatment costs.

which represents a long-standing and proven technology suitable for small treatment systems including high-rise buildings; and (3) an extensive VB-type (vertical bed) system, usually appropriate for small/medium houses. It should be stressed that the economic analysis described here does not intend to be comprehensive, but rather to serve as an example of comparison between various alternatives.

The following calculations were made according to the detailed principles described earlier. In addition, the cost of primary treatment (which includes an equalization tank) was calculated, as well as the cost of posttreatment (i.e., disinfection, either by chlorination or UV irradiation). For the sake of the calculation, it was assumed that each alternative includes pretreatment (equalization tank) and the final stage of disinfection.

The overall treatment cost (investment, operation, and maintenance) and space requirements of the three technologies are compared in Table 7.3. The cost of treatment is highly dependent on the unit's size (see the following discussion), so it is important to note the design flows of the facilities tested. The calculations were performed for three systems that are suitable for a design flow of 3.3 m³/day, which corresponds to treatment of greywater produced by 55 persons. This is equivalent to a building of 4–5 floors, with four apartments on each floor. As expected, treatment by MBR was found to be the most expensive, 2.5 times more expensive than the treatment using RBC and 10 times more than the treatment using VB. The space requirement of this facility was only 12% smaller than that of the RBC space requirement. The cheapest treatment was by VB, also as expected, although the space requirement of this facility is 3–3.5 times greater than that of RBC and MBR. In calculating the costs of these three systems, pretreatment that includes an equalization tank was taken into account. This tank ensures that water flow and pollutant concentrations are balanced at the entrance to treatment. The initial investment in this tank is low, and the maintenance required is primarily for periodic cleaning and emptying of the accumulated sludge. The total cost of disinfection, by chlorination or by UV radiation, was found to be very similar.

### 7.4.1 Composition of Treatment Cost

Greywater treatment cost consists of (1) the investment (specifically the return on investment) in the raw and treated greywater conveyance system and treatment system and (2) the cost of operation and maintenance. In each treatment technology, the ratio between the operation/maintenance costs and the construction costs is different and also changes as a function of the facility size (which is itself a function of the building size). Figure 7.4 presents the distribution of relative costs of each cost component as part of the total costs for the three representative treatment systems (design flow of 3.3 m$^3$/day; equivalent to 55 persons). The more intensive the treatment technology, the higher the proportion of energy cost relative to the capital cost payback (Equations 7.5–7.7 calculate the capital return). In the extensive treatment (VB), the cost of energy comprises only 2% of the total cost, compared with 8% and 13% in RBC and MBR, respectively. The main difference between RBC and MBR is in the relative share of the cost of capital return. In RBC, the capital cost reimbursement is 47% of the total cost, while in MBR it is 56%. As illustrated in the figure, the total cost (normalized for a treated cubic meter) with VB treatment is the lowest; however, the footprint of this technology is relatively high, and therefore it is less suitable for implementation in densely built areas where the cost of land is high. It is more suitable for detached houses or several houses in a rural or suburban area.

Figure 7.5 displays the total specific cost (cost per cubic meter of treated greywater), for the RBC and MBR systems as a function of the size of the building (VB was not included in this analysis because it is suitable only for small buildings). With the increase in the building size, a steep decrease occurs in the total cost, especially in small houses range (of four stories and less). In buildings where the treatment is performed by RBC, the cost of treatment decreases, beginning with six floors down to below $2\,\text{US\$}/\,m^3_{\text{treated}}$. In buildings of 22 floors or more, the cost stabilizes at less than $1\,\text{US\$}/\,m^3_{\text{treated}}$ and hardly drops any lower. Treatment by MBR is more expensive and declines more moderately. It falls below the $2\,\text{US\$}/\,m^3_{\text{treated}}$ line only in buildings higher than 22 floors (88 apartments).

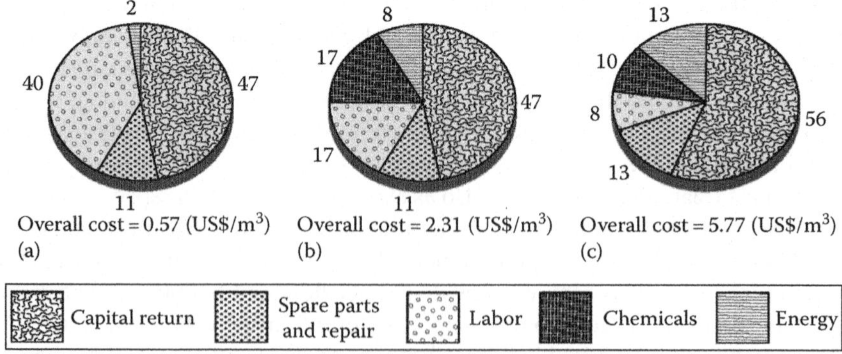


**FIGURE 7.4** Distribution of the relative share of the cost components of treatment systems (design flow rate 3.3 m$^3$/day). (a) VB, (b) RBC, and (c) MBR. Disinfection in MBR and RBC is performed with chlorine, in VB with UV radiation.

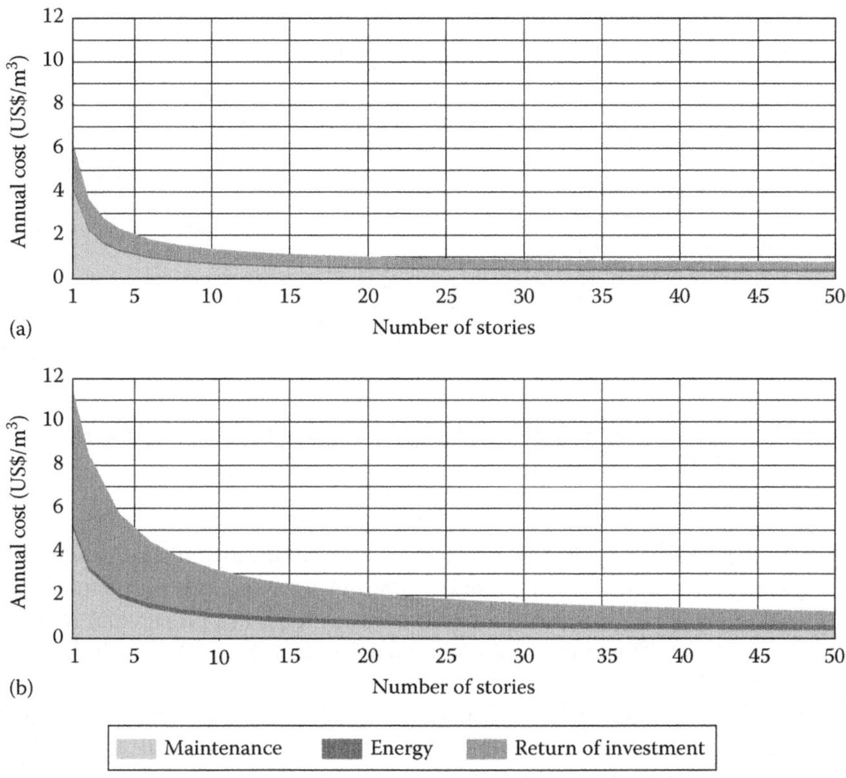


**FIGURE 7.5** Specific treatment cost (including all components) as a function of the number of floors in a building (four apartments *per* floor) for (a) RBC and (b) MBR.

To examine the effect of the number of apartments on the cost of treatment and reuse of greywater, the cost per flat was calculated for the MBR and RBC systems (Figure 7.6). Moreover, the savings potential in the annual water bill per flat (according to the price of water in Israel in 2012, 2.40 US$/m$^3$ [up to 2.5 m$^3$/(person × month)] and 3.50 US$/m$^3$ for any additional quantity) resulting from replacing potable water with treated greywater in toilet flushing is highlighted in the figure. Again, in keeping with the previous illustration, a sharp decline in the specific cost per apartment per year is seen, especially in small buildings, indicating economies of scale. In buildings where an RBC system is installed, the financial savings potential becomes greater than the overall treatment cost when the number of apartments in the building exceeds 12 (equivalent to three floors). In other words treatment and reuse of greywater begins to be economical for the individual consumer from this size up. For treatment with the MBR system, only in buildings of 10 floors (40 apartments) or more does the savings potential exceed the specific cost of treatment. A decrease in the annual interest rate, an increase in the price of potable water, and/or a reduction in the up-front cost of the systems (from construction or energy use) may change the differences between the MBR and RBC systems.

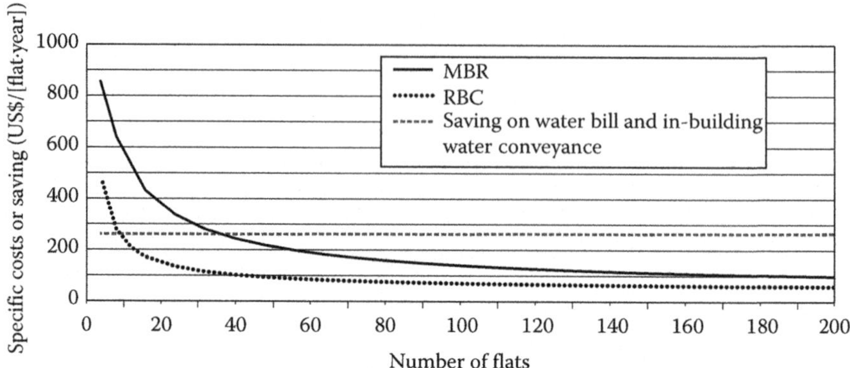


**FIGURE 7.6** Treatment with MBR or RBC: cost of treatment and reuse of greywater per flat and savings in water bill per flat.

Due to the relatively high cost of treatment with MBR, it would be better to build such systems for a cluster of high-rise buildings using a semicentral treatment facility. This can increase the number of apartments in the reuse project and reduce the cost per apartment. This option was examined for clusters of 10-story buildings (40 apartments each). In 2006 when water prices were significantly lower (60%–70% lower) than current water prices, beginning with four buildings of this size, the system became economic (Friedler and Hadari, 2006). With today's (2014) water prices, MBR-based systems become economic starting at a much lower number of apartments (40 apartments, as compared with 160 apartments in 2006).

It is noteworthy that in recent years, more high-rise apartment buildings have been built around the world. Israel has also designed and constructed residential buildings of 30 floors and more. For example, in Ramat Gan (a city on the Greater Tel Aviv metropolitan area), there is a plan to construct the "Elite" building of 70 floors, of which 40 floors are designated for residence (Dvir, 2009); the Akirov Towers in Tel Aviv have 34 floors each, a total of three towers of 360 apartments; and the two Yoo Towers (also in Tel Aviv) have 39 and 41 floors.

Figure 7.7 displays the cost of initial investment in the treatment system and reuse as part of the total cost of the apartment for an average apartment price in Israel in 2010 and for prices that are higher and lower by 50% of the average price. As expected, one can see that the relative cost of investment in an RBC system is lower than that in an MBR system. In addition, it can be seen that the relative investment costs of the RBC system is sensitive to size, especially for small buildings (less than 24 apartments, 6 floors). For buildings above this size, the relative cost is lower than 0.25% of the (average) apartment price. The influence of the building size on the relative investment costs of the MBR system is more moderate: for a building with six floors (24 apartments), the extra cost for constructing an MBR system for the whole building is 0.9% of the apartment price. Only when the building is more than 14 stories (56 apartments) does the extra cost decrease fall below 0.5%.

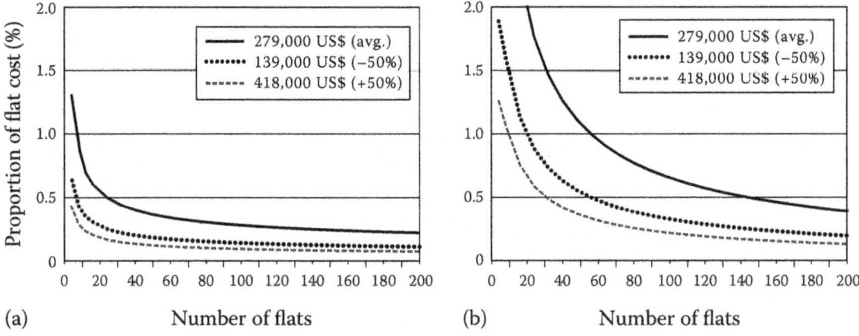


**FIGURE 7.7** Construction costs of the treatment and reuse system of greywater in multi-apartment residential buildings as part of the price of an apartment in Israel: (a) RBC system; (b) MBR system. Prices drawn from the ICBS data for 2010 (ICBS, 2011).

## 7.5 BENEFIT

An example of cost–benefit analysis for an individual consumer is described in this section as well as in the following. The analysis in the example refers to prices of water in Israel but could be easily adapted to other regions. The main (direct) benefit derived by the individual consumer from the reuse of treated greywater is savings in the water and sewage bill.

In the last 3 years, the water rate in Israel increased by about 25%. As of July 7, 2010, the rate for the *approved quantity* (up to 2.5 $m^3$/[person × month]) was 2.40 US$/$m^3$. For any additional quantity, the rate is 3.50 US$/$m^3$ (Israel Water Authority data, Figure 7.8). It should be noted that these rates include sewage fees that were

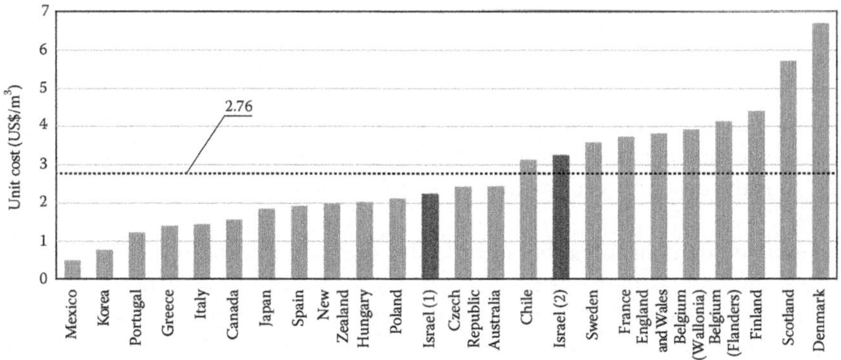


**FIGURE 7.8** Water prices in the OECD countries and in Israel (the average price in the OECD members is 2.76 $US/$m^3$). (From Friedler, E., The water saving potential and the socio-economic feasibility of urban on-site greywater reuse. Keynote speech, in *International Conference on Integrated Water Management*, Murdoch University, Western Australia, February 2011.)

not included in the price of water in the past. Notably, the water tariffs for domestic consumers in Israel are around the average for the 23 OECD countries, and the price of water is higher especially in countries that are not considered water scarce. The reason for this apparent abnormality is that in these (European) water-rich countries, the indirect costs of water use are included in the consumer price. It is important to note that on January 1, 2013, there was another increase in water rates in Israel that raised the price to 2.53 and 4.07 US$/m$^3$ for the lower and higher tariff, respectively. This rise was accompanied by an increase in the amount of water for which the lower rate is paid from 2.5 to 3.5 m$^3$/(person × month).

In addition to the benefits gained by individual consumers from a certain penetration threshold rate of greywater systems, the general public will also benefit from greywater reuse. This is due to the reduction in total water consumption and the corresponding reduction in the volume of wastewater conveyed to wastewater treatment plants (WWTPs). It should be noted that it is difficult to directly quantify some of the benefits of greywater reuse, so it is done indirectly. For example, the improvement in the environment as a result of greywater reuse can be expressed in several ways: prevention of environmental pollution caused by untreated wastewater, improved landscaping with watered gardens and a green cover, and the use of recycled water for artificial pools. It is difficult to quantify such benefits in monetary terms (Chen and Wang, 2009).

## 7.6 EXAMINATION OF THE ECONOMIC FEASIBILITY OF GREYWATER TREATMENT AND REUSE

### 7.6.1 Annual Savings versus Operation and Maintenance Costs

The net annual savings in costs as a result of greywater reuse can pay back the initial investment in the system (Equation 7.5):

$$R = R_{WS} - C_{O\&M} \tag{7.5}$$

where

$R$ is the annual net cost savings (US$/year)

$R_{WS}$ is the annual savings of water bill and sewage fee (US$/year)

$C_{O\&M}$ is the cost of operation and maintenance of treatment system (US$/year)

This equation demonstrates that in order for greywater reuse to be potentially economical, the savings in the water bill and sewage fee resulting from the water reuse must be higher than the operation and maintenance cost of the reuse system. This is a necessary condition, but not sufficient, because the initial investment also has to be repaid (see the following for more details). In Figure 7.6, one can see that reuse implemented with RBC-based system already meets this condition for a building that has 10 apartments (three to four floors). As expected, an MBR-based system meets this condition only for a building of 36 apartments (nine floors).

## 7.7 EXAMINING THE PAYBACK PERIOD OF THE INITIAL INVESTMENT IN THE SYSTEM

The payback period of the investment can be calculated using the following equation:

$$R = \frac{i \cdot (1+i)^n}{(1+i)^n - 1} \cdot C \tag{7.6}$$

where

- $R$ is the annual payment (US$/year)
- $i$ is the annual interest
- $n$ is the payback period (years)
- $C$ is the total cost of investment (US$)

Representing the payback period from this equation as a function of the other variables yields the following:

$$n = \frac{\log\left[1/(1-(C \cdot i/R))\right]}{\log(1+i)} \tag{7.7}$$

This shows that the equation is solvable only when the expression in the logarithm is positive ($R > C \cdot i$). This is logical, because the return on investment begins only when the annual savings ($R$) is higher than the annual interest repayment ($C \cdot i$). This is, as mentioned, a necessary but not sufficient condition. However, for the system to be economically feasible, the payback period ($n$) should be shorter than the expected life of the system (serviceable life). Since electromechanical equipment comprises most of the investment, the serviceable life of the system was determined to be 15 years. The annual interest was set at 3.5%.

Figure 7.9 presents the capital payback period of greywater reuse as a function of the number of apartments. The system's serviceable life (15 years) is marked by a dotted horizontal line. The figure shows that the payback period decreases from 15 years when the number of apartments exceeds 12 apartments for an RBC system and when the building has more than 36 apartments for an MBR system. In other words, a system does not cost the consumer when it is installed in buildings of this size or in larger buildings. It is important to note that the payback period of the RBC system decreases much more sharply with the size of the building than does the payback period of the MBR system. Subsequently, the size of the building that will ensure a repayment period of 5 years (1/3 of the system's serviceable life, a common repayment period in the industrial sector) was examined, and it was found that for the RBC system, a building of 24 apartments is required (six floors), while for an MBR system, a building of 100 apartments is required.

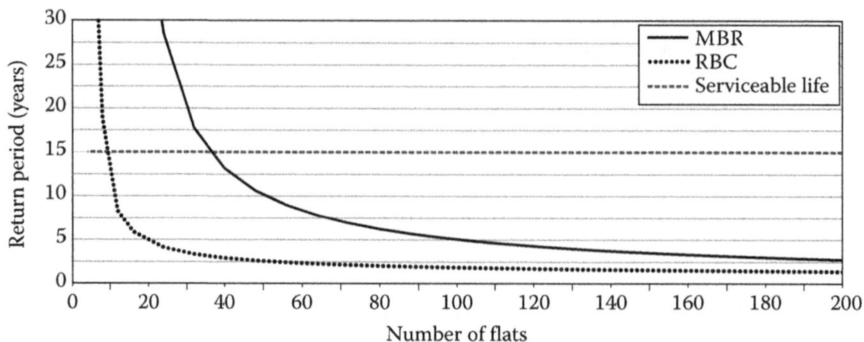


**FIGURE 7.9** Capital payback period of greywater reuse using RBC and MBR as a function of the number of apartments.

### 7.7.1 Subsidies and Incentives

As the aforementioned (in Section 7.2), it seems that the economic benefit from on-site (or local) greywater reuse on a regional and national scale is much more significant than the economic benefit to the private/individual user. This is especially true in areas suffering from a shortage of freshwater. Therefore, authorities (local and national) should incentivize individuals to install such reuse systems. Private/individual users can be encouraged to reuse greywater by various means such as subsidies per cubic meter of greywater being reused, loan funds with reduced interest to finance the construction of systems, and property tax reductions for buildings that reuse greywater. It should be noted that in some developed countries such as Australia and several U.S. states, incentives are given to citizens who install greywater reuse systems, either by a direct payment or through tax reduction.

In Israel, desalination facilities are built to overcome the national shortage of water. Israel already desalinates about 300 MCM per year in 2013, and the volume is expected to increase to 700 MCM per year by 2030. Greywater reuse could (if becoming a common practice) reduce the need for some of this desalination capacity and thus save in costs to the public. Hence, the state should find a way to subsidize greywater reuse at a rate that can outdo the total direct cost to the state of desalinated water.

## 7.8 SUMMARY

This chapter analyzed the economic feasibility of establishing on-site greywater reuse in new construction in the urban sector. The economic feasibility for the individual user is a precondition for the expansion of the integration of these systems in urban construction.

We distinguished between two separate entities in the urban sector: the individual user and the general public. For the economic analysis, an extreme model of cost sharing was selected in which the individual customer bears all the costs (and enjoys reduced water bill and sewage fee), while the general public does not bear the burden of the cost involved in greywater reuse. However in reality, the public also enjoys the benefit arising from the installation of the reuse systems by individual users thanks to positive outcomes from a general reduction of water consumption and sewage discharge.

Under this model, in which the individual user alone bears all expenses, the cost of investment in a greywater reuse system in Israel using RBC drops below 0.25% of the cost of an average Israeli apartment (279,000 US$) in a building of 24 apartments or more (six floors). The cost of investment in a greywater reuse system using MBR falls below 0.5% of the cost of an average Israeli apartment only in buildings of 14 stories or more (56 apartments). This is a small addition to the expenditure on housing, and it underlines the attractiveness to the homeowner and the contractor who builds the building: for a negligible increase in cost (which would be borne by the customer anyhow), the building gains a valuable *environmentally friendly* image.

Considering the marginal cost of water (plus sewage fee) of 3.5 US$/m$^3$, the size threshold for an RBC-based system to become economically feasible is four stories (equaling 12 apartments). Beyond that size, the individual user starts making a profit. In an MBR-based system, economic viability begins at a 9-story building (36 apartments).

The payback period of the investment was determined to be highly sensitive to the price of water and the size of the building, especially in reuse systems of smaller-sized buildings. This is because the specific cost of the treatment system (which is very sensitive to changes in size) is the main component of the initial investment cost, while the specific investment in collection and transmission pipelines (which is not sensitive to size) is only a fraction of the total initial investment. In the Israel example, it was found that for an RBC system, the payback period drops below 5 years when the building has over 24 apartments, while for MBR, the payback period decreases to below 5 years when the building has more than 100 apartments.

As noted, in accordance with the extreme economic model tested, the general public benefits from greywater reuse thanks to reduced water consumption and reduced sewage discharges by individual users, but the latter bears all the costs. Because the benefit gained can be significant from a regional and national perspective, authorities may want to incentivize residents to install on-site/local greywater reuse systems. The authorities can encourage the installation by providing a subsidy per each recycled cubic meter, by providing a property tax rebate, or through other economic mechanisms. As mentioned earlier, there are a number of countries that already encourage greywater recycling by providing subsidies or tax breaks.

It is important to note that, with the growing prevalence of greywater reuse systems, the construction and installation costs of such systems will be reduced, and they will become more attractive to the public of potential users. This fact is true for many systems including RBC but is especially true for MBR systems. The latter is an innovative technology that is controlled by a limited number of international manufacturers, especially in the small sizes.

The economic analysis in this chapter demonstrates that the use of on-site systems for greywater reuse is a feasible solution to reduce urban water consumption, not only for environmental motives and water use efficiency but also from an economic perspective.

This chapter dealt only with the direct cost–benefit analysis greywater reuse systems. A broader analysis should also include indirect cost and benefits such as the use of natural resources, use of raw materials, reduction in energy consumption (and its significance for air pollution), and increase in the availability of water for gardens in arid regions. Such an analysis is beyond the scope of this book but would make a logical next step in this line of analysis.

# Summary

The aim of this book is to bind together unbiased scientifically based information on greywater and its reuse. It was demonstrated that along with its various challenges, greywater reuse is one mechanism that can contribute to significant water savings on various scales (e.g., individual, regional, and national).

An engineer who wants to design a greywater reuse system often faces conflicting information online and is unable to separate false from the reliable information. In practice, this situation has created a knowledge vacuum, and many thousands of systems recycle greywater without official guidance or adequate information. Most of the available information comes from companies. Amateur gardeners who try to market greywater reuse systems provide little information (Gross et al., 2008). Moreover, many people market greywater treatment systems in a very wide price range. Without a binding standard, there is no way of knowing whether these systems are effective and what the quality of water will be after treatment.

We have shown that sanitary and environmental challenges posed by the use of greywater have a range of solutions that meet economic viability. To implement these solutions effectively, a mechanism must be established to ensure the use of greywater is accompanied by a public education campaign. To date, several countries have adopted various regulations; the effectiveness of these regulations should be vigilantly followed as they are implemented. Field studies should be initiated to monitor the various systems, including an epidemiological follow-up of user families. According to the results of these studies, the need and direction for improving the regulations can be continuously assessed and adapted.

# References

ABS. 2007. *Environmental Issues: People's Views and Practices*. Canberra, Australian Capital Territory, Australia: Australian Bureau of Statistics.

ACT-Health. 2007. *Greywater Use Guidelines for Residential Properties in Canberra*. Weston Creek, Australian Capital Territory, Australia: Australian Capital Territory, Department of Health.

ADEQ. 2001. *Direct Reuse of Reclaimed Water*. Administrative Code, Title 18, Chapter 9, Article 7. Phoenix, AZ: Arizona Department of Environmental Quality.

AMSA. 2000. *Evaluation of Domestic Sources of Mercury*. Association of Metropolitan Sewerage Agencies, Washington, DC.

ANZECC. 2000. *Australian Water Quality Guidelines for Fresh and Marine Waters*. Canberra, Australian Capital Territory, Australia: Australian and New Zealand Environment and Conservation Council.

APHA. 2005. In: *Standard Methods for the Examination of Water and Wastewater*, Ed. A D Eaton, L S Clesceri, E W Rice, and A E Greenberg, 20th ed. Washington, DC: American Public Health Association.

Abdessemed D and Nezzal G. 2002. Treatment of primary effluent by coagulation–adsorption–ultrafiltration for reuse. *Desalination* **152**: 367–373.

Abu Ghunmi L, Zeeman G, van Lier J, and Fayyed M. 2008. Quantitative and qualitative characteristics of grey water for reuse requirements and treatment alternatives: The case of Jordan. *Water Science and Technology* **58**(7): 1385–1396.

Abu-Zreig M, Rudra R P, and Dickinson W T. 2003. Effect of application of surfactants on hydraulic properties of soils. *Biosystems Engineering* **84**(3): 363–372.

Acher A, Fischer E, Turnheim R, and Manor Y. 1997. Ecologically friendly wastewater disinfection techniques. *Water Research* **31**(6) (June): 1398–1404.

Adel M, Shmueli L, Bas Y, and Friedler E. 2012. Onsite greywater recycling and its potential future impact on desalination and wastewater reuse in Israel. In: *4th International Conference on Drylands, Deserts and Desertification*, Ben Gurion University, Sede Boqer, Israel, November 2012.

Aertgeerts A and Angelakis A. 2003. *State of the Art Report: Health Risks in Aquifer Recharge Using Reclaimed Water. EUR/03/5041122*. Copenhagen, Denmark: WHO Regional Office for Europe.

Al-Hamaideh H and Bino M. 2010. Effect of treated grey water reuse in irrigation on soil and plants. *Desalination* **256**: 115–119.

Al-Jayyousi O R. 2003. Greywater reuse: Towards sustainable water management. *Desalination* **156**(1–3): 181–192.

Alfiya Y, Green M, and Lahav O. 2007. Modeling the aeration efficiency of a passively aerated vertical-flow biological filter. *Journal of Environmental Engineering—ASCE* **133**(10) (October): 970–978.

Alfiya Y, Gross A, Sklarz M, and Friedler E. 2013. Reliability of onsite greywater treatment systems in Mediterranean and arid environments—A case study. *Water Science and Technology* **67**(6): 1389–1395.

Allen L, Christian-Smith J, and Palaniappan M. 2010. *Overview of Greywater Reuse: The Potential of Greywater Systems to Aid Sustainable Water Management*. Oakland, CA: Pacific Institute.

Almeida M C, Butler D, and Friedler E. 1999. At-source domestic wastewater quality. *Urban Water* **1**: 49–55.

Armstrong J and Armstrong W. 1990. Light-enhanced convective throughflow increases oxygenation in rhizomes and rhizosphere of phragmites-australis (cav) trin ex steud. *New Phytologist* **114**: 121–128.

Ashbolt N J, Petterson S E, Roser D J, Westrell T, Ottoson J, Schönning C, and Stenström T A. 2004. Microbial risk assessment tool to aid in the selection of sustainable urban water systems. In: *Proceedings from 2nd IWA Leading-Edge Conference on Sustainability in Water Limited Environments*, Sydney, New South Wales, Australia, November 2004.

Ayers R S and Westcot D W. 1985. Water quality for agriculture. Food and Agriculture Organization of the United Nations.

Bahman S. 2010. Whitepaper on graywater. San Francisco, CA. American Water Works Association. In: Ed. Allen L et al., *Overview of Greywater Reuse: The Potential of Greywater Systems to Aid Sustainable Water Management.* Oakland, CA: Pacific Institute.

Bar-Ilan Y, Pearlmutter D, and Tal A. 2010. *Policy Mechanisms for Promoting Energy Efficiency in Buildings in Israel.* Haifa, Israel. The Technion, Center for Urban and Regional Studies Press.

Barko J W, Gunnison D, and Carpenter S R. 1991. Sediment interactions with submersed macrophyte growth and community dynamics. *Aquatic Botany* **41**: 41–65.

Becker N, Lavee D, and Katz D. 2010. Desalination and alternative water-shortage mitigation options in Israel: A comparative cost analysis. *Journal of Water Research and Protection* **2**: 1042–1056.

Belmont M A and Metcalfe C D. 2003. Feasibility of using ornamental plants (zantedeschia aethiopica) in subsurface flow treatment wetlands to remove nitrogen, chemical oxygen demand and nonylphenol ethoxylate surfactants—a laboratory-scale study. *Ecological Engineering* **21**: 233–247.

Birks R, Colbourne J, Hills S, and Hobson R. 2004. Microbiological water quality in a large in-building, water recycling facility. *Water Science and Technology* **50**(2): 165–172.

Birks R and Hills S. 2007. Characterisation of indicator organisms and pathogens in domestic greywater for recycling. *Environmental Monitoring and Assessment* **129**(1–3): 61–69.

Bondrenko N, Oron G, Gross A, and Ronen Z. 2006. Greywater Reuse: Using BOD for Water Quality Characterization. Faculty of Engineering Sciences, Unit of Environmental Engineering, Jacob Blaustein Institute for Desert Research (BIDR). Ben-Gurion University of the Negev.

Boyd C E. 1995. *Bottom Soils, Sediment, and Pond Aquaculture.* New York: Chapman & Hall.

Brennan M J and Patterson R A. 2004. Economic analysis of greywater recycling. In: *1st International Conference on Onsite Wastewater Treatment and Recycling Organised by Environmental Technology Centre*, Murdoch University, Perth, Western Australia, Australia, February 11–13, 2004.

Brisson J, and Chazarenc F. 2009. Maximizing pollutant removal in constructed wetlands: Should we pay more attention to macrophyte species selection? *Science of the Total Environment* **407**: 3923–3930.

Brix H. 1994. Functions of macrophytes in constructed wetlands. *Water Science and Technology* **29**: 71–78.

Brix H. 1997. Do macrophytes play a role in constructed treatment wetlands? *Water Science and Technology* **35**: 11–17.

Brix H. 1998. Denmark. In: *Constructed Wetlands For Wastewater Treatment In Europe.* Eds., Vymazal J, Brix H, Cooper P F, Green M B, and Haberl R. Backhuys Publishers, Leiden, pp. 123–156.

Brown R and Palmer A. 2002. *Water Reclamation Standard Laboratory Testing of Systems Using Grey Water.* Berkshire, England: Building Services Research & Information Association.

BSI. 2010. Greywater Systems BS 8525-1&2:2010. British Standards Institute.

Bungay H R, Whalen W J, and Sanders 3rd W M. 1969. Microprobe techniques for determining diffusivities and respiration rates in microbial slime systems. *Biotechnology and Bioengineering* **11**: 765.

Butler D, Friedler E, and Gatt K. 1995. Characterising the quantity and quality of domestic wastewater inflows. *Water Science and Technology* **31**(7): 13–24.

Burrows W D, Schmidt M D, Carnevale R M, and Schaub S A. 1991. Nonpotable reuse: Development of health criteria and technologies for shower water recycle. *Water Science Technology* **24**(9): 81.

CAMRA. 2012. Dose response assessment. http://wiki.camra.msu.edu/index.php?title=Dose_response_assessment. Accessed February 17, 2012.

CBSC. 2009. Chapter 16A "Nonpotable Water Reuse Systems" as appearing in the 2007 California Plumbing Code, Title 24, Part 5, Chapter 16A, Part 1.

CBSC. 2012. Chapter 16A "Nonpotable Water Reuse Systems" as appearing in the 2007 California Plumbing Code, Title 24, Part 5, Chapter 16A, Part 1.

CDWR. 1997. Graywater Systems. California Administration Code. State of California, Department of Water Resources, Division of Planning and Local Assistance.

Campisano A and Modica C. 2008. Experimental investigation on water saving by the reuse of washbasin grey waters for WC flushing. In: *11th International Conference on Urban Drainage*, Edinburgh, Scotland, U.K., 2008.

Carlander A, Schönning C, and Stenström T A. 2009. Energy forest irrigated with wastewater: A comparative microbial risk assessment. *Journal of Water and Health* **7**(3): 413–433.

Carpenter S. 2009. Gray water debate in Sacramento steams up. *Los Angeles Times*, March 17.

Casanova L M, Little V, Frye R J, and Gerba C P. 2001. A survey of the microbial quality of recycled household greywater association. *Journal of the American Water Resources* **37**(5): 1313–1319.

Chen R and Wang X C. 2009. Cost-benefit evaluation of a decentralized water system for wastewater reuse and environmental protection. *Water Science and Technology* **59**(8): 1515–1522.

Christova-Boal D, Eden R E, and McFarlane S. 1996. An investigation into greywater reuse for urban residential properties. *Desalination* **106**: 391–397.

Čížková-Končalová H J, Květ J, and Lukavská J. 1996. Response of phragmites australis, glyceria maxima, and typha latifolia to additions of piggery sewage in a flooded sand culture. *Wetlands Ecology and Management* **4**: 43–50.

Coagan T A, Bloomfield S F, and Humphrey T J. 1999. The effectiveness of hygiene procedures for prevention of cross-contamination from chicken carcasses in the domestic kitchen. *Letters in Applied Microbiology* **29**: 354–358.

Connnel D W. 1974. *Water Pollution: Causes and Effects in Australia*. St. Lucia, Queensland, Australia: The University of Queensland Press.

Cordy G E, Duran N L, Bouwer H, and Kolpin D W. 2004. Do pharmaceuticals, pathogens, and other organic waste water compounds persist when waste water is used for recharge? *Ground Water Monitoring & Remediation* **24**(2): 58–69.

Crittenden J C, Trussell R R, Hand D W, Howe K J, and Tchobanoglous G. 2005. *Water Treatment—Principles and Design*, 2nd ed. Hoboken, NJ: John Wiley & Sons.

De-Shalit A. 1995. From the political to the objective: The dialectics of Zionism and the environment. *Environmental Politics* **4**(1): 70–87.

Dental K S, Allen E H, Srinivasarao C, and Divincenzo J. 1994. *Effects of Surfactant on Sludge Dewatering and Pollutant Fate*. Newark, DE: University of Delaware, Report to Water Resources Center Project 6.

Diaper C, Dixon A, Bulier D, Parsons S A, Strathern M, and Strutt J. 2001. Small scale water recycling systems—Risk assessment and modelling. *Water Science and Technology* **43**(10): 83–90.

Diaper C, Jefferson B, Parsons S A, and Judd S J. 2001. Water-recycling technologies in the UK. *Journal of the Chartered Institution of Water and Environmental Management* **15**(4): 282–286.

Diaper C, Toifl M, and Storey M. 2008. *Greywater Technology Testing Protocol*. CSIRO: Water for a Healthy Country National Research Flagship, Vol. 1835-095X. Victoria, Australia.

Dixon A, Butler D, and Fewkes A. 1999a. Water saving potential of domestic water reuse systems using greywater and rainwater in combination. *Water Science and Technology* **39**(5): 25–32.

Dixon A M, Butler D, and Fewkes A. 1999b. Guidelines for greywater re-use: Health issues. *Water and Environment Journal* **13**(5): 322–326.

Domènech L and Saurí D. 2010. Socio-technical transitions in water scarcity contexts: Public acceptance of greywater reuse technologies in the Metropolitan Area of Barcelona. *Resources, Conservation and Recycling* **55**(1): 53–62. doi:10.1016/j.resconrec. 2010.07.001.

Doneen L D. 1971. *Irrigation Practice and Water Management. Food and Agriculture Organization of the United Nations. Water Resources Development and Management Service*. Rome: FAO Irrigation and Drainage Paper; No. 1, 84pp.

Donner E, Eriksson E, Revitt D M, Scholes L, Lützhøft H-CH, and Ledin A. 2010. Presence and fate of priority substances in domestic greywater treatment and reuse systems. *Science of the Total Environment* **408**(12): 2444–2451.

Drechsel P, Christopher A S, Raschid-Sally L, Mark R, and Akiça B, Eds. 2010. *Wastewater Irrigation and Health: Assessing and Mitigating Risk in Low-Income Countries*. London, UK. Earthscan, IDRC, IWMI.

Drobotko V G, Rashba E Y, Aizenman B E, Zelepukha S I, Novikova S I, and Kaganskaya M B. 1958. Antimicrobial activity of alkaloids obtained from *valeriana officinalis, chelidonium majus, nuphar luteum* and *asarum europium. Antibiotiki 22, Chem. Abstr.* **53**: 12589d.

Droste R L. 1997. *Theory and Practice of Water and Wastewater Treatment*. Hoboken, NJ: John Wiley & Sons.

Dwyer J P. 1990. The pathology of symbolic legislation. *Ecology Law Quarterly* **17**: 233–316.

EPA. 2000. *Constructed Wetlands Treatment of Municipal Wastewater*, Cincinnati, OH.

EPHC. 2008. *Australian Guidelines for Water Recycling*. Augmentation of Drinking Water Supplies, Canberra, Australia.

EU. 1991. Council Directive of 21 May 1991 concerning urban waste water treatment (91/271/EEC).

EU. 2010. Is greywater safe for irrigation?

Elmitwalli T A and Otterpohl R. 2007. Anaerobic biodegradability and treatment of grey water in upflow anaerobic sludge blanket (UASB) reactor. *Water Research* **41**(6): 1379–1387.

Eriksson E. 2002. *Potential and Problems Related to Reuse of Water in Households*. Lyngby, Denmark: Technical University of Denmark.

Eriksson E, Andersen H R, and Ledin A. 2008. Substance flow analysis of parabens in Denmark complemented with a survey of presence and frequency in various commodities. *Journal of Hazardous Materials* **156**(1–3): 240–259.

Eriksson E, Andersen H R, Madsen T S, and Ledin A. 2009. Greywater pollution variability and loadings. *Ecological Engineering* **35**(5): 661–669.

Eriksson E, Auffarth K, Henze M, and Ledin A. 2002. Characteristics of grey wastewater. *Urban Water* **4**(1): 85–104.

Eriksson E and Donner E. 2009. Metals in greywater: Sources, presence and removal efficiencies. *Desalination* **248**(1–3): 271–278.

Exner M E, Perea-Estrada H, and Spalding R F. 2010. Long-term response of groundwater nitrate concentrations to management regulations in Nebraska's central Platte valley. *Scientific World Journal* **10**: 286–297.

Farooq S and Ansari Z. 1983. Wastewater reuse in Muslim countries: An Islamic perspective number. *Environmental M*anagement **7**(2): 119–123.

Farré M, Pérez S, Kantiani L, and Barceló D. 2008. Fate and toxicity of emerging pollutants, their metabolites and transformation products in the aquatic environment. *Trends in Analytical Chemistry* **27**(11): 991–1007.

Fernandes T M A, Schout C, De Roda Husman A M, Eilander A, Vennema H, and van Duynhoven Y T H P. 2007. Gastroenteritis associated with accidental contamination of drinking water with partially treated water. *Epidemiology and Infection* **135**(5): 818–826. doi:10.1017/S0950268806007497.

Fernández Cirelli A, Ojeda C, Castro M, and Salgot M. 2008. Surfactants in sludge-amended agricultural soils: A review. *Environmental Chemistry Letters* **6**(3): 135–148.

Fewkes A and Ferris S A. 1982. The recycling of domestic waste water. A study of the factors influencing the storage capacity and the simulation of the usage patterns. *Building and Environment* **17**(3): 209–216.

Fewtrell L and Kay D. 2007. Quantitative microbial risk assessment with respect to *Campylobacter* spp. in toilets flushed with harvested rainwater. *Water and Environment Journal* **21**(4): 275–280.

Fraser L H, Carty S M, and Steer D. 2004. A test of four plant species to reduce total nitrogen and total phosphorus from soil leachate in subsurface wetland microcosms. *Bioresource Technology* **94**: 185–192.

Friedler E. 2004. Quality of individual domestic greywater streams and its implication for on-site treatment and reuse possibilities. *Environmental Technology* **25**(9): 997–1008.

Friedler E. 2008. The water saving potential and the socio-economic feasibility of greywater reuse within the urban sector—Israel as a case study. *International Journal of Environmental Studies* **65**(1): 57–69.

Friedler E. 2011. The water saving potential and the socio-economic feasibility of urban on-site greywater reuse. Keynote speech. In: *International Conference on Integrated Water Management*, Murdoch University, Western Australia, February 2011.

Friedler E and Alfiya Y. 2010. Physicochemical treatment of office and public buildings greywater. *Water Science and Technology* **62**(10): 2357–2363.

Friedler E and Butler D. 1996. Quantifying the inherent uncertainty in the quantity and quality of domestic wastewater. *Water Science and Technology* **33**(2): 65–78.

Friedler E and Galil N I. 2003. On-site greywater reuse in multi-storey buildings: Sustainable solution for water saving. In: *Efficient 2003—2nd International Conference on Efficient Use and Management of Urban Water Supply*, Tenerife, Canary Islands, Spain.

Friedler E and Gilboa Y. 2010. Performance of UV disinfection and the microbial quality of greywater effluent along a reuse system for toilet flushing. *Science of the Total Environment* **408**(9) (April 1): 2109–2117.

Friedler E and Hadari M. 2006. Economic feasibility of on-site greywater reuse in multi-storey buildings. *Desalination* **190**(1–3): 221–234.

Friedler E, Katz I, and Dosoretz C G. 2008. Chlorination and coagulation as pretreatments for greywater desalination. *Desalination* **222**(1–3): 38–49.

Friedler E, Kovalio R, and Ben-Zvi A. 2006. Comparative study of the microbial quality of greywater treated by three on-site treatment systems. *Environmental Technology* **27**(6): 653–663.

Friedler E, Kovalio R, and Galil N I. 2005. On-site greywater treatment and reuse in multi-storey buildings. *Water Science and Technology* **51**(10): 187–194.

Friedler E, Lahav O, Jizhaki H, and Lahav T. 2006. Study of urban population attitudes towards various wastewater reuse options: Israel as a case study. *Journal of Environmental Management* **81**: 360–370.

Friedler E, Yardeni A, Gilboa Y, and Alfiya Y. 2011. Disinfection of greywater effluent and regrowth potential of selected bacteria. *Water Science and Technology* **63**(5): 931–940.

GWI. 2010. *New Bill Allows for Grey Water Reuse*. Global Water Intelligence, Oxford, UK.

Gander M, Jefferson B, and Judd S. 2000. Aerobic MBRs for domestic wastewater treatment: A review with cost considerations. *Separation and Purification Technology* **18**(2) (March 6): 119–130.

Geller G. 1997. Horizontal subsurface flow systems in the german speaking countries: Summary of long-term scientific and practical experiences; recommendations. *Water Science and Technology* **35**: 157–166.

Gerba C P, Castro-Del Campo N, Brooks J P, and Pepper I L. 2008. Exposure and risk assessment of Salmonella in recycled residuals. *Water Science and Technology* **57**(7): 1061–1065.

Gerba C P, Rose J B, Haas C N, and Crabtree K D. 1996. Waterborne rotavirus: A risk assessment. *Water Research* **30**(12): 2929–2940.

Gerba C P, Straub T M, Rose J B, Karpiscak M M, Foster K E, and Brittain R G. 1995. Water quality study of graywater treatment systems. *Water Resources Bulletin* **31**(1): 109–116.

Gersberg R M, Elkins B V, Lyon S R, and Goldman C R. 1986. Role of aquatic plants in wastewater treatment by artificial wetlands. *Water Research* **20**: 363–368.

Gethke K, Herbst H, Keysers C, and Pinnekamp J. 2007. Grey water reuse in hotel and catering industry. In: *6th IWA Specialist Conference on Wastewater Reclamation and Reuse for Sustainability*, Antwerp, Belgium, October 9–12, 2007.

Ghisi E and Ferreira D F. 2007. Potential for potable water savings by using rainwater and greywater in a multi-storey residential building in southern Brazil. *Building and Environment* **42**(7): 2512–2522.

Gilboa Y and Friedler E. 2008. UV disinfection of RBC-treated light greywater effluent: Kinetics, survival and regrowth of selected microorganisms. *Water Research* **42**(4–5): 1043–1050.

Godfrey S, Labhasetwar P, and Wate S. 2009. Greywater reuse in residential schools in Madhya Pradesh, India—A case study of cost-benefit analysis. *Resources, Conservation and Recycling* **53**(5): 287–293.

Godfrey S, Labhasetwar P, Wate S, and Jimenez B. 2010. Safe greywater reuse to augment water supply and provide sanitation in semi-arid areas of rural India. *Water Science and Technology* **62**(6): 1296–1303.

Grady C P Jr., Daigger G T, and Lim H C. 1999. *Biological Wastewater Treatment*, 2nd ed. New York: Marcel Dekker.

Green M, Friedler E, and Safrai I. 1998. Enhancing nitrification in vertical flow constructed wetlands utilising a passive air pump. *Water Research* **32**(12): 3513–3520.

Gross A, Azulai N, Oron G, Ronen Z, Arnold M, and Nejidat A. 2005. Environmental impact and health risks associated with greywater irrigation: A case study. *Water Science and Technology* **52**(8): 161–169.

Gross A, Kaplan D, and Baker K. 2007. Removal of chemical and microbiological contaminants from domestic greywater using a recycled vertical flow bioreactor (RVFB). *Ecological Engineering* **31**: 107–114.

Gross A, Shmueli O, Ronen Z, and Raveh E. 2007. Recycled vertical flow constructed wetland (RVFCW)—A novel method of recycling greywater for irrigation in small communities and households. *Chemosphere* **66**(5): 916–923.

Gross A, Wiel-Shafran A, Bondarenko N, and Ronen Z. 2008. Reliability of small scale greywater treatment systems and the impact of its effluent on soil properties. *International Journal of Environmental Studies* **65**: 41–50.

HCC. 2011. Hobart City Council: Reusing greywater within the Hobart City Council Municipality in sewer reticulated areas. http://www.hobartcity.com.au/Environment/Environmental_Health/Greywater_Reuse. Accessed December 2, 2012.

Haas C N, Rose J B, and Gerba C P. 1999. *Quantitative Microbial Risk Assessment*. New York: John Wiley & Sons.

Hammer D A and Bastian R K. 1989. Wetland ecosystems: Natural water purifiers? In: *Constructed Wetlands for Wastewater Treatment: Municipal, Industrial and Agricultural*. Ed. Hammer D A. Lewis Publishers: Chelsea, MI, pp. 5–20.

Hardin G. 1968. The tragedy of the commons. *Science* **162**: 1243–1248.

Haslam S M. 1971. Community regulation in phragmites-communis trin. 1. Monodominant stands. *Journal of Ecology* **59**: 65–73.

Hattersly J G. 2000. The negative health effects of chlorine. *The Journal of Orthomolecular Medicine* **15**(2): 89–102.

Helvetas. 2005. (Schweizer Gesellschaft für Internationale Zusammenarbeit). In *Water Consumption in Switzerland* [*in German: Wasserverbrauch in der Schweiz*], 3p. Helvetas, Zurich, Switzerland.

Henkel J, Cornel P, and Wagner M. 2009. Free water content and sludge retention time: Impact on oxygen transfer in activated sludge. *Environmental Science and Technology* **43**(22): 8561–8565.

Hernández-Leal L, Temmink B, Zeeman G, and Buisman C. 2010. Comparison of three systems for biological grey water treatment. *Water* **2**(2): 155–169.

Hernández-Leal L, Zeeman G, Temmink H, and Buisman C. 2007. Characterisation and biological treatment of greywater. *Water Science and Technology* **57**(5): 2059–2064.

Hills S, Smith A, Hardy P, and Birks R. 2001. Water recycling at the millennium dome. *Water Science and Technology* **43**(10): 287–294.

Ho C-FH, Pitt P, Mamais D, Chiu C, and Jolis D. 1998. Evaluation of UV disinfection systems for large-scale secondary effluent. *Water Environment Research* **70**(6): 1142–1150.

Homedes N. 1996. *The Disability-Adjusted Life Year (DALY) Definition, Measurement and Potential Use*. HCD, World Bank.

Horan N J. 1990. *Biological Wastewater Treatment Systems, Theory and Operation*. New York: John Wiley & Sons.

Hourlier F, Massé A, Jaouen P, Lakel A, Gérente C, Faur C, and Le Cloirec P. 2010. Membrane process treatment for greywater recycling: Investigations on direct tubular nanofiltration. *Water Science and Technology* **62**(7): 1544–1550.

Huelgas A and Funamizu N. 2010. Flat-plate submerged membrane bioreactor for the treatment of higher-load graywater. *Desalination* **250**(1): 162–166.

Hurlimann, A. 2011. Household use of and satisfaction with alternative water sources in Victoria Australia. *Journal of Environmental Management* **92**: 2691–2697.

Hurlimann A and Dolnicar S. 2010. When public opposition defeats alternative water projects—The case of Toowoomba Australia. *Water Research* **44**(1): 287–297.

Inbar Y. 2007. New Standards for treated wastewater reuse in Israel. In: *Environmental Security. Proceedings of the NATO Advanced Research Workshop "Wastewater Reuse—Risk Assessment, Decision-Making and Environmental Security"*, Istanbul, Turkey, October 2006. Ed. M K Zaidi. Springer: NATO Science for Peace and Security Series C, pp. 291–296.

Israel Ministry of Environment. 2000. Reducing wastewater salinity. *Israel Environment Bulletin* **23**(4): 15–17.

Israeli Water Authority. 2009. *Gardens and Landscape Planing*, Beer, R., ed., Israeli Water Authority, pp. 61.

Itayama T, Kiji M, Suetsugu A, Tanaka N, Saito T, Iwami N, Mizuochi M, and Inamorim Y. 2006. On site experiments of the slanted soil treatment systems for domestic gray water. *Water Science and Technology* **53**(9): 193–201.

IWA. 2000. Constructed Wetland for Pollution Control. IWA specialist group on use of macrophytes in water pollution control, London, UK.

Jamrah A, Al-Futaisi A, Ahmed M, Prathapar S, Al-Harrasi A, and Al-Abri A. 2008. Biological treatment of greywater using sequencing batch reactor technology. *International Journal of Environmental Studies* **65**(1): 71–85.

Jefferson B, Laine AL, Stephenson T, and Judd SJ. 2001. Advanced biological unit processes for domestic water recycling. *Water Science and Technology* **43**(10): 211–218.

Jefferson B, Palmer A, Jeffrey P, Stuetz R, and Judd S. 2004. Grey water characterisation and its impact on the selection and operation of technologies for urban reuse. *Water Science and Technology* **50**(2): 157–164.

Jeffrey P and Jefferson B. 2003. Public receptivity regarding 'in-house' water recycling: Results from a UK survey. *Water Science and Technology: Water Supply* **3**(3): 109–116.

Jenssen P D and Vrale L. 2003. Greywater treatment in combined biofilter/constructed wetlands in cold climate. In: *Ecosan—Closing the Loop. Proceedings of 2nd International Symposium Ecological Sanitation*, Lübeck, Germany, April 2003. pp. 875–881.

Jeppesen B. 1996. Domestic greywater re-use: Australia's challenge for the future. *Desalination* **106**(1–3): 311–315.

Kadewa W W, Le Corre K, Pidou M, Jeffrey P J, and Jefferson B. 2010. Comparison of grey water treatment performance by a cascading sand filter and a constructed wetland. *Water Science and Technology* **62**(7): 1471–1478.

Kadlec R H and Knight R L. 1996. *Treatment Wetlands*. Boca Raton, FL: CRC Press.

Katz-Gerro T. 2009. New middle class and environmental lifestyle in Israel. In: *Globalizing Lifestyles, Consumerism, and Environmental Concern: The Case of the New Middle Class*, Ed. H Lange and L Meier. Dordrecht, the Netherlands: Springer, pp. 197–215.

Kennedy M D, Kamanyi J, Salinas Rodriguez S G, Lee N H, Schippers J C, and Amy G. 2008. Water treatment by microfiltration and ultrafiltration. In: *Advanced Membrane Technology and Applications*, Ed. N N Li, A G Fane, W S W Ho, and T Matsuura. Hoboken, NJ: John Wiley & Sons.

Kim J, Song I, Lee S, Kim P, Lee M, and Choung Y. 2009. Novel pilot plant-scale graywater treatment system using titanium ball, membrane and advanced oxidation process. *Desalination and Water Treatment* **10**(1–3) (October): 153–157.

Kim R-H, Lee S, Jeong J, Lee J-H, and Kim Y-K. 2007. Reuse of greywater and rainwater using fiber filter media and metal membrane. *Desalination* **202**(1–3): 326–332.

Kovlio R. 2004. Study of three systems for treatment and reuse of greywater for toilet flushing. MSc Thesis, Technion–Israel Institute of Technology, Israel.

Kraume M, Scheumann R, Baban A, and El Hamouri B. 2010. Performance of a compact submerged membrane sequencing batch reactor (SM-SBR) for greywater treatment. *Desalination* **250**(3): 1011–1013.

Langergraber G. 2005. The role of plant uptake on the removal of organic matter and nutrients in subsurface flow constructed wetlands: A simulation study. *Water Science and Technology* **51**: 213–223.

Langergraber G, and Simunek J. 2005. Modeling variably saturated water flow and multicomponent reactive transport in constructed wetlands. *Vadose Zone Journal* **4**: 924–938.

Lantzke I R, Heritage A D, Pistillo G, and Mitchell D S. 1998. Phosphorus removal rates in bucket size planted wetlands with a vertical hydraulic flow. *Water Research* **32**: 1280–1286.

Lazarova, V, Savoye P, Janex M L, Blatchley E R, and Pommepuy M. 1999. Advanced wastewater disinfection technologies: State of the art and perspectives. *Water Science and Technology* **40**: 203–213.

LEEDS. 2010. Leeds Domestic Waterusers Association Water Conservation Plan, Adapted April 15, 2010.

Lesjean B and Gnirss R. 2006. Grey water treatment with a membrane bioreactor operated at low SRT and low HRT. *Desalination* **199**(1–3): 432–434.

Lewis E L. 1980. The practical salinity scale 1978 and its antecedents. *IEEE Journal of Oceanic Engineering* **5**(1): 3–8.

Li F, Behrendt J, Wichmann K, and Otterpohl R. 2008. Resources and nutrients oriented greywater treatment for non-potable reuses. *Water Science and Technology* **57**(12): 1901–1907.

Li F, Gulyas H, Wichmann K, and Otterpohl R. 2009. Treatment of household grey water with a UF membrane filtration system. *Desalination and Water Treatment* **5**(1–3): 275–282.

Lin C-J, Lo S-L, Kuo C-Y, and Wu C-H. 2005. Pilot-scale electrocoagulation with bipolar aluminum electrodes for on-site domestic greywater reuse. *Journal of Environmental Engineering* **131**(3): 491–495.

Lindstrom C R. 2000. *Graywater, Facts about Graywater—What It Is, How to Treat It, When and Where to Use It.* Cambridge, MA: Electronic Manuscript.

Little V. 2000. *Residential Graywater Reuse: The Good, The Bad, The Healthy.* Tucson, AZ: The Water Conservation Alliance of Southern Arizona (Water CASA).

Loh M and P Coghlan. 2003. Domestic water used study in Perth, Western Australia 1998–2001. Perth, Western Australia, Australia: Water Corporation, 36.

Lowe K S, Rothe N K, Tomaras J M B, DeJong K, Tuchholke M B, Drewes J D, McCray J E, and Munakata-Marr J. 2007. *Influent Constituent Characteristics of the Modern Waste Stream from Single Family Sources: Literature Review.* Environmental Science and Engineering Division, Colorado, Water Environment Research Foundation.

Lucía Hernández L, Temmink H, Zeeman G, and Buisman C J N. 2010. Comparison of three systems for biological greywater treatment. *Water* **2**(2): 155–169.

Luederitz V, Eckert E, Lange-Weber M, Lange A, and Gersberg R M. 2001. Nutrient removal efficiency and resource economics of vertical flow and horizontal flow constructed wetlands. *Ecological Engineering* **18**: 157–171.

Maas E V. 1990a. Salt tolerance of plants. In: *Handbook of Plant Science in Agriculture*, Ed. B R Christie. Boca Raton, FL: CRC Press, pp. 57–75.

Maas E V. 1990b. Crop salt tolerance. Chapter 13. In *Agricultural Salinity Assessment and Management* Ed. K K Tanji, American Society for Civil Engineering: New York.

Maas E V and Hoffman G J. 1997. Crop salt tolerance-current assessment. *Journal of the Irrigation and Drainage Division. American Society for Civil Engineering* **103**(IR2): 115–134.

Maimon A. 2010. Safe reuse of greywater for irrigation in private households. MSc thesis, Albert Katz International School for Desert Studies (AKIS), Sede boqer campus, Israel: Ben Gurion University of the Negev.

Maimon A, Tal A, Friedler E, and Gross A. 2010. Safe on-site reuse of greywater for irrigation—A critical review of current guidelines. *Environmental Science & Technology* **44**(9) (May 1): 3213–3220. doi:10.1021/es902646g.

Mandal D, Labhasetwar P, Dhone S, Dubey A S, Shinde G, and Wate S. 2010. Water conservation due to greywater treatment and reuse in urban setting with specific context to developing countries. *Resources Conservation and Recycling* **55**(3): 356–361.

Mara D and Horan N. 2003. *Handbook of Water and Wastewater Microbiology.* San Diego, CA: Elsevier, Academic Press.

Mara D D, Sleigh P A, Blumenthal U J, and Carr R M. 2007. Health risks in wastewater irrigation: Comparing estimates from quantitative microbial risk analyses and epidemiological studies. *Journal of Water and Health* **5**(1): 39–50.

March J G, Gual M, and Orozco F. 2004. Experiences on greywater re-use for toilet flushing in a hotel (Mallorca, Island, Spain). *Desalination* **164**(3): 241–247.

Masi F, El Hamouri B, Abdel Shafi H, Baban A, Ghrabi A, and Regelsberger M. 2010. Treatment of segregated black/grey domestic wastewater using constructed wetlands in the Mediterranean basin: The zer0-m experience. *Water Science and Technology* **61**(1): 97–105.

Meinzinger F and Oldenburg M. 2009. Characteristics of source-separated household wastewater flows: A statistical assessment. *Water Science and Technology* **59**(9): 1785–1791.

Memon F A and Butler D. 2006. Domestic water consumption trends and techniques for demand forecasts. In: *Water Demand Management*, Ed. D Butler and F A Memon. London, U.K.: IWA, pp. 1–25.

Meng Q E and Ganczarczyk J. 2004. Full scale comparison of heterotrophic and nitrifying RBC biofilms. *Environmental Technology* **25**: 165–171.

Merz C, Scheumann R, El Hamouri B, and Kraume M. 2007. Membrane bioreactor technology for the treatment of greywater from a sports and leisure club. *Desalination* **215**(1–3): 37–43.

Metcalf and Eddy. 2003. *Wastewater Engineering: Treatment and Reuse*. Ed. G Tchobanoglous, F L Burton, and H D Stensel. New York: McGraw-Hill.

Misra R K and Sivongxay A. 2009. Reuse of laundry greywater as affected by its interaction with saturated soil. *Journal of Hydrology* **366**(1–4): 55–61.

Molle P, Liénard A, Boutin C, Merlin G, and Iwema A. 2005. How to treat raw sewage with constructed wetlands: An overview of the French systems. *Water Science and Technology* **51**(9): 11–21.

Morel A and Diener S. 2006. Greywater Management—in Low and Middle-Income Countries. Sandec (Water and Sanitation in Developing Countries) at Eawag (Swiss Federal Institute of Aquatic Science and Technology). Vol. 14/06.

Munch C, Kuschk P, and Roske I. 2005. Root stimulated nitrogen removal: Only a local effect or important for water treatment? *Water Science and Technology* **51**: 185–192.

Ne'eman-Avramovich A and Katz-Gerro A. 2007. Social bases of environmental attitudes and behavior in Israel. *Megamot* **44**(4):736–758 (in Hebrew).

Nelson M, Cattin F, Rajendran M, and Hafouda L. 2008. Value-adding through creation of high diversity gardens and ecospaces in subsurface flow constructed wetlands: Case studies in algeria and australia of wastewater gardens systems. In *Proceedings of 11th International Conference On Wetland Systems For Water Pollution Control, vol. 1. Institute of Environmental Management And Plant Sciences*. Eds., Billore S, Dass P, and Vymazal J. Vikram University, Ujjain, pp. 344–356.

NRMMC and EPHC. 2006. *National Guidelines for Water Recycling: Managing Health and Environmental Risks (Phase 1)*. Natural Resource Management Ministerial Council, Environment Protection and Heritage Council, Australian Health Ministers' Conference.

NSF. 2011. NSF/ANSI 350-2011, Onsite Residential and Commercial Water Reuse Treatment Systems.

NSW. 2006. The Local Government (General) Amendment (Domestic Greywater Diversion) Regulation 2006.

NSW. 2008. *Management of Private Recycled Water Schemes*. New South Wales, Australia: Water for Life.

NSW-DEUS. 2007. *NSW Guidelines for Greywater Reuse in Sewered Single Household Residential Premises*. New South Wales, Australia: Department of Energy, Utilities and Sustainability.

NSW-Health. 2011. Certificate of accreditation for domestic greywater treatment system Nubian GT600 version 1.3 DGTS.

Narkis N, Armon R, Offer R, Orshansky F, and Friedland E. 1995. Effect of suspended solids on wastewater disinfection efficiency by chlorine dioxide. *Water Research* **29**(1): 227–236.

Nazer D W, Siebel M A, van der Zaag P, Mimi Z, and Gijzen H J. 2010. A financial, environmental and social evaluation of domestic water management options in the West Bank, Palestine. *Water Resources Management* **24**(15): 4445–4467.

Nolde E. 1999. Greywater reuse systems for toilet flushing in multi-storey buildings—Over ten years experience in Berlin. *Urban Water* **1**(4): 275–284.

Nolde E. 2005. Greywater treatment systems in Germany: Results, experiences and guidelines. *Water Science and Technology* **51**(10): 203–210.

OFWAT. 2010. *Water Meters—Your Questions Answered, Information for Household Customers*. OFWAT, The Water Services Regulation Authority, 28pp.

OasisDesign, 2009. History of Greywater Regulation. http://www.oasisdesign.net/greywater/law/history/index.htm. Accessed October 5, 2009.

Oesterholt F, Martijnse G, Medema G, and van der Kooij D. 2007. Health risk assessment of non-potable domestic water supplies in the Netherlands. *Journal of Water Supply: Research and Technology. AQUA* **56**(3): 171–179.

Ogoshi M, Suzuki Y, and Asano T. 2001. Water reuse in Japan. *Water Science and Technology* **43**(10): 17–23.

Ottoson J and Stenstrom T A. 2003. Faecal contamination of greywater and associated microbial risks. *Water Research* **37**(3): 645–655.

Palma L D, Merli C, Paris M, and Petrucci E. 2003. A steady state model for the evaluation of disk rotational speed influence on RBC kinetics: Model presentation. *Bioresource Technology* **86**: 193–200.

Parkin R T. 2007. Microbial risk assessment. In: *Risk Assessment for Environmental Health*, Ed. M G Robinson and W A Toscano. San Francisco, CA: Jossy-Bass.

Parks J L and Edwards M. 2005. Boron in the environment. *Critical Reviews in Environmental Science and Technology* **35**(2):81–114.

Paulo P L, Begosso L, Pansonato N, Shrestha R R, and Boncz M A. 2009. Design and configuration criteria for wetland systems treating greywater. *Water Science and Technology* **60**(8): 2001–2007.

Paz S, Ayalon O, and Haj A. 2012. The potential conflict between traditional perceptions and environmental behavior: compost use by Muslim Farmers. *Environment, Development and Sustainability*. DOI: 10.1007/s10668-012-9421-1.

Pedersen A, Woelfe-Erskine C, and Hill-Hart J. 2007. Policy recommendations for Montana. Greywater action. http://greywateraction.org/content/policy-recommendations-montana. Accessed February 17, 2012.

Penn R, Hadari M, and Friedler E. 2012. Evaluation of the effects of greywater reuse on domestic wastewater quality and quantity. *Urban Water Journal* **9**(3): 137–148.

Pepper IL, Gerba CP, and Brusseau ML. 2006. *Environmental and Pollution Science*. San Diego, CA: Elsevier, Academic Press.

Petticrew E L and Kalff J. 1992. Water-flow and clay retention in submerged macrophyte beds. *Canadian Journal of Fisheries and Aquatic Sciences* **49**: 2483–2489.

Pidou M, Avery L, Stephenson T, Jeffrey P, Parsons S A, Liu S, Memon F A, and Jefferson B. 2008. Chemical solutions for greywater recycling. *Chemosphere* **71**(1): 147–155.

Pidou M, Memon F A, Stephenson T, Jefferson B, and Jeffrey P. 2007. Greywater recycling: Treatment options and applications. *Proceedings of the Institution of Civil Engineers: Engineering Sustainability* **160**(3): 119–131.

PlanetArk. 2007. Greywater fact sheet series. Planet Ark, March 7. http://products.planetark.org/documents/doc-163-greywater-information-guide.pdf. Accessed December 2, 2012.

Po M, Kaercher J, and Nancarrow B E. 2003. Literature review of factors influencing public perceptions of water reuse. CSIRO Land and Water.

Prathapar S A, Jamrah A, Ahmed M, Al Adawi S, Al Sidairi S, and Al Harassi A. 2005. Overcoming constraints in treated greywater reuse in Oman. *Desalination* **186**(1–3) (December 30): 177–186.

Prillwitz M and Farwell L. 1995. *Graywater Guide*. San Francisco, CA: Department of Water Resources.

Qian Y L and Mecham B. 2005. Long term effects of recycled wastewater irrigation on soil chemical properties on golf course fairways. *Agronomy Journal* **97**: 717–721.

Radcliff J. 2004. *Water Recycling in Australia*. Parkville, Victoria, Australia: Australian Academy of Technological Sciences and Engineering.

Ramon G, Green M, Semiat R, and Dosoretz C. 2004. Low strength graywater characterization and treatment by direct membrane filtration. *Desalination* **170**(3): 241–250.

Raude J, Mutua B, Chemelil M, Kraft L, and Sleytr K. 2009. Household greywater treatment for peri-urban areas of Nakuru Municipality, Kenya. *Sustainable Sanitation Practice* **10**(1): 10–15.

Rebhun M, Heller-Grossman L, and Manka J. 1997. Formation of disinfection byproducts during chlorination of secondary effluent and renovated water. *Water Environment Research* **69**(6): 1154–1162.

Reed S C, Crites R W, and Middlebrooks E J. 1995. *Natural Systems for Waste Management and Treatment*. McGraw-Hill, New York, NY.

Rhoades JD and Loveday J. 1990. Salinity in irrigated agriculture. In: *Irrigation of Agricultural Crops*, Eds B A Steward, and D R Nielsen, ASCE (Monograph 30). American Society of Agronomists, pp. 1089–1142.

Ripple W. 1883. Capacity of storage reservoirs for water supply. *Minutes of Institution of Civil Engineers* **71**: 270–278.

Rittmann B E and McCarty P L. 2001. *Environmental Biotechnology: Principles and Applications*. New York: McGraw-Hill.

Roesner L, Criswell M, Stromberger M, and Klein S. 2006. Long term effects of landscape irrigation using household graywater—Literature review and synthesis. Prepared for WERF, published with SDA.

Ronen Z, Guerrero A, and Gross A. 2010. Greywater disinfection with the environmentally friendly hydrogen peroxide plus (HPP). *Chemosphere* **78**(1): 61–65.

Rose J B, Sun G S, Gerba C P, and Sinclair N A. 1991. Microbial quality and persistence of enteric pathogens in greywater from various household sources. *Water Research* **25**: 37–42.

Rosenthal A. 1990. State agricultural pollution regulation, a quantitative assessment. *Water Environment and Technology* **2**(8): 50–58.

Ruiz-Rueda O, Hallin S, and Baneras L. 2009. Structure and function of denitrifying and nitrifying bacterial communities in relation to the plant species in a constructed wetland. *Fems Microbiology Ecology* **67**: 308–319.

SADH. 2004. *Application for Alternative on-site Wastewater/Waste ControlSystem Installation*. South Australia, Australia: Department of Health, Environmental Health Service.

SADH. 2006. *Installation of Permanent Onsite Domestic Greywater Systems*. South Australia, Australia: Department of Health.

SADH. 2007. *Manual Bucketing & Temporary Diversion of Greywater*. South Australia: Department of Health.

Salt D E, Blaylock M, Kumar N P B A, Dushenkov V, Ensley B D, Chet I, and Raskin I. 1995. Phytoremediation—a novel strategy for the removal of toxic metals from the environment using plants. *Bio-Technology* **13**: 468–474.

Santa-Barbara. 1989. *Appendix C: Change to SB County Building Code Ordinance 3665*. Santa Barbara, CA.

Scheumann R and Kraume M. 2009. Influence of hydraulic retention time on the operation of a submerged membrane sequencing batch reactor (SM-SBR) for the treatment of greywater. *Desalination* **246**(1–3): 444–451.

Schumacher E F. 1973. *Small is Beautiful: Economics as if People Mattered*. New York: Harper & Row.

Scott M J and Jones M N. 2000. The biodegradation of surfactants in the environment. *Biochimica et Biophysica Acta (BBA)—Biomembranes* **1508**: 235–251.

Seidel K. 1964. Abbau von bacterium coli durch höhere wasserpflanzen. *Naturwiss* **51**: 395.

Seidel K. 1976. Macrophytes and water purification. In: *Biological Control Of Water Pollution*, Eds. Tourbier J and Pierson R W J. University of Pennsylvania Press: Pennsylvania, PA, pp. 109–123.

Shelef O, Golan-Goldhirsh A, Gendler T, and Rachmilevitch S. 2011. Physiological parameters of plants as indicators of water quality in a constructed wetland. *Environmental Science and Pollution Research* **18**: 1234–1242.

Shelef O, Gross A, and Rachmilevitch S. 2012. The use of *Bassia indica* for salt phytoremediation in constructed wetlands. *Water Research* **46**: 3967–3976.

Shelef O, Gross A, and Rachmilevitch S. 2013. Role of plants in a constructed wetland: Current and new perspectives. *Water* **5**: 405–419.

Shelef O, Lazarovitch N, Rewald B, and Golan-Goldhirsh S, Rachmilevitch S. 2010. Root halotropism: Salinity effects on *Bassia indica* root. *Plant Biosystems* **144**: 471–478.

Shin H-S, Lee S-M, Seo I-S, Kim G-O, Lim K-H, and Song J-S. 1998. Pilot-scale SBR and MF operation for the removal of organic and nitrogen compounds from greywater. *Water Science and Technology* **38**(6): 80–88.

Shuval H, Lampert Y, and Fattal B. 1997. Development of a risk assessment approach for evaluating wastewater reuse standards for agriculture. *Water Science and Technology* **35**: 15–20.

Sinclair M, O'Toole J, Forbes A, Carr D, and Leder K. 2010. Health status of residents of an urban dual reticulation system. *International Journal of Epidemiology* **39**(6) (December 1): 1667–1675.

Sklarz M Y, Gross A, Ines M, Soares M, and Yakirevich A. 2010. Mathematical model for analysis of recirculating vertical flow constructed wetlands. *Water Research* **44**(6) (March): 2010–2020.

Smith I D, Bis G N, Lemon E R, and Rozema L R. 1997. A thermal analysis of a sub-surface, vertical flow constructed wetland. *Water Science and Technology* **35**: 55–62.

Soffer Y, Ben-Aim R, and Adin A. 2005. Membrane fouling and selectivity mechanisms in effluent ultrafiltration coupled with flocculation. *Water Science and Technology* **51**(6–7): 123–134.

Sorrell B K, and Boon P I. 1992. Biogeochemistry of billabong sediments. 2. Seasonal-variations in methane production. *Freshwater Biology* 27: 435–445.

Šostar-Turk S, Petrinić I, and Simonič M. 2005. Laundry wastewater treatment using coagulation and membrane filtration. *Resources Conservation and Recycling* **44**(2): 185–196.

Stottmeister U, Wiessner A, Kuschk P, Kappelmeyer U, Kastner M, Bederski O, Muller R A, and Moormann H. 2003. Effects of plants and microorganisms in constructed wetlands for wastewater treatment. *Biotechnology Advances* **22**: 93–117.

Stumm W and Morgan J J. 1996. *Aquatic Chemistry: Chemical Equilibria and Rates in Natural Waters*. John Wiley & Sons, New York.

Sturgis S. 2008. North Carolina OKs greywater use. *Raleigh EcoNews*, March 13.

Tal A. 1998. Enforceable standards to abate agricultural pollution: The potential of regulatory policies in the Israeli context. *Tel Aviv University Studies in Law* **14**: 223–286.

Tal A. 2006. Seeking sustainability: Israel's evolving water management strategy. *Science* **313**: 1081–1084.

Tang S L, Yue D P T, and Li X Z. 2006. Comparison of engineering costs of raw freshwater, reclaimed water and seawater for toilet flushing in Hong Kong. *Water and Environment Journal* **20**(4): 240–247.

Tanner C C, Clayton J S, and Upsdell M P. 1995. Effect of loading rate and planting on treatment of dairy farm wastewaters in constructed wetlands. 2. Removal of nitrogen and phosphorus. *Water Research* **29**: 27–34.

Tas-Environment. 2002. *Environmental Guidelines for the Use of Recycled Water in Tasmania*. Tasmania, Australia: Environment Division.

Tencer Y, Idan G, Strom M, Nusinow U, Banet D, Cohen E, Schroder P, Shelef O, Rachmilevitch S, Soares I. et al. 2009. Establishment of a constructed wetland in extreme dryland. *Environmental Science and Pollution Research* **16**: 862–875.

Thomas H and Burden R. 1963. *Operations Research in Water Quality Management*. Cambridge, MA: Division of Engineering and Applied Physics, Harvard University.

Travis M J, Weisbrod N, and Gross A. 2008. Accumulation of oil and grease in soils irrigated with greywater and their potential role in soil water repellency. *Science of the Total Environment* **394**(1): 68–74.

Travis M J, Weisbrod N, and Gross A. 2012. Decentralized wetland-based treatment of oil-rich farm wastewater for reuse in an arid environment. *Ecological Engineering* **39**: 81–89.

Tufvesson A. 2009. *Greywater Treatment and Technology*. World Plumbing Info.

UK Environment-Agency. 2011. *Greywater for Domestic Users: An Information Guide*, May 2011.

UN Water. 2007. *Coping with Water Scarcity—Challenge of the 21st Century*. World Water.

VICEPA. 2006. *Reuse Options for Household Wastewater*. Victoria, Australia: Environmental Protection Agency.

VICEPA. 2008. *Code of Practice—Onsite Wastewater Management*. Victoria, Australia: Environmental Protection Agency.

Venkataraman R and Ramanujam T K. 1998. A study of microbiology film layer in rotating biological contactors. *Bioprocess Engineering* **18**: 181–186.

Vieira P, Almeida M C, Baptista J M, and Ribeiro R. 2007. Household water use: A Portuguese field study. *Water Science and Technology: Water Supply* **7**(5–6): 193–202.

Vymazal J. 2005. Removal of enteric bacteria in constructed treatment wetlands with emergent macrophytes: A review. *Journal of Environmental Science and Health, Part A* **40**(6–7) (June 1): 1355–1367. doi:10.1081/ESE-200055851.

Vymazal J. 2010. Constructed wetlands for wastewater treatment. *Water* **2**(3) (August 27): 530–549. doi:10.3390/w2030530.

Vymazal J, Brix H, Cooper P F, Green M B, and Haberl R. 1998. *Constructed Wetlands for Wastewater Treatment in Europe*. Leiden, the Netherlands: Backhuys Publishers, 366pp.

Wand H, Vacca G, Kuschk P, Kruger M, and Kastner M. 2007. Removal of bacteria by filtration in planted and non-planted sand columns. *Water Research* **41**: 159–167.

Wathugala A G, Suzuki T, and Kurihara Y. 1987. Removal of nitrogen, phosphorus and cod from waste-water using sand filtration system with phragmites-australis. Water *Research* **21**: 1217–1224.

WDH. 2011. *Guidance for Performance, Application, Design, and Operation & Maintenance, Tier Two and Three*, Greywater Subsurface Irrigation Systems Chapter 246–274 WAC.

WEF and ASCE. 1998. Water Environment Federation (Manual of practice No. 8) and American Society of Civil Engineers (Manuals and Reports on Engineering Practice No. 76), Design of Municipal Wastewater Treatment Plants.

Weis J S and Weis P. 2004. Metal uptake, transport and release by wetland plants: Implications for phytoremediation and restoration. *Environment International* **30**: 685–700.

WHO. 2006a. *Guidelines for the Safe Use of Wastewater, Excreta and Greywater*. WHO, Geneva, Switzerland.

WHO. 2006b. *Overview of Greywater Management Health Considerations*. Amman, Jordan. World Health Organization, Regional Office for the Eastern Mediterranean, Centre For Environmental Health Activities.

WHO. 2008. *Guidelines for Drinking-Water Quality*, Vol. 1, 3rd ed. incorporating 1st and 2nd addenda. World Health Organization, Geneva, Switzerland.

Wood A. 1995. Constructed wetlands in water pollution control: Fundamentals to their understanding. *Water Science and Technology* **32**: 21–29.

WST. 2008. Manual bucketing of greywater—Plumbing regulation advisory note.

Washington-State. 2011. *Greywater Reuse for Subsurface Irrigation*.

Weissenbacher N and Müllegger E. 2009. Combined greywater reuse and rainwater harvesting in an office building in Austria: Analyses of practical operation. *Sustainable Sanitation Practice* **10**(1): 4–9.

Westman W E. 1972. Some basic issues in water pollution control legislation: Contrasts between technological and ecological perspectives on the regulation of effluents underlie current debates in water pollution legislation. *American Scientist* **60**(6): 767–773.

Wiel-Shafran A, Gross A, Ronen Z, Weisbrod N, and Adar E. 2005. Effects of surfactants originating from reuse of greywater on capillary rise in the soil. *Water Science and Technology* **52**(10–11): 157–166.

Wiel-Shafran A, Ronen Z, Weisbrod N, Adar E, and Gross A. 2006. Potential changes in soil properties following irrigation with surfactant-rich greywater. *Ecological Engineering* **26**(4): 348–354.

Winward G P, Avery L M, Frazer-Williams R, Pidou M, Jeffrey P, Stephenson T, and Jefferson B. 2008. A study of the microbial quality of grey water and an evaluation of treatment technologies for reuse. *Ecological Engineering* **32**(2): 187–197.

Wu F and Farland W H. 2007. Risk assessment and regulatory decision making in environmental health. In: *Risk Assessment for Environmental Health*, Ed. M G Robinson and W A Toscano. San Francisco, CA: Jossy-Bass, pp. 31–53.

Yang L, Chang H T, and Huang M N L. 2001. Nutrient removal in gravel- and soil-based wetland microcosms with and without vegetation. *Ecological Engineering* **18**: 91–105.

Ying G G. 2006. Fate, behavior and effects of surfactants and their degradation products in the environment. *Environment International* **32**: 417–431.

Zurita F, De Anda J, and Belmont M A. 2009. Treatment of domestic wastewater and production of commercial flowers in vertical and horizontal subsurface-flow constructed wetlands. *Ecological Engineering* **35**: 861–869.

Zuma B M, Tandlich R, Whittington-Jones K J, and Burgess J E. 2009. Mulch tower treatment system Part I: Overall performance in greywater treatment. *Desalination* **242**(1–3): 38–56.

## ORIGINALLY IN HEBREW

Be'er R, Shalem A, Balaban I and Pauker R. 2009. *Planning Low Water Consumption Gardens and Landscape*, 2nd ed. Water Authority, the Department to Promote Water Savings (Hebrew).

Dvir N. 2009. The sky is the limit: Towers in Israel. *City Mouse*. (Hebrew) http://www.mouse.co.il/CM.articles_item,1042,209,38083,.aspx. Accessed February 2, 2012.

Friedler E., Resnitsky L., 2004. *Study of Boron, Sodium and Chlorides Contribution of Domestic Dishwashers to Municipal Wastewater in Israel and Its Implication on Wastewater Reuse—Present Situation and Long Term Forecasting*. Submitted to the Israeli Ministry of Environment, 99pp. (Hebrew).

Germaz M, Friedler E. and Katz A. 2010. Greywater recycling. In: *Environmental Policy to Manage the Water Sector*, Ed. G Rosenthal and R Arad. Tel Aviv, Israel: The Environmental Organizations, pp. 57–66 (Hebrew).

Goldberger S. 2008. *Sanitary Conditions to Recover Greywater for Gardening and Toilet Flushing—The Engineer's Guidelines*. Israeli Ministry of Health (Hebrew).

Halperin R. and Aloni A. 2003. *Guidelines for Wastewater Reuse in the City, on Vacation and in the Industry*. A Report to the Israeli Ministry of Health. (Hebrew)

Horowitz N. 2009. *Rational Use of Natural Resources Water—Allowing Reuse of Wastewater*. Knesset, Israel: Environmental Protection Bill.

ICBS. 2010. Selected data on housing, by type of settlement. *Israel Statistic Yearbook* (5.35). Israel Central Bureau of Statistics (Hebrew).

ICBS. 2011. *Average Prices of Flats Owned by Tenants* (Table 6.2). Israel Central Bureau of Statistics (Hebrew).

Inbar I. 2003. *The Regulation Committee—Effluent Quality Standard* (Inbar Committee), Summary Report. February 2003 (Hebrew).

Israel Meteorological Service (Hebrew). http://ims.gov.il/IMS/Meteorologika/evaporation+Tub/monthly+data/. Accessed February 17, 2012.

Israel Union for Environmental Defense. 2009. The water crisis: The greywater bill. *The Voice of the Environment, Israel Union for Environmental Defense Quarterly*. September 2012. http://www.adamteva.org.il/_Uploads/dbsAttachedFiles/Sep09Heb.pdf (Hebrew). Accessed September 25, 2012.

Kovalio R. 2003. Comparative investigation of three treatment and reuse systems of greywater for toilet flushing. Thesis for the Degree of Master of Science in Environmental Engineering. The Technion—Israel Institute of Technology (Hebrew).

Mazuri D. 2011. The Greywater Bill was Endorsed for First Reading. NRG, Ma'ariv (Hebrew). http://www.nrg.co.il/online/1/ART2/248/105.html. Accessed April 28, 2012.

Neaman-Abramowitz A and Katz-Gerro T. Social bases of concern for the environment and environmental behavior in Israel. *Megamot* **44**(4): 736–758 (Hebrew).

Ronen I. 2009. *The Possibility of Using Greywater in the Private Sector. The Knesset: The Center of Research and Information*. A Report Written to the Interior and Environment Committee of the Israeli Knesset, July 5, 2009, Jerusalem (Hebrew).

Rozin A. 1997. *Recommendations on an Irrigation Policy with Treated Wastewater to prevent the Destruction of the Soil Structure*. The Joint Committee of the Division of Soil Conservation and Drainage and Field Services in the Agricultural Extension Service (SHAHAM), Ministry of Agriculture (Hebrew).

Shani A. 2008. The Water Authority Chairman, Personal Communication, December 2008 (Hebrew).

Shomera for a Better Environment. Shomera's initiative for greywater recycling (Hebrew). Retrieved: http://www.shomera.org/hebgraywater.htm. Accessed April 28, 2012.

SII (The Standards Institute of Israel), 1999. IS 438—Cleaning powders, Environmental requirements and labeling requirements: Laundry powders, 3pp. (Hebrew).

SII (The Standards Institute of Israel), 2006. IS 1417—Dishwashing powders for dishwashing machine: Environmental quality assurance requirements and labeling requirements, 8pp. (Hebrew).

Tal U. 2008. *Data on the Water Crisis in Israel*. The Israeli Knesset: Research and Information Center, March 17, 2008, Jerusalem (Hebrew).

Tartman E. 2008. *Wastewater Recycling Bill*. The Israeli Knesset 2008. P/17/3975 (Hebrew).

*The Gardner's Journal*. 2000. *The Gardening and Landscaping Organization in Israel* (Hebrew).

The Israeli Water Authority. 2012. *A Long-term National Master Plan for the Water Sector*. Part A—Policy Document, 4th ed., 70pp. (Hebrew).

USGBC. 2008. Water reuse. V3 - LEED 2008 WEc1, U.S. green building council. http://www.usgbc.org/credits/homes/v2008/wec1.

# Index

## U



## V



## W



## X